Fairness and Competence in Citizen Participation

# Technology, Risk, and Society

*An International Series in Risk Analysis*

VOLUME 10

*Editors*

Jeryl Mumpower, *State University of New York, Albany, USA*
Ortwin Renn, *Center of Technology Assessment, Baden-Württemberg, Germany*

*The titles published in this series are listed at the end of this volume.*

# FAIRNESS AND COMPETENCE IN CITIZEN PARTICIPATION

## Evaluating Models for Environmental Discourse

*Edited by*

ORTWIN RENN
*Center for Technology Assessment, Stuttgart, Germany*

THOMAS WEBLER
*Antioch New England Graduate School, Keene, New Hampshire, U.S.A.*

and

PETER WIEDEMANN
*National Research Center, Jülich, Germany*

KLUWER ACADEMIC PUBLISHERS
DORDRECHT / BOSTON / LONDON

Library of Congress Cataloging-in-Publication Data

Fairness and competence in citizen participation : evaluating models for environmental discourse / edited by Ortwin Renn, Thomas Webler, and Peter Wiedemann.
p. cm. -- (Technology, risk, and society ; v. 10)
Includes index.
ISBN 0-7923-3517-1 (alk. paper)
1. Environmental policy--Citizen participation. I. Renn, Ortwin. II. Webler, Thomas. III. Wiedemann, Peter M., 1948- . IV. Series.
GE170.F35 1995
363.7'0525--dc20 95-12720

ISBN 0-7923-3517-1

---

Published by Kluwer Academic Publishers,
P.O. Box 17, 3300 AA Dordrecht, The Netherlands.

Kluwer Academic Publishers incorporates
the publishing programmes of
D. Reidel, Martinus Nijhoff, Dr W. Junk and MTP Press.

Sold and distributed in the U.S.A. and Canada
by Kluwer Academic Publishers,
101 Philip Drive, Norwell, MA 02061, U.S.A.

In all other countries, sold and distributed
by Kluwer Academic Publishers Group,
P.O. Box 322, 3300 AH Dordrecht, The Netherlands.

*Paperback cover design by*
*Daniel Courtney,*
*Worcester, Massachusetts*

*Printed on acid-free paper*



Printed in the Netherlands

We dedicate this book to all the citizens who, at the bidding of researchers, have taken uncertain chances by agreeing to experiment with new forms of public participation. They have contributed enormously to the development of better models for citizen participation. Without them this book would not have been possible.

This book is also dedicated to Susan Hadden, one of the contributing authors. Susan did not live to see this in print. She was a leader in the study of public participation, a woman devoted to science and to environmental and social justice. Her untimely death is a loss to us all.

# Contents

# *List of Contributors*

Professor Audrey Armour, Faculty of Environmental Studies, 355 Lumbers Building, 4700 Keele Street, North York, Ontario M3J 1P3. Canada

Dr. Joanne Bayer-Linnerooth, International Institute for Applied Systems Analysis, A-2361 Laxenberg, Austria

Dr. Michael Baughman, Intertech Consultants, 4220 S. Maryland Parkway, Suite 219A, P.O. Box 93537, Las Vegas, Nevada 89193. USA

Dr. Frank Claus, Institut für Kommunikation und Umweltplanung, Altfriedstrasse 16, D4600 Dormund, Germany

Dr. Ned Crosby, Director, Jefferson Center, 364 Century Plaza, 111 3rd Avenue South, Minneapolis, MN 55404. USA

Professor Thomas Dietz, George Mason University, Fairfax, Virginia, 22030. USA

Professor Peter Dienel, Bergische Universität Wuppertal, Fachbereich 1, Postfach 10 01 27, D-5600 Wuppertal, Germany

Dr. Claus Frank Institute for Communication and Environmental Planning (IKU), Altfriedstasse 16; 4600 Dortmund 18, Germany.

Dr. Daniel Fiorino, Associate Director of the Office of Policy Analysis, US EPA, 401 M Street, SW, Washington, D.C. 20460. USA

Professor Susan Hadden, LBJ School of Public Affairs. Drawer Y, University Station, University of Texas Austin, TX 78713-7450. USA

Professor Jack Kartez, Hazard Reduction Recovery Center, Texas A&M University, College Station, TX 77843-3137. USA

Professor Howard Kunreuther, Risk and Decision Processes Center, 1332 Steinberg Hall – Dietrich Hall, Wharton School, Philadelphia, PA 19104-6366. USA

Dr. Frances Lynn, University of North Carolina, CB 7410315 Pittsboro Street, Chapel Hill, NC 27599-7410. USA

Professor Cees Midden, Eindhoven University of Technology, Netherlands Department of Philosophy and Social Sciences, P.O. Box 513, 5600 MB, Eindhoven, Netherlands

Professor Jeryl Mumpower, Center for Policy Research, University at Albany – SUNY Milne 300, Albany, NY 12222. USA

Dr. Werner Nothdurft, Institut für Deutsche Sprache, Universität Mannheim, Postfach 10 16 21, D-6800 Mannheim, Germany

Felix Oberholzer, Institut für wirtschaftliche Forschung, Universität Zürich, Kleinstrasse 15, CH-8008 Zürich, Switzerland

Professor Ortwin Renn, Center of Technology Assessment, Industriestraße 5, D-70565 Stuttgart, Germany

Dr. Hansjörg Seiler, Polyprojekt "Risiko und Sicherheit", ETH Zentrum, CH-8092 Zürich, Switzerland

Professor Anna Vari, 1145 Budepest, Varsö u. 29, Hungary

Dr. Thomas Webler, 132 Farley Road, Wendell, Massachusetts, 01379. USA.

Dr. Peter Wiedemann, Technik, Umwelt, und Gellschaft (TUG) Forschungszentrum Jülich, Postfach 1913, D-5170 Jülich, Germany

# Foreword

*Ortwin Renn*
*Thomas Webler*
*Peter Wiedemann*

In late July of 1992 the small and remote mountain resort of Morschach in the Swiss Alps became a lively place of discussion, debate, and discourse. Over a three-day period twenty-two analysts and practitioners of public participation from the United States and Europe came together to address one of the most pressing issues in contemporary environmental politics: How can environmental policies be designed in a way that achieves both effective protection of nature and an adequate representation of public values? In other words, how can we make the environmental decision process competent and fair? All the invited scholars from academia, international research institutes, and governmental agencies agreed on one fundamental principle: For environmental policies to be effective and legitimate, we need to involve the people who are or will be affected by the outcomes of these policies. There is no technocratic solution to this problem. Without public involvement, environmental policies are doomed to fail.

The workshop was preceded by a joint effort by the three editors to develop a framework for evaluating different models of public participation in the environmental policy arena. During a preliminary review of the literature we made four major observations. These came to serve as the primary motivation for this book. First, the last decade has witnessed only a fair amount of interest within the sociological or political science communities in issues of public participation. Most of the recent literature has focused either on specific case studies or manuals for conducting practical applications. Second, the traditional forms of public participation – hearings and advisory committees – certainly have advantages, but are not able to fulfill popular demands for widespread and meaningful citizen involvement in environmental decision making. Third, there is a perceived need for new models of participation that enhance decision making competence and legitimacy through more meaningful participation by citizens. Several analysts and practitioners have experimented with novel models. Fourth, a systematic framework for evaluation on any but the most abstract level is completely absent. This is needed if those who implement participation are to appropriately match purpose to method. Such an evaluation would have to define and defend normative ends that public participation aims to achieve. This cannot be done without taking a position in the debate about how much citizens should be involved in governance. The evaluative framework would have to provide assessments of each model's scope of application, potential benefits and drawbacks, most-suited problem type, and transferability to other cultural contexts.

In early 1992, we met several times to develop a common framework for reviewing and critiquing participation models. As explained in Chapter 3, this framework is highly influenced by critical theory, in particular the theory of communicative action developed by the German sociologist and philosopher Jürgen Habermas. Thomas Webler, who was basically responsible for this part of the work, adopted Habermas's theory of communication and applied it to the concrete situation of citizen involvement in environmental decision making. Habermas's ideas of the ideal speech situation and communicative competence became the guiding principles of an evaluative framework. Since Habermas is rather vague about providing clear instructions of how to define an ideal speech situation or communicative competence, we developed a list of criteria and indicators in the spirit of his work. At the same time, however, we restricted our scope to citizen participation in environmental decision making and we departed from Habermas in two ways. First, we included rational strategic actions by individuals and parties as an unwanted, but nevertheless realistic behavioral assumption in any type of discourse. Second, we were more skeptical than Habermas about reliance on the lifeworld as a reservoir of rules and selection principles for redeeming cognitive claims. Although our framework is clearly inspired by critical theory, we believe that the resulting criteria and indicators have a validity of their own. Even those participants at the Morschach workshop who were not sympathetic to critical theory could agree with our criteria.

Next we identified models of participation that represented a wide variety of formats and contacted leading proponents and critiques of these models. We invited a proponent and a critical reviewer of each model to Morschach. Both the reviewer and the proposer received our framework with the criteria for review. We asked the proponents to make an argument for how well these criteria were met by their models. To balance the analysis, we asked reviewers to make more critical assessments using our criteria.

During the workshop, both the proponents and the reviewers made claims and counterclaims and the whole group discussed each argument in detail. Workshop sessions scheduled for two hours ran several hours over time and – most surprisingly for the editors – all participants attended all sessions without any exception. The result of the workshop was twofold. Proponents acknowledged the limitations and problems of their models, but were also reassured about the specific merits that their approach encompassed. Many critical remarks turned out to be based on misunderstandings or the application of the method to a problem unsuited for such a method. In essence, the workshop succeeded in meeting the most important objective: to define the legitimate scope of each model and to evaluate its relative advantages and drawbacks.

After the workshop, the participants composed their papers based on the discussions and the commonly agreed directions. Papers were reviewed by the editors and two independent reviewers. In addition, the whole manuscript was reviewed by two outside reviewers chosen by the editorial team at Kluwer. The authors were asked to use the reviews while writing a second draft.

This book would have not been possible without the support and effort of many people. We are especially grateful to all the contributors who played along in this unique attempt to submit their own thoughts to a common framework. We would also like to thank our reviewers, Dr. Cvetkovich and Dr. Dietz and the anonymous reviewers hired by Kluwer for their valuable comments and criticisms. Thanks are also due to Kathy Plunkett at the State University of New York for her invaluable assistance during the final editing and production of this volume. Finally, we want to express our appreciation for the publishing team at Kluwer, most notably to Ms. Strata from Boston, who has been an ardent supporter of this book project and assisted us in many ways. The project was financed by the Humboldt Foundation (Bonn, Germany) and the Polyproject "Risk and Safety of Technological Systems" of the Swiss Institute of Technology (Zürich, Switzerland). We are very grateful for the generous financial support by these two sponsors. Without it we would have not been able to pursue the elaborate procedure culminating in this unusual book.

Summer, 1994

Ortwin Renn (Stuttgart, Germany)
Thomas Webler (Wendell Massachusetts, USA)
Peter Wiedemann (Jülich, Germany)

# Preface

## Democracy and Science

*Thomas Dietz*

By 2050, the human population of the earth will double. The per capita level of affluence and resource demand will increase by 50 percent. The result will be a tripling of human activity in the biosphere. Stresses on human, biological and physical systems, already at critical levels, will be greatly exacerbated. In addition, new technologies will make our ability to manipulate biological and physical systems more profound. Evolving artificial life may complement artificial intelligence. We will be able to program genomes and hybridize natural and engineered systems. We are remaking the biosphere through the intended and unintended consequences of our actions in ways unprecedented in human history.

How will we respond to these challenges? How can we shape decent lives for ourselves, our ancestors and the other species that share the planet with us? To meet these challenges, we must improve our ability to answer two questions. One is descriptive: How does the world work? The other is prescriptive: What should we do? These questions have motivated human intellectual activity since our ancestors acquired language. Our success at answering the first question is a major reason to be both fearful and optimistic about the future. It is our understanding of the physical, biological and social worlds that allows the profound and powerful manipulations that characterize the 20th century.

Deciding what is better or worse is the subject of the second question. Our ability to use descriptive knowledge towards good ends depends on answering the prescriptive questions: towards what ends should we direct our efforts? The importance of this problem has never been more evident. But our intellectual progress on this problem has not been impressive. The tradition of democracy is our best hope. Democracy is an ancient mode of societal decision making that has its roots in the fundamental elements of human adaptation – communication and social learning. Democracy is usually traced to the Greeks, but is actually a heritage from the earliest human social groups. It is at the core of our cultural and biological heritage, and was the mode of decision making for food foraging societies that constitute 99 percent of human history. Humans have evolved to coordinate group action by discussion and shape individual action through social learning and reflection. While the forms of democratic process have been modified throughout history, and must be modified again, the basic concept has very deep roots.

However sound the democratic tradition, it is under pressure in the contemporary world. One source of pressure is the power of science and technology to transform the world. We can produce changes that are not easily related to the daily experience of most people and thus not linked to our moral sense. Most of us lack the moral context to understand the implications of artificial life, genetic engineering or other new technologies and thus have trouble making informed ethical judgments about them. Weber suggested that science has "disenchanted" the world, but in another way, science "re-enchants" by casting a spell of incomprehensibility on much that is important to our lives. Put simply, the implications of our technology are hard to understand.

Another pressure comes from the scale at which democracy must operate. When the U.S. government was founded two centuries ago, the average Representative spoke for about 75,000 people, and reached decisions in a body of only 66 peers. Now, the average member of the House of Representatives speaks for over half a million citizens and must carry out debate with 364 colleagues. The American example of increasing complexity is mirrored throughout the world. The discourse that underpins democratic process is attenuated by problems of scale. And, while in some sense all politics is local – it relates critically to the lived experience of the citizens – all politics is now global. Actions taken in one locale often have profound implications across the planet. So the context in which decisions must be assessed is much more complex.

For several decades, a handful of scholars and activists have proposed ways democracy can cope with these profound problems. Jürgen Habermas provides the most thorough and systematic approach to the issue. Habermas offers a theory of society and of human action. He argues that the rational way to make collective decisions is through fair and competent discourse. But Habermas' opus, while theoretically profound, is not articulate about practice. I have been among a handful of scholars, working at the intersection of theory and practice, who have proposed methods for putting Habermas into practice, in contexts ranging from impact assessment to planning.

This volume is the most important work to date in that tradition. And it is more than that – this is the best extant examination of experiments in democratic process. Instead of cynical complaints about the failures of democracy, the editors and authors take up the challenge of proposing new forms of democratic process and examining how they work in practice. Habermas' theory provides an organizing principle to compare democratic strategy and tactics, allowing thoughtful comparisons across methods and contexts. The result is a volume that is essential reading for anyone interested in the future of democratic process and decision-making.

The discussion focuses appropriately on the toughest challenge facing democracy: problems of environment and technology. Environmental and technological issues are at the intersection of science and politics, involve subtle and uncertain risks, large time and spatial scales and a myriad of conflicting interests and values. It is precisely in this area that traditional democratic process seems to falter. The resulting policies are often inequitable and poorly aligned with science.

Those concerned with environmental and technological policy should read this volume carefully. So should those interested in Habermas' theory and those concerned with the future of democracy. This collection is not the end of the debate on these issues, but the place from which all future work must start.

December 19, 1993

Thomas Dietz
Professor of Sociology at the George Mason University
President: Society for Human Ecology

*Chapter 1*

# *A Need for Discourse on Citizen Participation: Objectives and Structure of the Book*

*Ortwin Renn*
*Thomas Webler*
*Peter Wiedemann*

## Introduction

More and more the decision makers and affected parties engaged in solving environmental problems are recognizing that traditional decision making strategies are insufficient. Often heavily shaped by scientific analysis and judgment, these kinds of decisions are vulnerable to two major critiques. First, because they de-emphasize the consideration of affected interests in favor of "objective" analyses, they suffer from a lack of popular acceptance. Second, because they rely almost exclusively on systematic observations and general theories, they slight the local and anecdotal knowledge of the people most familiar with the problem and risk producing outcomes that are incompetent, irrelevant, or simply unworkable.

Citizen involvement in decision making has been widely acknowledged as a potential and partial solution to these problems. Precisely which citizens should be involved in which way, when, and to what degree, however, has been the subject of a tremendous amount of argument, posturing, experimentation, and theorizing. One of the main objectives of this book is to set out a normative theory of citizen participation in environmental decision making and to illustrate how such a theory could inform the design and implementation of novel approaches to public involvement. In this sense the book addresses the challenge for environmental problem solvers to transcend the traditional means of policy and decision making and to consider adopting novel approaches to conflict resolution and citizen participation. Such procedures may be philosophically preferable, but they may also help to improve the effectiveness of environmental policies and enhance the potential for affected citizens to become part of the decision making process rather than being the victims of the decisions made by anonymous agencies or institutions.

## Defining Public Participation

Although other forms of political activity are important as well, in this volume we concentrate on how intentionally designed structures for participation can be improved. Toward this end, we adopt the definition of public participation as: *forums for exchange that are organized for the purpose of facilitating communication between government, citizens, stakeholders and interest groups, and businesses regarding a specific decision or problem.*[1] This definition explicitly excludes activities such as protest, expert workshops, and serving as governmental officials. It is inclusive of public hearings, public meetings, focus groups, surveys, citizen advisory committees, referendums and initiatives, and negotiation, among other models.

We use the term “model” to distinguish formats for participation that can be implemented in a variety of imaginable problem contexts from those that are unique experiences. Models represent different institutional forms for public participation and they are described by their characteristic structures and procedures. In selecting models for study in this volume, we did not necessarily intend to endorse some while condemning others. Several models do not appear in this volume which still might be promising. We would simply mention recent promising work in electronic democracy (Strauss and Clark 1980), in surveys (Fischer et al. 1991), social impact assessment (Daneke et al. 1983; Freudenburg 1983; Dietz 1987), risk communication (Chess et al. 1991; Stern 1991; Renn 1992), and public hearings (Checkoway 1981).

### *Scope*

The focus of this book is on participatory models for environment decision making. Environmental problems preoccupy modern societies for a number of reasons (Dunlap 1987; Fietkau et al. 1982; Kessel and Tischler 1984). The dangers these problems pose to human health and safety are certainly among the most noticeable, but they are sometimes outweighed by the inequities and ethical and moral implications. Environmental problems and solutions also demand difficult choices of society, particularly balancing the public

---

1 The terms "public participation" and "public involvement" will be used interchangeably. Using the word "public" in the singular conveys the untrue impression that "the public" is somehow homogeneous, when in actuality, it is a vast and heterogeneous group of individuals. The phrase "the public interest" is additionally misleading. There is no such thing as a single "public interest." Rather, society consists of individuals with unique sets of interests, some of which are identical to those of others, and collectivities of people which form shared interests. Shared interests are not merely the intersection of individuals' particular interests.

regulation of technology and land-use patterns against economic growth and change. Finally, there is the pervasive, but unknown threat of regional or global ecological catastrophe that encourages some voices to call for more prudence.

Environmental problems are closely linked to those of protecting human health and safety through the language of risk. The threat posed by low probability, high consequence risk sources, the inequities associated with large-scale technologies, the potential for global environmental deterioration, and the decreased marginal value of economic commodities have also contributed to growing sensitivity and awareness toward technological risks (Hohenemser et al. 1983). This concern highlights, according to the German sociologist Ulrich Beck, a gradual change of the predominant social conflict in this century (Beck 1986; cf. Schnaiberg 1986). The primary organizing principle in the early 20th century was the distribution of wealth. After the Second World War, and particularly in the 1960's, this changed to the distribution of power in politics and economics; in more recent times the major conflict is about the distribution and the tolerability of risks for social groups, regions, and future generations. Professionalization of risk analysis and institutionalization of conflict resolution mechanisms are reinforced by the characteristics of environmental politics and risk issues in the political arena.

## *Environmental discourse*

This shift of focus in social organization implies new forms of collective decision making and conflict resolution. Moreover, it underlines the importance of communication among the project proponents and government and the affected population as a necessary, though not sufficient, step towards a social equilibrium (Luhmann 1986). In addition, the capability of societal institutions to tame powerful natural sources for economic purposes and reduce the concomitant risks of potential side effects to human health and the natural environment depends largely on consent-producing mechanisms among institutions and groups (Lake 1980) and the formation of specialized risk or danger cultures within these institutions (Rip 1985).

Discourse is the term we have adopted to highlight the style and nature of citizen participation that can aptly handle these kinds of decision problems (Habermas 1984; Gumpert 1982). The word implies equality among the participants, peer review as a means for verifying understandings (i.e. holding knowledge claims up to public scrutiny), and an orientation toward resolving conflicts in consensual rather than adversarial ways. The question of how to incorporate relative frequencies in the decision process, i.e. how to balance options with different compositions of magnitude and probability of impacts, has not been adequately addressed and assimilated by the political decision system. Numerous strategies to cope with this new challenge have evolved over time. They include technocratic decision making through expert committees, or ignoring probabilistic information altogether. The incorporation of probability assessments in decision making requires new rationales for evaluating policy options and necessitates a revision of institutional routines (Freudenburg 1988). This book takes up the line of argument that discourse with citizens as equal partners, who are informed about the cognitive aspects of the problem and consequences, is one good way to realize better decision making.

## Problems of Evaluation: How to Avoid Vicious Cycles

### *Consensus and conflict approaches to social values*

Value free evaluation is an oxymoron. By definition, "objective" evaluation does not exist. Evaluating or designing new models of participation must deal with the question of how values can be selected and how they can be traded off against each other. Two major types of theories have emerged in sociology to explain how values are selected and prioritized.

Consensus theories put forth that society is maintained through shared opinions about norms and values, which are expressed in what Durkheim called a *collective conscience* (Etzioni 1993). Social order is based on tacit agreement, and changes emerge through gradual evolution. Conflict originates with people who deviate from social norms, endangering society's order and efficiency.

Conflict theories explain that the ruling groups impose their values upon others and that there is no collective agreement on values (Coser 1956, 1967; Blalock 1989). Social order, in this perspective, is maintained only through patterns of domination, and change is something that comes suddenly and which revolutionizes society. Consensus is coercive because it convinces people to surrender aspects of their individual interests in exchange for the betterment of the majority.

Neither theory has been able to defeat the other, despite the fact that both make strong arguments. Consensus theories have been criticized for justifying dictatorships, totalitarian regimes, tyranny of the majority, devaluation of the individual, and mechanistic and technocratic approaches to governing (Bachrach 1967). Conflict theories have been criticized for ignoring the importance of order for social reproduction: of the individual, the society, and the culture (Bernard 1983).

Confusion about the goals of citizen participation accompanies every practical application. Goals do not appear in clearly defined consensual form, neither in practice nor in theory (Rosener 1978a; Rosenbaum, 1979:81). Part of the reason for the lack of agreement on goals is that the parties involved tend to define them in ways beneficial to their interests. Furthermore, they sometimes find it advantageous to be vague about announcing their interests. Administrators, citizens, stakeholders, and experts may all desire participation, but for quite different reasons. Thus, they may have different ideas about how the process should be conducted.

What is the purpose of public participation? Is it to provide a means by which the collective conscience is revealed? Or is it a means to facilitate conflict and power redistribution? Not surprisingly, there are supporters of both views. Locked into an irreconcilable debate that extends at least as far back to differences between Plato and Aristotle, agreeing on the way to evaluate public participation seems impossible. Which theory should provide the values for the evaluation? The answer to that question seems to be: It depends whose side you are on. As Judy Rosener put it,

> There should be little doubt that knowing who is doing the perceiving is crucial to any understanding of the effectiveness of citizen participation (1978a:458).

## *Past approaches to evaluation: The citizens' view*

One way that participation is commonly evaluated is from the perspective of the different participants (see Potter and Norville, 1983). Conflict theories and consensus theories are used to support different positions. Critics concerned with the plight of local citizens remark that, first and foremost, public participation as it has been performed has failed to fulfill citizens' needs and demands. Sherry Arnstein's "Ladder of Participation" takes as its main focus the powerfulness of citizens in different participation models (1969). Arguments based in consensus theory point to the citizenry as the source for society's collective values. For example, Barry Checkoway and Jon Van Til (1978) note that participation provides society with a means for social change to occur gradually through public involvement in establishing social values and needs. Daniel Mazmanian (1976) notes that it provides administrators with the means by which they can acquire political support. Conflict theories are more often used to justify the position of citizen values. For example, Nelson Rosenbaum (1978) argues that participation is supposed to give citizens a voice in decisions that affect them, so that they can protect their own interests.

According to these explanations, public participation has failed because citizens feel the existing participatory opportunities are inadequate (Langton 1978; DeSario and Langton 1987; Brion 1988; Portney 1984; Heiman 1990a; Fiorino 1990). They argue that this is clearly evident in the low attendance rates and the plenteous attempts by activists to influence the participatory process, sometimes before it has even begun.

## *Past approaches to evaluation: The administrators' view*

Others point out that participation has failed to fulfill government administrators' expectations. They are charged with finding solutions to problems. Cuthbertson (1983) contends that participation is organized only to inform decision makers. According to other accounts, the prime motivation for administrators to organize public participation is to acquire the public support necessary to implement policy (Ethridge 1987; Cupps 1977; Rosenbaum 1978). In the very act of seeking this support, however, officials often encounter--and may even contribute to the development of--public opposition. This opposition is not easily transformed into policy adjustments or alternatives. In short, administrators cannot do their jobs.

A consequence of this failure is that administrators are unable to implement policies, problems go unattended and system organization is endangered (through loss of faith among the citizenry that the government can protect their interests) – a kind of legitimation crisis ensues. When the publics' frustrations are channeled into opposition movements, the legitimacy of the current mode of political organization is challenged. If this opposition is not tempered, for instance, by incorporating their demands into the agreement, then it will continue to grow and may eventually challenge the existing order (Habermas 1975). The inability of governments to site hazardous waste facilities may be a good example of this situation (Webler, Rakel, Ross 1992). Consensus theorists find this problematic, while conflict theorists find this inevitable or desirable. Creighton (1983) exemplified the conflict perspective when he noted that participation could serve the interests of change and produce better decisions by expanding the scope of conflict.[2]

## *Evaluation: Escaping from the subject*

Viewing the problem in terms of how it affects government regulators, stakeholders, interest groups, or private citizens focuses attention on one particular agent. Certainly, this is a valid perspective to take. After all, individuals are involved in public participation partly to protect their interests. It is unavoidable that people will evaluate the process in terms of how well their interests fared in the final outcome. However, when operating under this orientation, it is difficult not to see the decision making process as a zero-sum game – a process which must have winners and losers. Since the interests of

---

2 Schattschneider (1960) argued that expanding the scope of conflict is a necessary step toward reaching an agreement on "the public interest."

the participants are nearly always diverse, it is always possible to produce a negative evaluation by orienting the evaluation around the interests of the participant who lost. Philosophically, it is difficult to justify one evaluation over another. The problem is putting one group's values and norms above all others, thereby de-emphasizing due process and the shared interests among the group that may stand in opposition to the dominant social values and norms.

In efforts to escape from the traps of subject-centered evaluations, researchers have phrased the goals of public participation in different ways. For example, Kweit and Kweit (1987) suggest that participation should help to "improve" decisions through using "comprehensive information." Another view, which could be used to argue for or against either consensus or conflict theory, is that participation is essential to having a responsive and responsible government (Daneke et al. 1983).

But these definitions do not avoid the problems of the subject, they merely disguise them. When one delves more deeply into the wording, troubles become apparent. For example, Kweit and Kweit have hidden the value dispute in the word "improve." Improved according to whose interests? Likewise with Daneke's use of the phrase "responsive and responsible." To whom should it be made responsible and responsive? Perhaps the only statement made about subject-centered evaluation that everyone can accept is that it is easy to agree on a method of evaluation when there is agreement on the means and the ends of the participation (Rosener 1978a). Unfortunately, this is rarely the case.

It is unfortunate that evaluations of public participation have been so tightly linked to serving the values and interests of individuals and groups. What has resulted is a standoff between government administrators on the one hand, who offer participation opportunities with no real power to citizens, and locally-impacted citizens on the other hand, who block projects through other means such as protest and complaint. Opinions on who should be served by participation clash in largely unproductive ways. Although it would be impossible to free public participation entirely of that burden, it should be possible to move *away from an subject-centered view of participation and closer to shared values and interests.* This can be achieved without adopting all of the unwanted baggage of consensual theories or conflict theories. In other words, it is not necessary to conceptualize public participation as either an activity that stabilizes society (thereby serving the interests of the ruling elite), or as a mechanism to accelerate social change (thereby empowering citizens).

The non-subject centered perspective is based on a belief that both consensual theories and conflict theories have important contributions to make, although neither theory can dominate. It is certainly true that people have some interests and values in common, thus they organize themselves into labor unions, interest groups, corporations, and communities. At the same time, there is a great deal of conflict among people as they compete for scarce resources and power. Both of these elements are present in society and public participation is one of the realms where they occur. By participating with others in democratic decision making, people learn that they have things in common. By banding together, they can achieve their goals and meet their needs. They also discover that other groups and individuals have interests and values in direct contradiction to their own. Resolving the conflict is not always the main purpose of participation, but acknowledging it is important.

There is a small record in the literature on non-subject centered approaches to evaluating public participation. Ray Kemp used Habermas's definition of the ideal speech situation as a measure against which to compare the discourse that occurred in the public inquiry process to permitting the British uranium reprocessing plant (Kemp 1985). Daniel Fiorino developed performance criteria for public participation from the theory of participatory democracy and evaluated several generic models of participation (1989). Recently, Frank Laird has supplemented Fiorino's criteria with another set from the theory of pluralist democracy and evaluated the same models (1993). The works in this volume continue in these traditions.

The survival of both consensus and conflict theories is a testament to their validity. But the positive contributions that both theories make to the understanding of social life is limited to the macro-level. Social theories of consensus, such as structural functionalism, and social theories of conflict, such as the radical conflict theory of Marx (Marx 1971; Evans 1972), study society as a whole. This contrasts with micro-level social theories, such as theories of social interaction by Weber and Mead, which seek to explain society through the individual (Weber 1962; Mead 1934). It is perfectly legitimate to study public participation as a macro-level phenomenon, as almost all of the literature does. But this seems to disregard the obvious fact that, to the participants, public participation is interaction among individuals. Theories of social interaction do not offer to replace conflict theories or consensus theories, as change and agreement also occur at the interpersonal level. They merely approach the understanding of society from the opposite direction.

## *Webler's approach to evaluation*

In Chapter 3, Webler follows up on the works of Kemp, Fiorino, and Laird to evaluate participation according to a consistent set of normative criteria. In this endeavor, he builds on the recent work of Jürgen Habermas, a German philosopher and dominant figure in the tradition of critical theory (Habermas 1970; 1984; 1987). Webler argues that Habermas's theory of society can provide the normative foundations as well as the analytical tools needed for evaluation. Adopting this approach does not commit one to seeing public participation merely as a tool that stabilizes the elite, or as a tool that topples the elite. At the micro-level, that is, in any single application of participation, social conflict will not be resolved nor will the functionality of society be determined. Those macro-level outcomes emerge out of infinite numbers of local experiences. The significant feature at the micro level is people interacting with each other. Interaction is oriented toward individual and shared goals of the actors through a coordinated process of discourse.

One of the most important reasons for selecting Habermas's theory is the credence it gives to individual autonomy. In the tradition of critical theory, Habermas believes that individuals ought to be free of all forms of domination. Once they are free, people are able to enter into social relations that encourage personal development as well as social and cultural reproduction. The key is critical self reflection. Habermas promotes introspection among free and autonomous beings so that they will think about the type of society *that they want*, before committing to new relations. Public participation can and should be a means to realize the critical awareness.

Accordingly, public participation should be guided by shared societal values about what type of society participants would like to have and expectations about how people should interact. Habermas's argument is that, to some extent, these values and norms are already in place. (To this extent, Habermas could be labeled a consensual theorist.) But he also acknowledges that, in the interest of freedom, participants must be able to challenge and change these pre-established norms and values when they feel that they are no longer useful (here he is a conflict theorist). It would be misleading to overemphasize the extent to which Habermas feels universalistic norms and values can be identified. He is cautious about claiming any but the most general universalities. For the most part, his normative theory frames out an unconstrained model of discourse, that is, one wherein values and norms themselves are discussed and agreed upon in a setting free of any kind of coercion.

## *Competence and fairness in citizen participation*

One of the simplest and yet essential results that emerge from the review of the public participation literature (Chapter 2) is that public participation must be fair and competent. These are principles which are widely agreed upon. We do not naively believe that any participation process can be fair to everyone's interests. However, if it is open and the rules are clear and consistent, at least the element of mystery (which is an open door for abuse) can be eliminated. We also do not believe that any single process will always produce the best decision. But by making decision quality a central aspect to the process design, decisions that are more favorable to the plurality of interests are more likely to be encouraged than if one interest group dominates the outcome.

In practice, interaction during public participation includes much more than Habermas's "rational discourse." People may participate with no intent to reach an agreement. They may want to use the opportunity as a stage to make statements about their unwillingness to cooperate, for example. Without being naive about the strategic activities which occur in discourse, we note that when people do decide it is in their best interests to cooperate, they will need a process that functions well for them. Such processes should be fair and competent.

## Participation Models Selected for the Book

### *Emphasis on discourse*

We identified models of participation that represented a wide variety of formats and contacted leading proponents and critiques of these models. Based on the literature and direct communication with other scholars, we decided to focus on the following models:

- Citizen Advisory Committees
- Citizen Panels (Planning Cells)
- Citizens Juries
- Citizen Initiatives
- Negotiated Rule Making
- Mediation
- Compensation and Benefit Sharing
- Dutch Study Groups

We deliberately excluded the most common traditional methods, such as public hearings or inquiries, social surveys, arbitration, or scientific advisory groups. The literature suggests that these methods, in spite of their widespread use, do not meet a sufficient number of our criteria to be seriously considered as candidates for ideal discourse. This does not exclude them from playing an important role as an additional element of a more complex hybrid model of participation, as we argue in the last chapter. Also excluded from our analysis is the method of referendum. Although it clearly meets our criteria for fairness, it does not constitute a discourse by itself. We suggest that referendums are an excellent way of legitimating a final decision after it has been subjected to a thorough review by citizens according to one of the above mentioned approaches.

The eight models analyzed in this book represent a wide variety of approaches. They are tools presently in use or available for use. It may seem awkward to think of mediation, negotiated rule making, or compensation as models for public participation. However, we have decided to treat them as such because that are increasingly being used to supplement or replace other forms of participation in environmental decision making, especially siting issues. Treating them as peers to citizen panel approaches, advisory committees, initiatives, and study groups also brings to the forefront the inabilities of the former approaches to fulfill democratic criteria while revealing their capabilities to produce outcomes that are often more acceptable or competent than the latter, more democratic procedures.

The Dutch Study Circles discussed in Chapters 18 and 19 has had only one empirical application, but we believe this still counts as a model because there was nothing to the Dutch context to prevent it from being administered in another country or problem context. Models do of course have limitations. The Dutch model may not be transferable to some locales, or it might not work for certain kinds of problems, but one of the main objectives of this book is to document these limitations in a systematic way that enables direct comparisons to be made between the available models.

## *Structure and content*

The structure of the book follows the procedure of the Morschach workshop. The next chapter represents a brief review of participation in Europe and the United States. The third chapter introduces our common framework for reviewing and evaluating models of citizen participation. The following sixteen chapters are devoted to the eight models selected for this volume. They are arranged in pairs: the first paper by the proponent, the second by the reviewer. Our intention in structuring the book as a sequence of optimistic proposals and serious critiques was to avoid the tendency to overemphasize the advantages or shortcomings of models.

The evaluation of the models according to our predefined list of criteria and indicators has proven more difficult than originally foreseen. Although the authors agreed with our criteria in principle, they felt that certain models were more associated with one set of criteria than others. Only a hybrid model could adequately address all the criteria in sufficient depth. They also objected to a formal evaluation of each model by the reviewer since each reviewer would interpret the scales differently (problem of calibration) and would use his or her subjective judgment for the task. Subjective judgment would make a comparison among the evaluations of the models meaningless. The judgment was made that each reviewer should use the criteria to describe and evaluate verbally each model and that the editors were left with the task to come up with a more formal evaluation based on their (hopefully) coherent subjective judgments.

The final chapter, which is co-authored by the editors, contains a synopsis that includes a comparative review of all the models and makes some suggestions for hybrid combinations. The editors assess all eight of the models according to the standard discursive criteria as well as on other important issues. The purpose is to give potential users of these models a basis for matching method with purpose.

Evaluations of public participation are necessary in order to improve our understandings of citizen participation. Improved understandings should, consequently enable participation processes to be conducted more transparently - as opposed to a muddling through approach, which may produce satisfactory outcomes, but does not facilitate knowledge transfer, decision transparency, or repeatability. These three goods are needed if participation models are to be widely used in diverse problem areas with reliable results. We hope that the readers will find this book makes a contribution to fulfilling this need.

## References

Arnstein, S., "A Ladder of Citizen Participation," *Journal of the American Institute of Planners*, 35 (1969), 216-224.

Bachrach, P., *The Theory of Democratic Elitism: A Critique* (Little Brown: Boston 1967).

Beck, U. *Die Risikogesellschaft. Auf dem Weg in eine andere Moderne* (Suhrkamp: Frankfurt/Main, 1986).

Bernard, T. *The Consensus-Conflict Debate: Form and Content in Sociological Theories* (Columbia University Press: New York 1983).

Blalock, H.M. *Power and Conflict. Toward a General Theory* (Sage: Newbury Park 1989).

Brion, D., "An Essay on LULU, NIMBY, and the Problem of Distributive Justice," *Environmental Affairs*, 15 (1988), 437-503.

Brooks, H., "The Resolution of Technically Intensive Public Policy Disputes," Science, Technology, & Human Values, 9 (Winter 1984), 39-50.

Checkoway, B., "The Politics of Public Hearings," *Journal of Applied Behavioral Science,* 17, No. 4 (1981), 566-582.

Checkoway, B., and J. Van Til, "What Do We Know About Citizen Participation? A Selective Review of Research," in: S. Langton (ed.) *Citizen Participation in America* (Lexington Books: Lexington 1978).

Chess, C., A. Saville, M. Greeberg, and Tamuz, M.; *From Crisis to Credibility: Behind the Scenes of the Risk Communication Program of Sybron Chemicals, Inc.* Environmental Risk Communication Research Program (Rutgers University: New Brunswick, NJ, 1991).

Coser, L.A., *The Function of Social Conflict* (Free Press: Glencoe 1956).

Coser, L.A., *Continuities in the Study of Social Conflict* (Free Press: Glencoe 1967).

Creighton, J. L., "The Use of Values: Public Participation in the Planning Process," in: G. A. Daneke, M. W. Garcia, and J. Delli Priscoli (eds.), *Public Involvement and Social Impact Assessment* (Westview Press: Boulder 1983), pp. 143-160.

Cupps, D. S., "Emerging Problems of Citizen Participation," *Public Administraion Review*, 37 (1977), 478-487.

Cuthbertson, I. "Evaluating Public Participation: An Approach for Governmental Practitioners." In: Gregory Daneke, Margot Garcia, and Jerome Delli Priscoli (eds.) *Public Involvement and Social Impact Assessment* (Westview: Boulder 1983), pp. 101-111.

Daneke, G. A., M. W. Garcia, and J. Delli Priscoli (eds.), *Public Involvement and Social Impact Assessment* (Westview Press: Boulder 1983).

DeSario, J., and S. Langton, "Toward a Metapolicy for Social Planning," in: J. DeSario and S. Langton (eds.), *Citizen Participation in Public Decision Making* (Greenwood Press: Westport 1987), pp. 205-221.

Dietz, T., "Theory and Method in Social Impact Assessment," *Social Inquiry*, 57 (1987), 54-67.

Dunlap, R. E., "Public Opinion on the Environment in the Reagan Era," *Environment*, 29, No. 6 (July/August 1987), 7-11 and 32-37.

Ethridge, M. "Procedures for Citizen Involvement in Environmental Policy: An Assessment of Policy Effects," in: Jack DeSario and Stuart Langton (eds.) *Citizen Participation in Public Decision Making* (Westport CT: Greenwood Press 1987), pp. 115-132.

Etzioni, A., *The Spirit of Community. Rights, Responsibilities and the Communitarian Agenda* (Crown: New York 1993).

Evans, M., "Karl Marx and the Concept of Political Participation," in: G. Parry (ed.), *Participation in Politics* (Manchester University Press: Manchester 1972), pp. 127-150.

Fietkau, H.-J., H. Kessel, and W. Tischler, *Umwelt im Spiegel der öffentlichen Meinung* (Campus: Frankfurt/Main 1982).

Fiorino, D., "Environmental Risk and Democratic Process: A Critical Review," *Columbia Journal of Environmental Law*, 14, No. 2 (1989), 501-547.

Fiorino, D., "Citizen Participation and Environmental Risk: A Survey of Institutional Mechanisms," *Science, Technology, & Human Values*, 15, No.2 (1990), 226-243.

Fischer, G.W., M.G. Morgan, B. Fischhoff, I. Nair, and L.B. Lave, "What Risks are People Concerned About?" *Risk Analysis* **11**(2) (1991), 303-314.

Freudenburg, W., "The Promise and the Peril od Public Participation in Social Impact Assessment," in: G.A. Daneke, M.W. Garcia and J. Delli Priscoli (eds.), *Public Involvement and Social Impact Assessment* (Westview: Boulder 1983), pp. 227-234.

Gumpert, J.J., *Discourse Strategies* (Cambridge University Press: Cambridge 1982).

Habermas, J., ''Towards a Theory of Communicative Competence,'' *Inquiry,* 13 (1970), 363-372.

Habermas, J., *Legitimation Crisis* (Beacon Press: Boston 1975).

Habermas, J., *Theory of Communicative Action. Vol. 1: Reason and the Rationalization of Society* translated by Thomas McCarthy, (Beacon Press: Boston, 1984).

Habermas, J., *Theory of Communicative Action. Vol. 2: System and Lifeworld* translated by Thomas McCarthy, (Beacon Press: Boston, 1987).

Heiman, M., "Using Public Authorities to Site Hazardous Waste Management Facilities: Problems and Prospects," *Policy Studies Journal,* 18, No. 4 (Summer 1990a), 974-985.

Heiman, M., "From 'Not in My Backyard' to 'Not in Anybody's Backyard!' Grassroots Challenge to Hazardous Waste Facility Siting," *Journal of American Planning Association,* 56, No. 3 (1990b), 359-362.

Hohenemser, C., R. W. Kates, and P. Slovic, "The Nature of Technological Hazard," *Science,* 220 (1983), 378-384.

Kemp, R., "Planning, Political Hearings, and the Politics of Discourse," in: J. Forester (ed.), *Critical Theory and Public Life* (MIT Press: Cambridge 1985), pp. 177-201.

Kessel, H., and W. Tischler, *Umweltbewußtsein. Ökologische Wertvorstellungen in westlichen Industrienationen* (Science Center: Berlin 1984).

Kweit, M. G., and R. W. Kweit, "The Politics of Policy Analysis: The Role of Citizen Participation in Analytic Decision Making," in: J. DeSario and S. Langton (eds.), *Citizen Participation in Public Decision Making* (Greenwood Press: Westport 1987), pp. 19-38.

Laird, F., "Participatory Analysis: Democracy and Technological Decision Making," *Science, Technology, and Human Values,* 18, Vol 3 (1993), 341-361.

Lake, L., *Environmental Mediation: The Search for Consensus* (Westview Press: Boulder 1980).

Laksmanan, J., "An Empirical Argument for Nontechnical Public Members on Advisory Committees: FDA as a Model," *Risk Issues in Health and Safety*, 1 (1990), 61-74.

Langton, S.," Citizen Participation in America: Current Reflections On State Of the Art," in: S. Langton (ed.), *Citizen Participation in America* (Lexington Books: Lexington 1978), pp. 1-12.

Luhmann, N. *Ökologische Kommunikation* (Westdeutscher Verlag: Opladen 1986).

Marx, K., Zur Kritik der Politischen Ökonomie, in: Karl Marx und F. Engels, *Ausgewählte Werke* (Progress Verlag: Moskau 1971 [1859]), pp. 186-190.

Mazmanian, D., "Participatory Democracy in a Federal Agency," in: John Pierce and Harvey Doerksen (eds.), *Water Politics and Public Involvement* (Science Publishers: Ann Arbor 1976).

Mead, G. H., *Mind, Self and Society* (University of Chicago Press: Chicago 1934).

Milbrath, L. W., "Citizen Surveys as Citizen Participation," *Journal of Applied Behavioral Science,* 17, No. 4 (1981), 478-496.

Portney, K., "Allaying the NIMBY Syndrome," *Hazardous Waste* 1:411 (1984).

Renn, O., "Risk Communication: Towards a Rational Dialogue with the Public," *Journal of Hazardous Materials,* 29, No. 3 (1992), 465-519.

Rip, A., "Experts in Public Arenas," in: H. Otway and M. Peltu (eds.), *Regulating Industrial Risks* (Butterworth: London 1985), pp. 94-110.

Rosenbaum, N., "Citizen Participation and Democratic Theory," in: S. Langton (ed.), *Citizen Participation in America* (Lexington Books: Lexington 1978), pp. 43-54.

Rosenbaum, N., "Elitism and Citizen Participation," in: S. Langton (ed.), *Citizen Participation Perspectives. Proceedings of the National Conference on Citizen Participation,* Washington, D.C., Sept. 28 – Oct. 1, 1978. (Tufts University Lincoln Filene Center for Citizenship and Public Affairs: Medford 1979).

Rosener, J., "Citizen Participation: Can We Measure Its Effectiveness?" *Public Administration Review,* 38 (1978a), 457-463.

Rosener, J., "Matching Method to Purpose: The Challenge of Planning Citizen Participation Activities." in Stuart Langton (ed.), Citi*zen Participation in America* (Lexington Books: Lexington: 1978b).

Schattschneider, E.E. *The Semi-Sovereign People* (Holt, Reinhart, and Winston: New York 1960).

Schnaiberg, A., "The Role of Experts and Mediators in Channeling Distributional Conflicts," in: A. Schnaiberg, N. Watts and K. Zimmermann (eds.), *Distributional Conflicts in Environmental Resource Policy* (Aldershot: Gower 1986).

Stern, P. C., "Learning Through Conflict: A Realistic Strategy for Risk Communication," *Policy Sciences,* 24 (1991), 99-119.

Strauss, D., and P. Clark, "Computer-Assisted Negotiation: Bigger Problems Need Bigger Tools," *Environmental Professional,* 2, No. 1 (1980), 75-87.

Weber, M., *Basic Concepts in Sociology,* translated and introduced by H.P. Secher (Philosophical Library: New York 1962).

Webler, T., H. Rakel, and R.J.S. Ross, "A Critical Theoretic Look at Technical Risk Analysis," *Industrial Crisis Quarterly* 6:23-38 (1992).

*Chapter 2*

# *A Brief Primer on Participation: Philosophy and Practice*

*Thomas Webler*
*Ortwin Renn*

## Public Participation in Western Democracies

In countries with the Anglo-Saxon tradition, public participation is synonymous with participatory democracy. People there associate the very concept of democracy with the activity of participating in government decision making. Although many do little more than vote, the term embraces much more. In some European countries, namely Germany, the Netherlands, and the Scandinavian countries, public participation practices evolved largely from the labor movement, in particular with regard to co-determination of corporate management. It is only recently that participation has spread to governmental activities (Guild 1979). Participation in Germany, for example, is largely realized through the institution of political parties. To be involved, even at a local level, one must first join one of the local political parties. In recent years, however, Germany as well as the several other European countries has experienced the emergence of many social movements which demand more direct democracy, in particular in decisions related to environmental quality and technological choices (Brand 1987: 31).

Public participation has been a major topic of debate and controversy in America, Canada, and all European countries from the beginning of the 19th century. The early development of democracy in the aftermath of the American and French revolutions was characterized by the gradual integration of citizens in the political system. In the beginning, voting was exclusively reserved for the rich and powerful. The fast change of constitutions during the French revolution from a class-oriented voting system (according to wealth), to an egalitarian system served as a model for most European nations. This was later transformed into radical one-party rule. After the restoration in the early 19th century, democratic reforms were gradually introduced to most European countries. These included incorporation of bills of rights into national constitutions, a division of power, equal access to voting privileges, permission of parties and labor unions, and other related measures. Progress in broadening the base for political power among all members of society took time. For example, womens' suffrage did not come until early to mid 20th century and even as late as the 1980's some Swiss cantons still formally denied women equality.

In addition to the continuous movement of citizens to fight for equal rights in the political arena, social attention turned to the economic system in the middle of the 19th century. Labor movements, particularly in northern Europe, demanded more power to influence decisions of corporations (Leminsky 1977:284ff). The quest for "co-determination" has been a predominant demand of the German and Swedish unions from the 1920's until today. In contrast to American unions, which have been primarily interested in issues of wages and working conditions, most European unions placed much emphasis on taking part in management decisions. In exchange for sharing power, they were willing to play a moderating role in labor-management conflicts. In these day, large corporations in Germany are required to have equal number of shareholders and representatives of labor on their boards of directors.

Direct involvement of citizens in political decision making beyond the conventional modes of voting, party involvement, and economic co-determination has been less pronounced in most European countries until fairly recently. Switzerland, with its frequent referendums and citizen initiatives and its still strong reliance on local governing is the obvious exception. The more centralized European countries (mostly in the south) reserve most political power to the national governmental, but leave room for individual action under the patronage of the local representative in the central government. The countries with strong federal traditions rely heavily on procedural rules when allowing citizens to take part in political decisions. Direct democracy is embedded in a complex framework of bureaucratic rules and regulations that ensure predictability of outcomes even on the expense of losing time.

Although social movements and citizens initiatives have been advocating more direct influence in political decision making since the early 1920's, their protest was not effective until the ecological movements of the 1970's. The road to more direct participation was partially prepared by the student movements of the late 1960's, even though their goals and objectives were not widely shared among the larger population. This situation changed dramatically with the emergence of ecologically motivated citizen initiatives. These new organizations became major political forces in West Germany, Sweden, Denmark, the Netherlands, and several other European countries. Many citizens throughout Europe, in particular the young generation, sympathized with or participated in these movements (Watts 1987: 51). Central to the self-identity of these new movements was the demand for more direct participation, particularly on the local level. New methods of political articulation such as large demonstrations, boycotts, non-violent protests, and blocking construction of environmentally controversial facilities (for example nuclear power plants) became familiar. The political cultures of most European countries were at first shaken by the popular protest but they adapted slowly to the new demands (Akademie der Wissenschaften zu Berlin 1992: 252ff).

In the United States formal participation was adopted to protect individuals from the infringements of government. Public participation of a consultative nature was already in effect in the early 1930's in the area of agricultural policy making (Daneke 1983: 12) and by the late 1930's with the Tennessee Valley Authority (Selznick 1966). Ever since the beginnings of the country, the need for public scrutiny of government has been widely recognized, but it was the rise of regulatory rulemaking in the federal agencies that produced the pressure for legal institutionalization of some participation activities. Formal avenues for participation in administrative decision making were first realized with the Administrative Procedures Act of 1944. Although the Act did not clarify the objectives of participation, it was an important milestone because it officially mandated norms for agency conduct (Daneke 1983). During the 1950's and 1960's a significant change in the government's objectives for public participation came about. Instead of promoting participation simply for reasons of protection of individual interests, some in the federal government suggested that participation was *essential* to good governance. Public participation was first *encouraged* by the government with the Revised Housing Act of 1954, and later by the Economic Opportunity Act of 1964 which sought "maximum feasible participation" in community development (Moynihan 1969). These laws were passed, in part with the enthusiastic belief that direct public participation in governance would improve living and economic conditions. But as problems of race relations, housing, education, and poverty persisted, and the Great Society era passed, enthusiasm for direct participation waned. Today, participation is still valued, but once again as a watchdog activity rather than a means to liberate citizens from poverty, exploitation, and injustice. Two pieces of legislation most profoundly drive and shape public participation today. They are The National Environmental Policy Act of 1969 (NEPA) and the Freedom of Information Act (FOIA) of 1974. Presently, nearly every federal agency is required to include some form of participation in their regulatory decision making procedures (DeSario and Langton 1987).

Driven by popular demand, most American and European governments have recently enhanced possibilities for direct citizen input in governmental decision making or became more lenient in extending the existing opportunities for public involvement. Switzerland, for example, introduced the right of organizations to litigate against licensing decisions even if their members were not directly affected. Other countries such as Sweden, Great

Britain, the Netherlands, Belgium, and Germany gave ecological organizations more opportunities to influence policies and to be either part of the neo-corporatist club or to become catalysts for political change. In the United States, environmental organizations have an opportunity to influence environmental decisions through the Environmental Impact Review process, which includes public review and comment. It is the litigative avenues, however, that have provided interest groups with their most formidable means to shape policy.

Another recent step forward in public participation has been increased public access to information. In the United States the Toxics Release Inventory (TRI) has empowered citizen organizations and environmental interest groups with the evidence they need to participate effectively. They have used this resource to effectively pursue their agendas. Similarly, the European Community promulgated directives to all member states that extended the right of the public to be better informed about hazards and to become more involved in local hazard management (Renn 1989).

As a result public participation has blossomed in certain domains on both sides of the Atlantic. Europe has become a major testing ground for new methods of citizen participation in the area of national policy formation, primarily because the new social movements were much more of a shock for the established parties and power players. Swedish Study Circles, Dutch Consensus Groups, and German Planning Cells are each an example of participation clearly meant to infuse public sentiments into national policy making (Nelkin and Pollak 1977). This has not been the case in the United States, where political protest and social movements have been present for as long as the country has existed. Participatory innovations there have been much more along the lines of immediate problem solving (as opposed to policy making). Alternate methods of conflict resolution, negotiated rulemaking, mediation, bargaining, and citizen juries are more consistent with the American adversarial regulatory process. Nevertheless, a cross-fertilization of European and American experiences with new forms of direct participation is occurring. Some of the most innovative European proposals to give citizens more power in influencing political decisions have been imported to the United States (for example, citizen panels and study circles). Likewise, the American dispute resolution tools were introduced in Europe recently and have already changed the closed-door policy making arena. It is this international sharing of participation experiences and techniques that was one of the major incentives to compile this volume.

## Participation: Elitist versus Egalitarian Theoretical Perspectives

The debate about the role of public participation centers around the elitist and egalitarian interpretations of democracy (Wiesendahl 1987). The competing theories behind these two interpretations are usually labeled participatory (or direct) democratic theory and liberal democratic theory. Based on the seminal works of Pareto, Michels, and Schumpeter, the elitist or liberal view claims that political elites compete for votes similarly to how entrepreneurs compete for customers. The populace has the right to determine which of the competing elites are allowed to govern, but the substance of political decisions is made within elite circles. In contrast the egalitarian view of democracy is inspired by the normative claim that each citizen be able to co-determine political decisions that affect his or her livelihood. Participatory democratic theorists emphasize that democratic institutions must be responsive to the social psychological character of the citizenry (Pateman 1970: 29). In between the two extremes are numerous combinations ranging from a pluralistic view in which political elites need the participation of key interest groups to legitimize their decisions (Dahl 1989) to the neo-corporatist view in which the key social actors such as industry, the unions, and technical associations negotiate commonly acceptable solutions behind closed doors (O'Riordan and Wynne 1987).

### *Ethical-normative arguments for participation*

In participatory democratic theory, public involvement is morally and functionally integral to the emergence and the sustenance of the two central values of democracy: *popular sovereignty* and *political equality* (N. Rosenbaum 1978). It is generally accepted that democracy is the outcome of an agreement among people who establish a sovereignty based upon their popular and mutual consent. All power within the sovereignty is allocated through this agreement. The ability of democracy to function is measured by the soundness of the decisions reached in the light of the needs of the community and by the scope of public participation in reaching them (Bachrach 1967: 3).

Rousseau argued that a sovereignty is composed of all citizens and requires input via public involvement to determine legitimate objectives. In his reasoning, public participation is justified out of necessity: citizens must engage in political affairs to keep the state alive, for only through interaction can the general will emerge from the plurality of particular wills (Rousseau 1968[1762]).

Preservation of popular sovereignty is not the only concern of the participatory democratic theorists. To guarantee political equality, democracy must engender a population of capable and socially responsible citizens (Bachrach 1967). At the core of both Rousseau's and John Stuart Mill's theories of democracy (Mill 1873) is a belief that citizens' moral and intellectual growth occurs through their involvement in political affairs. People "learn democracy" by becoming engaged in its workings. The result of the learning experience is an awakening to the realization that the public and private interests are linked.

Recent literature on public participation strongly echoes the value for popular sovereignty. Critics concerned with the plight of local citizens against the encroaching power of the state and large organized interests remark that, first and foremost, public participation must fulfill the demands and needs of the citizenry. Sherry Arnstein's (1969) "ladder of participation" advanced the normative argument that participation is better when it gives citizens power to influence decisions about their own governance. Noting that popular sovereignty is associated with a greater degree of democratization, there are many defenders of public participation as an enhancement of democracy (Lynn 1990; Schrader-Frechette 1990; Fiorino 1989; DeSario and Langton 1987; N. Rosenbaum 1978; Checkoway and Van Til 1978).

Defense of the value of political equality is also articulated. One of the most common arguments is that public participation should "level the playing field" so that citizens have an equal chance to influence decisions (Van Valey and Peterson 1987; Rosener 1982; Godschalk and Stiftle 1981). Nelson Rosenbaum (1978) repeated the Millsian argument that participation should give citizens a voice in decisions that affect them so that they can protect their individual interests. Others have noted the importance of participation in the personal development of individuals (Daneke 1983); their social development (W. Rosenbaum 1979), and in defining the collective will (Dienel 1989). From a different perspective, theorists from a Marxist persuasion have made similar arguments that citizens should be involved in governing to defend class interests, develop skills needed for self government, and help develop class consciousness (Kasperson 1974; Evans 1972).

## *Functional arguments for participation*

In contrast to ethical-normative arguments, which begin with a normative belief and then argue for its appropriateness, functional-analytic arguments are based on analysis of observations about how society works. Of course, these arguments also contain a normative premise – that stable functioning of society is of utmost importance – but the emphasis is on explaining how it functions and verifying theories with empirical data. Three distinct strains of functional-analytic arguments for and against participation can be distinguished. They are grounded in: the theory of democratic elitism (Pareto 1934), the theory of rational choice (Hume 1957), and the theory of structural functionalism (Parsons 1951). The former deserves the most attention because it stands in stark contrast to the classical democratic theoretical arguments presented above and because it addresses public participation in democratic decision making more directly than the others.

The theory of democratic elitism argues that liberalism can only be secured when an enlightened elite is in power *and* when it is susceptible to pressure by a counter elite. This theory reacts against the popular values of "classical democracy" as logically impossible and challenges the competency of citizens to participate meaningfully (Pateman 1970). It emerged from empirical observations of society in the early 19th century by Michels, Pareto, Mosca, and Schumpeter and was supported by sociological studies of American post-war society (Verba and Nie 1972; Greenstein 1963; Almond and Verba 1961; Dahl 1961 and 1989; and Berelson et al. 1954).

Advocates of this approach today do not necessarily argue that participation is bad for democracy, but there is a deep concern that too much participation may disrupt the social system operation (Burke 1968). Other arguments against too much participation are that it is economically inefficient (N. Rosenbaum 1978; Cupps 1977), technically incompetent (Aron 1979), unfair to political interests (Cupps 1977; Reagan and Fédor-Thurmon 1987), and it incites conflict and further unrest (Huntington 1970).

Perhaps because participation has become so widespread in contemporary society, many others now see it as necessary for system stability. This view is reflected in the common claim that participation can enhance the responsiveness and legitimacy of public institutions (Thomas 1990; Rosener 1982; Cupps 1977; Mazmanian 1976); in the argument that it leads to more efficiency than many traditional forms of administrative decision making (Laksmanan 1990; Krimsky 1979; Aron 1979); and in the observations about its practical usefulness of participation in helping to implement decisions and in reducing or resolving conflict (Fiorino 1990; Davis 1986; Elliot 1984; Grima 1983).

In summary, functional analytic justifications for public participation are based on its contribution to the social system's need to maintain itself. Problems that arise due to structural strains from social development and environmental change must be dealt with by coordinating human actions. Public participation is undertaken to attain the ends of system maintenance without satisfying the extraneous needs and desires of the individual participants. The level of success in accomplishing these ends can be measured with criteria developed from studying the historical direction of social system development.

## The Conventional Method of Public Participation: Public Hearings

The public hearing model is not covered in this book, as it does not constitute a novel and innovative proposal for a political discourse. However, since hearings have become so ubiquitous in most countries, they deserve comment. Public hearings, and public inquiries in Canada and Great Britain, are perhaps the cheapest, easiest, most common, and least studied form of participation. They are also one of the oldest in modern history. The earliest record dates to London in the year 1403 (Guild 1979). Public hearings and inquiries are legalistic. They can only be convened by government under legal statute. They have been used to investigate controversies, to provide advice to decision makers, and, most recently, to provide a forum for public discussion and a channel for public opinion to reach decision makers.

To a citizen, the thought of attending a public hearing immediately conjures up negative images. Citizens often picture the public hearing process as disempowering. Typically, attendance is slight (Milbrath 1981). To regulatory officials, experts, and project sponsors, the public hearing hall is a battle zone (Creighton, 1983). Legal obligations must be met, hopefully without raising the hackles of the local populace. A well-attended meeting is bad news (Connor 1993). Perhaps the worst attribute of public hearings is that, under some circumstances, they are a recipe for disaster. The *EPA Community Relations Handbook* notes: "The public hearing can easily become an adversarial confrontation" (USEPA, 1983: 4-14). This is probably due to both the structure of discourse within the public hearing process and the timing of their use in the decision making process.

Public influence through hearings is a matter of dispute and a much understudied question. The literature on the public hearing process is very sketchy, most is descriptive or prescriptive, with little analysis. Hearings have been criticized for being unfair. Hadden (1989) notes several failures. First, they are usually held late in the process when public impact can be, at best, minimal. Second, only a very small proportion of the population has an opportunity the speak at the hearing. Third, the structure of the event reveals its implicit communicative bias as experts stand on a stage above the citizens. Fourth, hearings are held primarily to satisfy legal requirements, rather than really promote public input. Checkoway (1981) found that low rates of public participation were due to weak pre-hearing procedures, poor and overly technical presentations of information, a bias of outcomes favoring participants with economic stakes, and minimal evidence that participation affects policy. Godschalk and Stiftle (1981) emphasized the undemocratic nature of public hearings. They studied who participated in hearings about water planning in North Carolina in the 1970's. Participants were more educated, politically active, and better informed then non-participating community members. In addition, over half the participants represented economic interests. This supported earlier work by Checkoway and Van Til (1978) who found that public hearings were dominated by organized interests with economic stakes.

To their favor, public hearings do offer citizens an opportunity to get first hand information about government and project proponents' intentions. They also offer administrators an opportunity to hear about contending interpretations and interests directly from the people. As forums held to strict legal requirements, hearings provide an excellent setting for citizens to get hard answers from the project sponsor, developer, or regulatory agency. They also provide a stage for other kinds of political posturing. In recent experience, citizens commonly used the public hearing forum to embarrass their opponents, to issue symbolic messages, and to make a show of power. By law, public hearings must be widely announced and open to all, regardless of the stake people have in the matter. Citizens are generally aware of when hearings are being held in their community and are familiar with the setting. These things, along with a strict time schedule, make it relatively easy for people to attend and make hearings an especially good way for people who only want to learn about a problem to get some answers. At the same time, administrators often use the hearing forum to announce their intentions and provide justifications.

Public hearings may also provide an effective means for people to influence policy, but more research is needed to support this counter-intuitive conclusion. Judy Rosener (1982), in a study of 1,816 public hearings in California, reportedly found a link between the contents of public statements in the hearings and the regulatory decisions, suggesting that citizens can influence the decision through speaking up at public hearings. Although citizens have demonstrated time and time again their proficiency at reversing governmental decisions made via the public hearing process, the sense of powerlessness felt by citizens attending public hearings has been well documented (e.g. see Levine 1982). Victories by ad hoc citizens groups usually transpire outside of the public hearing process. Notwithstanding Rosener's results, the fact that citizens often must resort to unconventional means (such as protest or legal action) to win these reversals is a sign that the public hearing process does not satisfactorily integrate public concerns in decision making.

## Problems of Participation

### *The paradox of participation*

Widespread negative experiences with public hearings have greatly contributed to the perception among citizens that business and government do not take their concerns seriously, but are more interested in taking the path of least resistance to achieve their desired ends. Citizens believe that governmental officials regard them as unwanted intruders in the decision making process.

Be that as it may, there are clearly supporters of public participation among the ranks of government. Not only have several laws been passed to institutionalize participation, but governmental officials and some businesses occasionally seek to enhance the extent and intensity of participation. (Although we note that the progressive cause is the mouse, the former the elephant.) Frustration on the part of the publics and piece wise concessions for more avenues of participation on the part of the government and business has led to the current condition that Stuart Langton termed "the participation paradox" (Langton 1978). He attributed it to three causes:

- a decline in mediating institutions,
- a rise of bureaucratic elite, and
- impact of the mass media.

To these should be added a fourth explanation:

- a rise in the awareness of and concern for environmental, social, economic, and physical impacts affiliated with regulatory decision-problems.[1]

There is really nothing paradoxical about this. In a nutshell, it is not the quantity of the public participation avenues that influences an individual's choice to participate, but the perception that the government sincerely wants them to participate in a meaningful way. We identify four major problems that participation faces today:

- *Participation as therapy.* Citizens feel cheated if they are asked to participate only to find out that the decision has already been made.
- *Insufficient knowledge.* Environmental managers are often uninformed about the citizens' concerns and neglect the experiences and preferences of the publics in when setting policy or making decisions.
- *Distrust.* Most people have limited trust in public institutions and limited confidence in the decision making process. As a consequence they demand to oversee the process and define objectives.
- *Conflicting rationalities.* The rationale managers use for making trade-offs between different cost and benefit dimensions is rarely compelling for

---

1 Before 1970, public participation occurred mainly around social issues such as housing, welfare, civil rights, and education. After 1970, public participation became increasingly prevalent in environmental management and technological risk-related decision making (DeSario and Langton, 1987). This conclusion was based on a study of the evolution of public participation from 1946 to 1981 (Langton, 1981).

citizens. This has led to accusations by citizens that environmental decision making is technocratic, while experts lament about public "irrationality."

## *Conflicting rationalities*

Environmental decision making difficulties are aggravated by differences between the social rationality of lay people and the bounded rationality of experts (Perrow 1984; Evers and Nowotny 1987). In many instances, decisions evoke conflicts between cost minimization and equity. To take an example from the field of risk, research shows that public perception of probabilities and risks differs considerably from professional analysis (Slovic 1987). For example, exporting waste to a community for disposal may minimize the overall public risk, but will likely be unacceptable to the host community. Whereas experts usually give equal weight to probabilities and magnitudes of a given risk, the intuitive risk perception of lay people reflects more concern for inequitable solutions. Both perspectives are rational. It should be a matter for the public policy making process to determine which is more appropriate, but usually decision makers are left to face the intrinsic problems of coping with these competing rationalities. Land-use practices, nuclear energy, genetic engineering, endangered species, and wilderness preservation are some of the areas which present severe challenges for participatory decision making. The aim should not to discount one rationale in favor or another, but to find ways to benefit from both.

## *The NIMBY syndrome*

The NIMBY (Not In My Back Yard!) label refers to fervent local citizen opposition to siting proposals or land-use activities with potential adverse impacts (Rosa 1988). Broad public support for environmental values (Dunlap 1987); dread of unknown and uncontrollable risks to personal health (Slovic, 1987); and the sudden increase in publicly available information (Hadden, 1989) have been posited as explanations for this reaction. NIMBY is a very formidable reaction. Today in the United States and Canada (and to an increasing degree in Europe as well), citizens have what amounts to a de facto veto over siting issues, despite legal provisions allowing legislative overrides of local opposition (Morell and Magorian, 1982; New York State Legislative Commission on Toxic Substances and Hazardous Waste, 1987).

Some argue that uncooperative citizens are selfish or irrational, and that society cannot afford to make everyone happy with every decision (Mazmanian and Morell 1990; Glaberson, 1988). Others find the local reaction rational and symptomatic of a crisis in modern democracy (Heiman 1990, Burns and Überhorst 1988). Regulatory agencies of the latter persuasion are experimenting with giving citizens what they demand: new opportunities for participation in decision making.

## *Complications of Participation*

Inviting the public to be part of the decision making process from the start improves the likelihood that the resulting decision will be considered appropriate. Unfortunately, this is not a silver bullet to end all conflict, it has its problems. First, publics expect and demand sufficient, if not absolute, protection of the environment and of human health and interests (Slovic et al. 1981). If an accident occurs or if negative side effects of pollution become visible, the legally responsible decision making body will be blamed for insufficient regulation or lack of proper implementation regardless of whether the public was involved in the decision making process. Second, public involvement may worsen environmental damage or impacts on interests. This can happen if the emotion or fashion of the moment is permitted to dictate a decision that flies in the face of firm contrary scientific evidence or professional judgment. Third, early public involvement may compromise the objective of efficient regulation. Actively participating citizens concerned with impacts may downplay costs and recommend expensive programs which achieve relatively small gains. Fourth, because the public consists of many groups with different value structures and preferences, consent among all the affected groups is difficult to achieve. Although everyone may agree on the overall goal of safety and environmental quality, the question of acceptable or tolerable impacts and the distribution of impacts may evoke substantial disagreement and conflict. Without a systematic procedure to reach consensus on values and preferences, the public's position often appears as unclear. Finally, participation may expand the scope of the conflict. If a decision making body avails itself to listening to the voices of the publics, some parties may take advantage of the opportunity to pursue a broader political or social objective.

Hence, regulatory agencies are confronted with a serious dilemma. The determination of acceptable or tolerable environmental impacts relies on subjective judgment and social values (Plough and Krimsky 1987). In pluralistic democratic societies, it is difficult for agencies to justify their selection and relative weighting of values vis-a-vis competing and conflicting values of various interest groups. Since values cannot be justified by referring to scientific measurements or legal requirements (although this is a popular strategy), the process of regulating depends heavily on public consent. A dynamic relationship based on reciprocity of expectations and promises exists between the government, the publics, the interest groups, and the project proponents. Administrative agencies do not merely respond to or measure public demands, they also play an active role in *establishing* norms of expected performance. It is precisely because there is no established mechanism for public norm formulation and consent that citizen participation is so contentious.

## References

Akademie der Wissenschaften zu Berlin, *Umweltstandards* (de Gruyter: Berlin 1992).

Almond, G. A., and S. Verba, *The Civic Culture* (Little Brown: Boston 1961).

Arnstein, S., "A Ladder of Citizen Participation," *Journal of the American Institute of Planners*, 35 (1969), 216-224.

Aron, J. B., "Citizen Participation at Government Expense," *Public Administration Review*, 39 (1979), 477-485.

Bachrach, P., *The Theory of Democratic Elitism: A Critique* ( Little Brown: Boston 1967).

Berelson, B., P. Lazarsfeld, and W. McPhee, *Voting* (University of Chicago Press: Chicago 1954).

Brand, K.-W., "Kontinuität und Diskontinuität in den neuen sozialen Bewegungen," in: R. Roth und D. Rucht (eds.), *Neue Soziale Bewegungen in der Bundesrepublik Deutschland* (Campus: Frankfurt am Main 1987), pp. 30-44.

Burns, T. R., and Überhorst, R., *Creative Democracy: Systematic Conflict Resolution and Policymaking in a World of High Science and Technology* (Praeger: New York 1988).

Burke, E., "Citizen Participation Strategies," *Journal of the American Institute of Planners*, 35 (1968), 287-294.

Checkoway, B., "The Politics of Public Hearings," *Journal of Applied Behavioral Science*, 17, No. 4 (1981), 566-582.

Checkoway, B., and J. Van Til, "What Do We Know About Citizen Participation? A Selective Review of Research," in: S. Langton (ed.) *Citizen Participation in America* (Lexington Books: Lexington 1978).

Connor, D., "A Generic Design for Public Involvement Programs," *Constructive Citizen Participation*, 21 (June & September 1993), 1-2.

Creighton, J. L., "The Use of Values: Public Participation in the Planning Process," in: G. A. Daneke, M. W. Garcia and J. Delli Priscoli (eds.), *Public Involvement and Social Impact Assessment* (Westview Press: Boulder 1983), pp. 143-160.

Cupps, D. S., "Emerging Problems of Citizen Participation," *Public Administration Review*, 37 (1977), 478-487.

Dahl, R. A., *Democracy and its Critics* (Yale University Press: New Haven 1989).

Dahl, R. A., *Who Governs? Democracy and Power in an American City* (Yale University Press: New Haven 1961).

Daneke, G. A., M. W. Garcia, and J. Delli Priscoli (eds.), *Public Involvement and Social Impact Assessment* (Westview Press: Boulder 1983).

Davis, C., "Public Involvement in Hazardous Waste Siting Decisions," *Polity,* 19. No. 2 (1986), 296-304.

DeSario, J., and S. Langton, "Toward a Metapolicy for Social Planning," in: J. DeSario and S. Langton (eds.), *Citizen Participation in Public Decision Making* (Greenwood Press: Westport 1987), pp. 205-221.

Dienel, P. C., "Contributing to Social Decision Methodology: Citizen Reports on Technological Projects," in: C. Vlek and G. Cvetkovich (eds.), *Social Decision Methodology for Technological Projects* (Kluwer Academic: Dordrecht 1989), pp. 133-151.

Dunlap, R. E., "Public Opinion on the Environment in the Reagan Era," *Environment,* 29, No. 6 (July/August 1987), 7-11 and 32-37.

Elliot, M., "Improving Community Acceptance of Hazardous Waste Facilities Through Alternative Systems for Mitigating and Managing Risk," *Hazardous Waste*, 1 (1984), 397-410.

Evers, A., and H. Nowotny, *Über den Umgang mit Unsicherheit. Die Entdeckung der Gestaltbarkeit von Gesellschaft* (Suhrkamp: Frankfurt/Main 1987).

Evans, M., "Karl Marx and the Concept of Political Participation," in: G. Parry (ed.), *Participation in Politics* (Manchester University Press: Manchester 1972), pp. 127-150.

Fiorino, D., "Citizen Participation and Environmental Risk: A Survey of Institutional Mechanisms," *Science, Technology, & Human Values*, 15, No.2 (1990), 226-243.

Fiorino, D., "Environmental Risk and Democratic Process: A Critical Review," *Columbia Journal of Environmental Law,* 14, No. 2 (1989), 501-547.

Glaberson, William, "Coping in the Age of 'NIMBY'," *The New York Times* June 19, Section 3:1 (1988).

Godschalk, D. R., and B. Stiftle, "Making Waves: Public Participation in State Water Planning," *Journal of Applied Behavioral Analysis,* 17 (1981), 597-614.

Greenstein, F. I., *The American Party System and the American People* (Prentice-Hall: Englewood Cliffs 1963).

Grima, A. P., "Analyzing Public Inputs to Environmental Planning" in: G. A. Daneke, M. W. Garcia and J. Delli Priscoli (eds.), *Public Involvement and Social Impact Assessment* (Westview Press: Boulder 1983), pp. 111-119.

Guild, Nicholas K. *Technology on Trial*: Public Participation in Decision Making Related to Science and Technology (Paris: OECD, 1979).

Hadden, S., *A Citizen's Right-to-Know: Risk Communication and Public Policy* (Westview Press: Boulder 1989).

Hadden, S., "Public Perception of Hazardous Waste," *Risk Analysis,* 11, No. 1 (1991), 47-57.

Heiman, M., "From 'Not in My Backyard' to 'Not in Anybody's Backyard!' Grassroots Challenge to Hazardous Waste Facility Siting," *Journal of American Planning Association,* 56, No. 3 (1990), 359-362.

Hume, D., *An Inquiry Concerning the Principles of Morals,* edited and introduced by C.W. Hendel (Bobbs-Merill: Indianapolis 1957 [1751]).

Huntington, S., "The Democratic Distemper," *The Public Interest,* 41 (1970), 9-38.

Kasperson, R. E., "Participating in Public Affairs: Theories and Issues," in: R. Kasperson and M. Breitbart (eds.), *Participation, Decentralization, and Advocacy Planning*. Resource Paper 25 (Association of American Geographers: Washington, DC., 1974), pp. 1-16.

Krimsky, S., "Citizen Participation in Scientific and Technological Decision Making," in: Stuart Langton (ed.), *Citizen Participation Perspectives. Proceedings of the National Conference on Citizen Participation*, Washington, DC., Sept. 28 – Oct. 1, 1978. (Tufts University Lincoln Filene Center for Citizenship and Public Affairs: Medford 1979).

Laksmanan, J., "An Empirical Argument for Nontechnical Public Members on Advisory Committees: FDA as a Model," *Risk Issues in Health and Safety*, 1 (1990), 61-74.

Langton, S.," Citizen Participation in America: Current Reflections On State Of the Art," in: S. Langton (ed.), *Citizen Participation in America* (Lexington Books: Lexington 1978), pp. 1-12.

Langton, S., "Evolution of a Federal Citizen Involvement Policy," *Policy Studies Review,* 1 (1981), 369-378.

Leminsky, G., "Bürgerbeteiligung, Mitbestimmung und Gewerkschaften – Eine vorläufige Übersicht," in: Hans Matthöfer (ed.), *Bürgerbeteiligung und Bürgerinitiativen. Argumente in der Energiediskussion.* Vol. 3 (Neckar Verlag: Villingen 1977), pp. 282-293.

Levine, Adeline Gordon *Love Canal: Science, Politics, and People* (Lexington, MA: Lexington Books, 1982).

Lynn, F. M., "Public Participation in Risk Management Decisions: The Right to Define, The Right to Know, and The Right to Act," *Risk Issues in Health and Safety*, 1 (1990), 95-101.

Mazmanian, D., "Participatory Democracy in a Federal Agency," in: John Pierce and Harvey Doerksen (eds.), *Water Politics and Public Involvement* (Science Publishers: Ann Arbor 1976).

Mazmanian, Daniel and David Morell, "The NIMBY Syndrome: Facility Siting and the Failure of Democratic Discourse," in: Norman J. Vig and Micheal E. Kraft, (eds.), *Environmental Policy in the 1990's. Toward a New Agenda* (CQ Press: Washington DC, 1990).

Milbrath, L. W., "Citizen Surveys as Citizen Participation," *Journal of Applied Behavioral Science,* 17, No. 4 (1981), 478-496.

Mill, J. S., *Considerations on Representative Government* (New York: Henry Holt and Company, 1873).

Morell, D. and Magorian, C. *Siting Hazardous Waste Facilities: Local Opposition and the Myth of Preemption* (Ballinger: Cambridge 1982).

Moynihan, Daniel *Maximum Feasible Misunderstanding* (Maximillian: New York, 1969).

Nelkin, D., and M. Pollak, "The Politics of Participation in the Nuclear Debate in Sweden, the Netherlands, and Austria," *Public Policy*, 25, No. 3, (1977), 333-357.

New York State Legislative Committee on Toxic Substances and Hazardous Wastes, *Hazardous Waste Facility Siting: A National Survey* (New York Legislative Committee: Albany, June 1987).

O'Riordan, T., and B. Wynne, "Regulating Environmental Risk: A Comparative Perspective," in: Paul R. Kleindorfer and Howard C. Kunreuther (Eds.), *Insuring and Managing Hazardous Risks: From Seveso to Bhopal and Beyond.* (Springer, Berlin 1987), pp. 389-410.

Pareto, V., *The Mind and Society: A Teatise on General Sociology.* Translated by A. Livingstone (Harcourt, Brace: New York 1934), Volumes 1-4.

Parson, T. E., *The Social System* (Free Press: Glencoe 1951).

Pateman, Carol *Participation and Democratic Theory* (Cambridge: Cambridge University Press, 1970).

Perrow, C., *Normal Accidents: Living with High Risk Technologies* (Basic Books: New York, 1984).

Plough, A., and S. Krimsky, "The Emergence of Risk Communication Studies: Social and Political Context," *Science, Technology, and Human Values,* 12 (1987), 78-85.

Reagan, M., and V. Fedor-Thurman, "Public Participation: Reflections on the California Energy Policy Experience," in: J. DeSario and S. Langton (eds.), *Citizen Participation in Public Decision Making* (Greenwood; Westport 1987), pp. 89-113.

Renn, O., "Risk Communication in the Community: European Lessons from the Seveso Directive," *Journal of the Air Pollution Control Association,* 40, No. 9 (1989), 1301-1308.

Rosa, E. A., "NAMBY PAMBY and NIMBY PIMBY: Public Issues in the Siting of Hazardous Waste Facilities," *Forum for Applied Research and Public Policy*, 3 (1988), 114-123.

Rosenbaum, N., "Citizen Participation and Democratic Theory," in: S. Langton (ed.), *Citizen Participation in America* (Lexington Books: Lexington 1978), pp. 43-54.

Rosenbaum, W., "Elitism and Citizen Participation," in: S. Langton (ed.), *Citizen Participation Perspectives. Proceedings of the National Conference on Citizen Participation*, Washington, DC., Sept. 28 – Oct. 1, 1978. (Tufts University Lincoln Filene Center for Citizenship and Public Affairs: Medford 1979).

Rosener, J., "Making Bureaucracy Responsive: A Study of the Impacts of Citizen Participation and Staff Recommendations on Regulatory Decision Making," *Public Administration Review,* 42 (1982), 339-345.

Rousseau, J. J., *The Social Contract.* Translated by Maurice Cranston, (Penguin: Harmondsworth UK, 1968 [1762]).

Selznick, Philip *T.V.A. and the Grass Roots* (New York: Harper and Row, 1966).

Shrader-Frechette, K., "Scientific Method, Anti-Foundationalism, and Public Policy," *Risk Issues in Health and Safety,* 1 (1990), 23-41.

Slovic, P., "Perception of Risk," *Science,* 236, No. 4799 (1987), 280-285.

Slovic, P., B. Fischhoff, and S. Lichtenstein, "Perceived Risk: Psychological Factors and Social Implications," in: *Proceedings of the Royal Society*, A376 (Royal Society: London 1981), pp. 17-34.

Thomas, J. C., "Public Involvement in Public Management: Adapting and Testing a Borrowed Theory," *Public Administration Review,* 50 (1990), 435-445.

United States Environmental Protection Agency, *Community Relations in Superfund: A Handbook* (US EPA: Washington, DC, 1983).

Van Valey, T. L., and J. C. Petersen, "Public Service Science Centers: The Michigan Experience," in: J. DeSario and S. Langton (eds.), *Citizen Participation in Public Decision Making* (Greenwood Press: Westport 1987), pp. 39-63.

Verba, S., and N. H. Nie, *Participation in America. Political Democracy and Social Equality* (Harper and Row: New York 1972).

Watts, N. S. J., "Mobilisierungspotential und gesellschaftliche Bedeutung der neuen sozialen Bewegungen," in: R. Roth and D. Rucht (Eds.), *Neue Soziale Bewegungen in der Bundesrepublik Deutschland* (Campus, Frankfurt/Main 1987), pp. 47-67.

Wiesendahl, E., "Neue soziale Bewegungen und moderne Demokratietheorie," in: R. Roth und D. Rucht (eds.), *Neue Soziale Bewegungen in der Bundesrepublik Deutschland* (Campus, Franfurt/Main 1987), pp. 364-384.

*Chapter 3*

# *"Right" Discourse in Citizen Participation: An Evaluative Yardstick*[*]

*Thomas Webler*

> Right thinking depends as much on right willing as right willing on right thinking.
>
> Max Horkheimer (1972:162).

> We should not expect a generally valid answer when we ask what is good for me, or good for us, or good for them; we must rather ask: what is equally good for all?
>
> Jürgen Habermas (1992: 248).

## Prelude

The official from the State Department of Environmental Protection locked the door to the County Agricultural Extension Service building and the five of us walked together out into the parking lot. After almost three and one-half hours of heated discussions with citizens in the meeting room, the fresh, cool air of the night was reinvigorating. I was thinking to myself, I'm glad that's over with! when the DEP official aggressively said, "Well that finishes any chance of our getting a landfill sited in this part of the state for the near future. The participation project was a complete failure. NIMBY strikes again." Between the lines I thought I heard him say, "Thanks for nothing."

As one of the research team organizing the public participation program, I naturally felt his resentment. After all, despite the fact that he had argued for giving citizens more say in the decisions, and despite the fact that he initiated this research into a novel approach to participation, he had also made it very clear to us how he wanted the process to end. In the same vein we were,

* My sincere thanks go to Ingrid Shockey, Neil Roughley, Ortwin Renn, Horst Rakel, Hans Kastenholz, Tom Dietz, and Bob Brulle for their encouragement, knowledge, and advice. I am indebted to Bob Brulle, George Cvetkovich, Tom Dietz, and Neil Roughley for their careful review and comments on the manscript. Errors and opinions remain my responsibility.

naturally, optimistic for an outcome that would satisfy everyone. Perhaps in the push to sell our model we led him to believe – or even expect – that the optimistic picture was all too easy to achieve. He has a certain right to be upset, I thought, because tomorrow he has to go back to the office and live with this result on a day-to-day basis. We, on the other hand, are driving home to another part of the country, probably never to return. Our biggest worry is what to write in our final report and whether or not we will ever get funding for another similar project.

In the van on the drive home we analyzed the event and looked for mistakes. We looked for people to blame. On the outset, there was plenty of blame to go around. If only the moderator had controlled the discussion a bit differently. If only that member of the research team did not argue with the expert. If only the state official had not given such a one-sided and lengthy presentation. If only we had pushed harder to have citizens from outside communities present. If only we had known about that other landfill siting debacle from last year. If only the citizens had been a little more patient, a little more open, and a little less hostile!

At one point during the long late-night ride I began to wonder, how are we justifying laying blame on people? Nothing that happened during the meeting was so unusual. What criteria are we using when we decide that somebody's actions were wrong? That not withstanding, even before there can be blame, we have to identify a mistake. Blame is laid on the person responsible for the error. But what was the mistake here? Why were we implicitly agreeing with the official that the process was a complete failure because it did not produce a sited landfill? What other bases could we use to evaluate public participation?

Should participation processes be evaluated only according to outcome? If so, what outcome – *whose* preferred outcome – should we use to base an evaluation? I thought back on the town, where the citizens were celebrating tonight. Surely everyone involved had their own expectations for the process. The research team wanted to reinvigorate democracy with a new instrument for collective decision making. It also wanted to provide its employer with a satisfactory outcome and better its chances for a research career in this area of work. The State wanted to site a landfill in a low-conflict and low-cost manner and also gain the trust and confidence of the citizenry. Residents of the targeted community, on the other hand, wanted to minimize their share of negative impacts (stigma, risk, annoyances, costs) and assert themselves against the traditional decision making authority of the State and experts. As a consequence, "preferred outcome" depends on the interests of each group involved in the event. To the citizens, the process was a success – they used it to keep the proposed landfill out (of course, they may end up paying more for waste disposal as a consequence, but this may not be as great a concern for them).

On the other hand one could argue that all participants, in addition to having their own interests, have shared interests and these are often more important (to each and everyone) than individual interests. For example, everyone has a concern for the negative health and environmental impacts of waste disposal. Public participation could be evaluated according to how well it serves these collective interests. In fact, this is often the argument of public officials who claim (undoubtedly in true honesty and with good reason) to be working for the good of all (i.e. the collective interest). Beyond interests, the Rouseauean idea of generalized will is relevant here. The difficulty associated with this approach is, of course, knowing with certainty the generalized will.[1]

Clearly there exists, among any group of people forming a community a *predisposition* for collective will, but it is not obvious that, beyond generalities like clean water, shelter and food, a detailed prescription for the good and just life exists *a priori* to self reflection, discourse, and understanding. It may help to distinguish between aspects of the collective will that can be determined objectively and are knowable through scientific inquiry (e.g. the need for clean drinking water), and other aspects that are socially constructed and knowable through intersubjective understanding (e.g. the interest in preserving a rural character of a region). However, it seems unpromising to attempt to evaluate participation against how well it serves the collective will given the fact that the collective will is changing and the participation itself shapes and modifies it. We would be hard pressed to characterize the collective will of this community based on what happened tonight.

Thinking back on the victorious citizens, perhaps subjective satisfaction should be used as a criterion for evaluating participation? After all, when people are happy with the process, the process has a good chance of being used for future problems. It stands to reason that a procedure people elect to use over and over again must have something going for it. Perhaps we could measure the happiness of everyone who participated, average the result, and use this as an overall evaluation of the process?

Even if we could get over the problems with quantifying such a feeling, however, and even if we could justify a weighting scheme,[2] what use would such a quantitative evaluation be? Does it make sense to compare two different participation models based on the relative satisfaction of the

---

1 The concept of generalized will should not be interpreted as the equivalent to Durkheim's collective conscious. It is not something that exists a priori, but is created through public discourse. Neither does the generalized will express the full range of consensus or dissent among the participants, but merely a limited consensus centered around the important aspects of the pressing decision.

2 Should everyone count equally? What about those with more at stake or more power? They may be more impacted by the process results, or they may have the power to stop the process next time.

participants of each model's application? This scheme seems susceptible to its own relativism. What basis is there for comparing satisfaction of models unless they are used in precisely the same context? For example, one problem context may include only choosing between bad and worse, another context may offer the choice between good or better. Overall satisfaction in the first case may be quantitatively high, but relatively low compared to the second.

Together with my colleagues in the van, I was reaching the conclusion that it is not enough to evaluate a single public participation exercise objectively, according to outcome (because of problems of knowing the generalized will), nor subjectively, according to the participants' assessments (because of the problem of aggregating individual preferences). To be able to make a judgment about how well a certain technique for participation works, we need some ideal against which to compare its performance – we need a procedural normative model of public participation.

## Looking for Agreement on Values

### *Fairness and competence as metacriteria*

A normative model of public participation is one that expresses and defends a vision about what public participation should accomplish and in what manner. This boils down to an argument that certain values should predominate. Not surprisingly, one finds in the literature little concise agreement, but rather scattered thoughts about what public participation should be (Rosener 1978). One way of making sense out of this confusion is to distinguish between ethical-normative and functional-analytic arguments (summarized in Chapter 2).

Ethical-normative and functional-analytic arguments for public participation contribute different aspects to a normative case for public participation. Taken together, they suggest that public participation should manifest the general goals of *fairness* and *competence.* Participatory democratic theorists emphasize that participation is consistent with political equality and popular sovereignty, because it provides every individual an equal and fair chance to defend his or her personal interests and values and to contribute to the definition of the collective will. Fairness is key to producing a forum where equality and popular sovereignty can emerge and personal competence can develop. When participation is fair, everyone takes part on an equal footing. As I will explain below, this means, not only are people provided equal opportunities to determine the agenda, the rules for discourse, and to speak and raise questions, but also equal access to knowledge and interpretations.

Functional-analytic arguments emphasize that participation is normatively right insofar as it contributes to sustaining social systems. This implies a need for competence. A systems-level definition of competence, however, is too abstract to enable a case-level evaluation of participation. For instance, in the landfill siting case mentioned above, it is more functional to have this landfill sited or not? Under a conservative interpretation of functionalism, one would argue for the siting, but considering other work on the functions of social conflict, one can imagine a situation wherein the landfill not being sited could result in a larger payoff to society.

Bringing this functionalist concern for competent social operation down to the level of social interaction can be aided by appealing to participatory democratic theory. Political equality and popular sovereignty also make an argument for competence. Namely, one should be capable of protecting his or her own interests while also being capable of contributing to the definition of the collective will. This refers to a competence in the achievement of personal development and in social interaction among individuals engaging in political life. Competence, in this sense, relates to psychological heuristics, listening and communication skills, self reflection, and consensus building.

What both arguments share is a recognition that competent shared social constructions of reality are essential. Functionality of the system order and protection and exercise of individual liberties both require competent understandings of the physical, social, and personal states of affairs. When the purpose of public participation is to produce a collective decision, competent understandings about terms, concepts, definitions, and language use; the objectified world of outer nature (nature and society); the social-cultural world of norms and values; and the subjective worlds of individuals are all essential. This is accomplished through the use of established procedures. How well the people in the discourse manage to apply these procedures is the measure of competence.

## *Public participation is social interaction*

Participatory and elitist democratic theories provide justification for very general normative principles of participation. Taken alone, these can and have been used to evaluate participation institutions or models (Fiorino 1990; Laird 1993). But because these theories explain political order at the macro level, they miss crucial elements at the micro level of participation.[3] Evaluating a

---

3 I believe Laird's claim to the contrary is off mark (1993:350). Consider, for example, just one of his criteria: improved understanding. Theories of democracy do not give any clue about which kinds of public participation produce better understandings. This requires a much more psychological approach, building on theories of small group interaction, among others. To be

participation exercise requires a much deeper understanding of how fairness and competence are realized in practice. The following observation is key to realizing this.

By and large, participation *is interaction among individuals through the medium of language*. Thus, it makes sense to ground a normative model of public participation in a theory of how language is used – also known as pragmatics. Toward this end the recent work of Jürgen Habermas appears promising (Habermas 1984, 1987). Habermas not only assembles the analytical tools and concepts necessary to the study of language use, but also sets the normative foundations for its evaluation. While there is much to be discussed about how effective he is at establishing and defending his normative contentions, his basic arguments about universal pragmatics (communicative competence) and the ideal speech situation are fundamentally consistent with the ethical-normative and functional-analytic arguments that public participation should be fair and competent.

## Approaching the Work of Jürgen Habermas[4]

### *Roots of cooperation*

Jürgen Habermas is a German philosopher and sociologist who comes out of the tradition of critical theory. In keeping with his roots in German philosophy, Habermas has sought to explain the changes in modernizing western societies as a process of rationalization – or the improvement and invention of techniques and skills we use to make sense out of the world. Habermas thinks of rationalization as occurring differently in each of three independent domains: science/technology, law/morality, and art/art criticism. Building on early critical theorists' apprehensions about the predominance of scientific rationality over all other forms of rationality, Habermas has diagnosed the current problem of modern societies as one-sided and uneven rationalization. For this problem he has prescribed a cure: we need to develop our non-scientific forms of thinking while also inventing ways to capitalize on all three kinds of knowing in cooperative decision making procedures. Using all three rationalities in a cooperative way is also a form of rationality, in fact

---

honest, this also points to a weakness of relying on Habermas, for psychological theories of small group interaction play little role in this work. Whether or not Habermas describes speech accurately, is a topic of much debate. Still, coming down from the level of political theory to Habermas's theory of social interaction is a movement in the correct direction.

4 Other reviews of this dimension of Habermas's recent work are given by (beginning with the briefest): Brulle 1992; Forester 1985; Kemp 1985; Pusey 1987; Bernstein 1985; McCarthy 1984; Braaten 1991; and White 1989.

it is the most important kind, a kind he calls *communicative rationality* (Habermas 1984:95).

Habermas believes that the roots of cooperation are found in the very structure of language. Built into language is the assumption that the speaker can defend his or her statements if needed. This amounts to an implicit commitment between any two persons talking with each other to cooperate. To prove his point Habermas points out that, without this assumption, all forms of language would fail: not only would it be difficult to reach agreement, but jokes would not be funny, lies would not work. Obviously, disagreements and misunderstandings do arise. In everyday speech we have certain conventions for coping with these problems. In situations where these conventions fail, as in many settings of environmental decision making, an institution for communication must be established.

In what follows I approach Habermas's theory of communicative action along a path that begins with the basic terms and concepts he uses to understand language. Then I trace how he answers his concern for renewing democracy with a prescription for discourse that aspires to an *unachievable* perfection called "ideal speech situation." The path then detours to a short discussion about using the ideal speech situation as a normative basis for evaluating public participation, before returning to introduce the core idea of communicative action. After this review, I return to the problem of finding criteria for competent and fair participation. The core of my argument is that the concept of the ideal speech situation, once it is refined to ensure competence, is suitable to use as a basis for a normative model of public participation.[5]

This work builds on that of others who have looked to Habermas's theories as a basis for including norms in policy making and planning. Thomas Dietz has worked to develop practical methods for small group interaction and social

---

5 Recently, Habermas has moved away from defining such conditions, because they tend to convey the idea that *all* discourse should attain this ideal, which is not possible or necessarily desirable (Habermas 1992:260). However, he has not retreated from the original argument that the ideal speech situation is a description of the idealizing presuppositions that all participants must hold before they can participate in communicative action. On this note, John Forester has asserted that the ideal speech situation "plays almost no significant role" in Habermas's work (Forester 1993:x). While this statement opens up intriguing possibilities, it is plainly incorrect (see: Habermas 1982:255-256). Forester must be referring to the conditions of the ideal speech situation that Habermas specified and later regretted, because the *concept* of the ideal speech situation, as opposed to the particular conditions, is essential to understanding the normative force of action. My evaluative framework is based on re-worked conditions of the ideal speech situation. Such an effort would seem to fly in the face of Habermas's retractions and Forester's assertion. But I believe this exercise is legitimate because I have narrowed the focus to only discourses about public participation in environmental decision making. I do not attempt to characterize all discourse in social life. This narrowed focus allows me to concentrate on one specific kind of discourse activity and a limited range of institutional settings.

impact assessment that counteract the instrumental rationality of cost benefit analysis with a form of value impact analysis that is discourse-based (Dietz 1984, Dietz and Pfund 1988). These tools have been applied successfully in Latin America and the Caribbean. Paul Stern and Thomas Dietz have suggested that Habermas's discourse ethics could form the basis for a method of contingent valuation in impact assessments (see Stern and Dietz forthcoming, Dietz 1987, 1988) and in understanding how values enter into collective decision making, especially with regard to global climatic change (Stern, Young, and Druckman 1992:192-196, Dietz 1992). Ray Kemp (1985) has taken an approach very similar to that taken here, using the rules of the ideal speech situation as normative criteria and applying them to evaluate a public hearing process. Finally, John Forester has recently published a theory of public planning based on Habermas's theory of communicative action (Forester 1993).

### *Process vs. outcome*

I take a procedural approach to evaluating the fairness and competence of citizen participation. Some people argue that it does not make sense to speak of a process as being fair or competent, only outcomes are fair or competent. I disagree with this position. While it is true that fair rules cannot guarantee fair play, or competent methods for knowledge selection cannot guarantee competent decisions, a person given a choice between two previously unknown procedures (no previous outcomes to evaluate) will not choose randomly, but make a judgment about expected performance. For new models of public participation or novel applications of old ones, procedural rules are the only basis for judgment. This question is not hypothetical. We are often faced with novel problems or the need to invent new tools. Evaluating outcomes of participation projects is certainly necessary, but it is also important that the decision making procedure be designed to engender fair and competent participation.

## Re-Politicization and the Ideal Speech Situation

### *Problems with democracy*

Jürgen Habermas has claimed that participatory democracy emerged in the *public sphere* [6] during the seventeenth and eighteenth centuries, as citizens

6 *Public sphere* is a term Habermas uses to denote the area of public life where intersubjective agreement on values can be reached in order to solve sociopolitical or practical questions.

became involved with the state in discussing practical issues and the role of the state (Habermas 1989; see also Kemp 1985). At this time, rules were developed to regulate hearing processes to resolve disagreements in (relatively) open, impartial, and rational ways. Habermas traced the history of the public sphere to its "disintegration" as a consequence of expanded state intervention in the late nineteenth century.

Habermas is concerned that the "scientization" of politics (increasing reliance on technological/scientific forms of rationality) has led to mere token public involvement, consequently jeopardizing society. When asked how to rectify these concerns, his answer is: "re-politicization of the public sphere," and he has laid out the requirements of discourse that can fulfill this prescription in a theoretical conception called *the ideal speech situation.* People who participate in such discourse must also meet certain requirements, these he outlines in a theory of *communicative competence.* Understanding how the concepts of the ideal speech situation and communicative competence can contribute to a normative model of public participation requires first acquiring a familiarity with how Habermas understands communication.

## *Speech act theory and universal pragmatics*

Habermas's theory of how language in everyday life is used to produce collective understandings and mutual agreements is called *universal pragmatics.* He uses speech act theory (Austin 1969) to explain what people *do* in discourse. (Hence the term: communicative *action.*) He focuses on four types of speech acts (assertions), each of which draws attention to a specific "world." A speech act makes a *validity claim,* that is, *the appeal implicit in a statement that makes the message meaningful.* This is the essential Truth-oriented component of the statement. Thus, communicative speech acts make validity claims to their *comprehensibility,* this is a contention that the utterance is linguistically proper. The location of this validity claim is in language. (Example: "A landfill is a site where trash is buried in a sanitary manner.") Constantive speech acts make claims to the *true or correct* depiction of existing states of affairs. (Ex. "The landfill over in Greenfield Township occupies a space of 46 acres.") Regulative speech acts claim *normative rightness* through appeals to legitimate interpersonal relationships. (Ex. "Operation of the landfill should be overseen by a committee of elected citizens.") Representative speech acts claim *sincerity* in reference to one's own subjectivity. (Ex. "I am concerned that the noise from the landfill will be unbearable.")

As part of the underlying normative agreement that makes speech possible, a speaker who makes an assertion implicitly presupposes that the validity claim(s)[7] can be verified to the satisfaction of all participants, in other words, *redeemed.* Speakers have no choice but to warrantee their validity claims with the unspoken promise of being able to offer convincing argument to anyone who challenges the assertion (Habermas 1984:302; 1979:1-5).

## *Types of discourse*

Different validity claims require fundamentally different means for *redemption,* that is, means by which the discourse participants collectively decide whether or not the speaker's claim is valid (and therefore acceptable or redeemable). Thus, he distinguishes between four *types of discourse* in which corresponding validity claims are discussed. These are called explicative, theoretical, practical, and therapeutic discourse. In explicative discourse references are made to language, terms, definitions, grammar. In theoretical discourse references are made to the objectified world of nature or society. In practical discourse references are made to social needs and the appropriate forms of social interaction. In therapeutic discourse references are made to the subjectivity of the speaker. Table 1 summarizes the connections between these terms.

## *Communicative competence*

*Communicative competence* is the ability to use language (more precisely: speech acts) to create understanding and agreement (Habermas 1970), that is, to communicate rationally. When communicative actions produce successful mutual understandings, they are judged communicatively competent. This requires people enter into a discourse with an attitude oriented toward reaching understanding. People must be committed to reflecting on their personal beliefs, values, preferences, and interests, they must be open to alternative definitions of reality, and they must listen to other peoples' arguments with an open mind.

There are four elements to communicative competence: *cognitive competence* – the ability of an individual to master the rules of formal logic; *speech competence* – mastery of linguistic rules; *pragmatic competence* – mastery of pragmatic rules; and *role competence* – mastery of rules for interaction (White 1989:29). The basic abilities that constitute competence at these levels are: the ability to focus on and comprehend the objective world;

---

7 Habermas argues that speakers engaged in communicative action simultaneously raise all four validity claims in every statement, but that one may be primary to the others.

*Table 1.* Structure of Habermas's theory of pragmatics. The four types of speech acts, their corresponding validity claims, and type of discourse.

| Speech Act | Validity Claims | Discourse |
|---|---|---|
| Communicative | Comprehensibility | Explicative |
| Constantive | | |
| - epistemic | True / Correct | Theoretical |
| - strategic | True / Effective ——▶ | x |
| Regulative | Normatively Right | Practical |
| Representative | Authentic<br>Truthful / Sincere | Therapeutic |

the ability to take up roles in the social world; and the ability to attain ego development (Braaten 1991:80). It is important to note that the attainment of each of these abilities is a communicative process. This means that the competent actor must not only know how to think logically, how to form grammatically correct sentences and phrase validity claims, but also how to participate in the redemption of those claims. Although Habermas sees all four of these aspects as integral to the idea of communicative competence, his contribution is to understand what it means to master the pragmatic rules of language and to interact through assuming roles competently.

Discerning among the four types of validity claims is central to communicative competence. The competent use of speech acts in each of these cases refers to the ability of the speaker to place him- or herself in a position relative to these worlds and to convey to others relevant aspects through the proper use of language. Habermas argues that, as people learn to speak, they *intuitively* develop the ability to do this, as well as to apply the appropriate redemption standards to the different validity claims.[8]

8 The extent to which this can occur, however, is related to the extent to which the culture is differentiated. This does not, of course, imply any judgment about the moral development of the culture.

## *The ideal speech situation*

The *ideal speech situation* is Habermas's attempt to describe the presuppositions that discourse participants must hold before communication without coercion can prosper.[9] He defines it in four conditions (Habermas 1973).[10] The first two conditions he calls "trivial:"

1. All potential participants of a discourse must have the same chance to employ communicative speech acts.
2. All discourse participants must have the same chance to interpret, claim or assert, recommend, explain, and put forth justifications; and problematize, justify, or refute any validity claim.

The second pair are non-trivial:

3. The only speakers permitted in the discourse are those who have the same chance to employ representative speech acts.
4. The only speakers permitted in the discourse are those who have the same chance to employ regulative speech acts.

The conditions for the ideal speech situation can be thought of as "rules for discourse" (White 1989:55), meaning that participants abiding by these rules will produce a rationally motivated agreement (or at least understanding), as opposed to one created through manipulation and coercion. Here, the word "rule" is not used in the sense of something externally imposed that determines the form of discourse – as rules of a game determine the play (Habermas 1991:91). Rather, "rule" here refers to the *unavoidable* presuppositions of speech that participants adopt, implicitly and intuitively. These are fundamental necessities for communicative action. They can be thought of as evaluative standards – standards that are unachievable because they represent a perfection that cannot be realized – and as ideal rules to which discourse participants should aspire.

---

9 There are differences of opinion about precisely what are the conditions for the ideal speech situation (see: Robert Alexy quoted in: White 1989:56; Stephan White 1989:56-7; Seyla Benhabib 1986:285-6; Ray Kemp 1985; and Pieter Jan M. Stallen 1985) and what status these "conditions" should have (Habermas 1984:25, 1991:92, 1992:160).

10 Translation by O. Renn and T. Webler. For each condition, he elaborates on what he means by the different types of speech acts.

## *Fairness in the ideal speech situation*

The ideal speech situation is a normative model that captures the ideal of fairness embedded in the ethical-normative arguments for popular sovereignty and political equality (see Chapter 2).[11] Clearly, the thrust of the ideal speech situation is a concern for fairness – people must presume each other to have equal chances to effect the formulation of the agreement. What is not so clear is how Habermas pictures the role communicative competence should play in the ideal model.

One way to begin is to notice that, while these *conditions* for the ideal speech situation embody fairness, the *concept* of the ideal speech situation must be grounded in the ideals of fairness *and* competence. Fairness and competence cannot be thought of as completely separable. When John Stuart Mill remarked that individuals are the best guardians of their own interests, he recognized that individuals have a legitimate right, based in their competence, to make judgments and take protective actions – namely to make political arguments about how their interests and values are affected by a proposed consensus. Habermas asserts that the basis of this right lies in the fact that knowledge is socially constructed. (How this social construction is accomplished is discussed below in the section on competence.)

A central premise of the theory of communicative action is that all validity is rooted in the *background consensus of the lifeworld,* that is, what we experience as individuals, what we construct as society, and what knowledge we have stored in culture (Habermas 1984:70). In the process of redeeming validity claims, people reflect upon, discuss, and renew this consensus. Habermas contends that claims to validity can be redeemed by those who have had a hand in building and preserving that body of meaning – the populace.[12] The consequence of putting this premise into practice is the realization of popular sovereignty.[13] Consequently, the ideal of popular sovereignty is quite close to Habermas's belief that validity is integral to the lifeworld and each

---

11 Ray Kemp argued that the ideal speech situation reflects the characteristics of the pre-disintegrated public sphere of British 19th century democracy: openness, impartiality, and justice (Kemp 1985:180, 189).

12 People may be more or less qualified to make validity judgments in different areas. The point is that even though everyone does not have the specific knowledge needed to make final judgments on all claims, everyone has the ability to recognize *what it would take* to prove a statement comprehensible, true, right, or sincere.

13 Fairness is both an ethical norm and a logical deduction based on this observation and an analytic study of how language is used. Logically, discourse should be fair because everyone has access (and contributes to shaping) this background consensus simply by virtue of being a member of society. Ethically, we cannot legitimately exclude anyone from this activity (except on legally established grounds). This legitimacy is cemented in the very nature of speech and is reaffirmed every time two people engage in speaking to each other.

individual has an *ipso facto* legitimate access to the underlying source of validity. In other words, an equal right to make determinations about validity claim redemption – and thereby, the social construction of reality.

## *Communicative rationality*

Habermas's normative theory of communication is part of a larger project to explain the evolution of modern society as a process of rationalization. Although he remains in the critical theory tradition,[14] Habermas has broken from his former colleagues in the Frankfurt School by reviving the project of the Enlightenment – the emancipation of people and society through the application of reason – via the adoption of a conception of rationalization based on an ideal form that he calls *communicative rationality*.[15] Habermas believes that this form of rationality, put into practice via discourses that aspire to (but can never be expected to achieve) the ideal speech situation, may be able to renew popular political life in a way that saves democracy and modern society from its own demise, which has been brought on by the wholesale unreflective use of instrumental (and strategic) reasoning.[16]

The approach taken by Habermas to develop the concept of communicative rationality is to start from action theory and then to switch over to a setting of communication. Just as in action theory, where peoples' actions are judged rational or not according to criteria linked to motives, Habermas proposes to evaluate communicative actions against the motive of reaching an non-coerced mutual understanding (Habermas 1979:1; 1984:8).

---

14 Critical theories are those that: (a) attempt to induce self-clarification by encouraging individuals to reflect upon how their actions are inconsistent with their interests; (b) seek to identify crisis potentials in society; and (c) are concerned with the emancipation of individual from arbitrary relations of power (Braaten 1991).

15 *Rationality* is the mode of making sense of the world. *Action* is evaluated according to how consistent it is with a rationality. Thus, communicative rationality is the basis by which communicative action is judged. *Competent action* is consistent with one's rationality.

16 For a critique of Habermas's view of instrumental and strategic action see Johnson 1991. Keep in mind that Habermas is not concerned with explaining action oriented toward the manipulation of others *(strategic action or instrumental action)*. He acknowledges that this kind of rationality is important to understanding how individuals act, especially in the economy and in political decision making (and many other aspects of our lives), but he observes that it is constrained by two primary forms: (a) social constraints imposed by others people in everyday life–communicative rationality, and (b) functional constraints associated with the institutions and structure of the system–formal rationality. This is not the same as saying that societal reproduction is endangered by strategic action per se. Of course, language is quite often used toward other strategic goals such as intentional confusion, misrepresentation, and manipulation, but these are all secondary forms of communicative intent that rely on the presupposition of language used for acquiring mutual understanding.

> "An assertion can be called rational only if the speaker satisfies the conditions necessary to achieve the [...] goal of reaching understanding about something in the world with at least one other participant in communication" (Habermas 1984:11).

The "conditions" are as follows: participants agree that things said can be called into question and resolved through mutual exploration and evaluation and that the basis for evaluating spoken statements is the stored knowledge, personal experiences, and shared reality accessible to the discourse participants. Ideally then, communicative rationality is used by participants of a discourse to select appropriate validity claims and help to construct a mutual understanding devoid of coercion.

Communicative rationality is a normative concept that Habermas believes is a universal normative principle.[17] He states that the whole purpose of communication is to build mutual understanding in order to construct a consensual social reality. Building on Mead, he argues that this is an aspect of our species' evolution, rooted in a biological predisposition and now solidified in the very structure of speech (Habermas 1984:256, 1987: 3ff). Our natural capacity for language contains the normative force of communicative rationality, which is: to communicate with the intention of reaching mutual understanding and agreement exclusively via the force of better argument. As a consequence, communicative rationality demands emancipated individuals coming together by their own free will.

## *Criteria for evaluating communicative action*

In elaborating on how communicative action can be realized, Habermas has tended to focus on practical discourse (about norms). He calls his theory of practical discourse *discourse ethics.* Discourse ethics does not replace the ideal speech situation, it simply delves more deeply into one kind of speech. A main result is Habermas's claim to a universal principle that applies to the redemption of all normative validity claims in communicative action. This is a criterion that can be used to evaluate discourses. The principle is: no norm can be considered valid unless all those affected can accept the consequences associated, to the extent those consequences can be known (Habermas 1991:65).

---

[17] Habermas defends this claim with the observation that no speech is possible that does not presuppose communicative rationality. As such, the ends of speech must be prior to personal ends. Despite what a person wishes to accomplish through speech (even if it is a strategic or manipulative end), he or she must acknowledge and renew *an underlying normative reciprocal agreement* to the ideal of reaching an non-coerced understanding–which is done merely in any act of uttering. See also Braaten 1991:66.

Constructing a normative model of public participation based on the ideal of communicative action requires more concrete criteria such as this.[18] Criteria will have to be derived from the normative principles outlined in the ideal speech situation. As it stands, however, the ideal speech situation specified by Habermas does not adequately capture the two metacriteria: fairness and competence. It is clear that fairness and competence are consistent with Habermas's core ideas (namely emancipation and communicative competence), but realizing these in a normative model for communication requires making a few alterations to the ideal speech situation.

## Alterations on Habermas: The Fair and Competent Ideal Speech Situation

### *New conditions for the fair ideal speech situation*

#### A critical review of Habermas's rules

As an evaluative yardstick for public participation discourses the ideal speech situation as stated by Habermas is unsatisfactory. To reconcile the problems, I first distinguish rules that encourage fairness and then rules that encourage competent actions. I then unite these to make up a revised set of rules for the fair and competent ideal speech situation.

Habermas's rules for the ideal speech situation do not relate in a clear way to what people actually do in discourse. One of his strongest contributions is the analytical distinction he draws between initiating assertions and challenging and debating others' assertions. Taking advantage of these valuable analytical deconstructions of speech is immensely helpful in describing what people do in public participation settings and in outlining evaluative norms for judging participation.

A second problem is that Habermas's fourth rule, which includes the making of normative claims, really has procedural and substantive aspects that should be distinguished. Procedural normative claims are proposals for resolving disputes. Since it is likely that consensus on all issues will not be reached in a discourse and since action may be necessary or desirable, there must be a way to resolve or bracket normative disputes so that coordinated actions can be taken.[19] To be fair, decisions about which procedures should

---

18 One such set of criteria have been proposed by Alexy (1978) and discussed by Habermas (1991: 87ff).

19 For instance, they could decide that differences about content of the agenda are to be resolved by a committee operating with two-thirds majority vote.

be used to resolve disputes (voting, sending to committee, postponing, etc.) must be made according to non-coerced consensus *before disputes arise.* Remedying the rules for the fair ideal speech situation to accommodate these objections is rather simple:

1. Anyone who considers him- or herself to be potentially affected by the results of the discourse must have an equal opportunity to attend the discourse and participate.

2. Every discourse participant must have an equal opportunity *to make validity claims to comprehensibility, truth, normative rightness, and sincerity.* (That is, to participate in explicative, theoretical, practical, or therapeutic discourse, including the translation of expressive claims – see below.)

3. Every discourse participant must have an equal opportunity *to challenge the comprehensibility, truth, rightness, or sincerity validity claims made by others.*

4. Every discourse participant must have an equal opportunity *to influence the choice of how the final determination of validity will be made and to determine discourse closure* (i.e., to decide how to decide when there is no consensus).

The first three rules now refer to the activities of participating, initiating, and discussing assertions (validity claims through speech acts). The fourth rule refers to the activity of redeeming the asserted validity claims through an agreed-upon procedure.

## Objections and justifications

One point where this formulation differs noticeably from Habermas is the inclusion of expressive and communicative statements in the discussion stage (No. 3). Habermas argues that these claims cannot be subject to critique and redemption in the same way as can cognitive or normative claims. He points out that, by definition, communicative validity claims cannot be discussed, since what is at issue is the comprehensibility of speech itself. In the extreme case (you speak Arabic and I speak Japanese) this is certainly true. But in citizen participation, the kinds of communicative validity claims that arise are mostly points of definition (technical jargon) and clarification ("Could you say that again in another way?"). These kinds of communicative validity claims can be (and are) discussed.

Expressive claims originate in the subjective, a world to which others have no direct access, Habermas therefore, posits that these cannot be redeemed as other claims. Monitoring the consistency between what a speaker says and does is the only suggestion he offers for how to redeem sincerity validity claims. However, I suggest that, without challenging the integrity of the individual, competent translation of *some* expressive claims into their factual or normative constituents is realistic.[20] Translation makes expressive claims partly testable. If an expressive claim is found to be based upon factually untrue contentions or normatively unacceptable depictions, then the speaker may retract the expressive statement. To assure that the autonomy of the individual is not sacrificed, it is crucial that the translation procedure be *agreed to* and the translation *approved by* the person making the subjective statement. In other words, the translation process is a means for *the speaker* to reflect upon and reconfirm the authenticity and sincerity of his or her own feelings, based on a guided empathetic review of their logical consistency and possible belief origins.

### Attendance: defining the affected population

The need to attend is primary to achieving fairness and every participation process must deal with the problem of identifying who has a legitimate right to participate. This is often the same as the problem of defining the potentially affected population – individuals or groups whose interests or values may be affected by the problem or the proposed decision action. Impacts can occur at the physical, psychological, social, economic, or value levels. In many instances, the affected population is the same as the citizens of a particular community (they live in close proximity and are legally responsible for the consequences). Other times the impacts extend beyond political boundaries (such as waste transport through neighboring communities). In still other cases, such as value-laden disputes (for example, national energy policy, national park management, endangered species protection), distance from the site may not be relevant in determining the "affected population."

Organizers need to make certain that everyone who could be affected is notified, even if the delineation does not coincide with political boundaries. When excluding people is a practical necessity, it should be done in a way that is fair to all – such as random selection.[21] Both objective and subjective processes should be used to identify these people and groups. Objective procedures (risk analysis, exposure assessment, social impact analysis, value tree analysis) reveal the causal links that expose physical, social, economic, or

---

[20] For example, the speaker might be asked, "Can you tell us what it is about the proposed landfill that you feel it so threatening?" If the speaker can reflect and discover the key fears (e.g. "noise, hazardous waste leaking, ...) then expressive claims can be restated in cognitive or normative terms.

[21] Of course, to be done correctly, random selection would chose a number of people dependent on the distribution of interest and value positions among the population, so that no position is underrepresented. This is much easier said than done.

value impacts. Subjective processes are those which allow people to decide for themselves if they will participate. Particularly in the United States, people expect that they can be involved in issues of their governance if they feel so inclined. While opening the doors to anyone who wants to participate may pose logistical problems for the discourse, it is essential to the legitimacy of the process.

## *Bringing competence into the normative model*

### Promoting competence by discourse rules

This leads to my major point of disagreement with Habermas. He has developed the concept of communicative competence, but has not laid out a set of ideal conditions by which we are to assess how competently people perform in a discourse. As a consequence, in applications of the theory of communicative action (see Forester 1985), the idea of competence seems to have become lost in the overwhelming concern for neutralizing power relations in discourse and achieving fairness.

Conditions for a competent speech situation should list the inescapable presuppositions regarding competence that a person inherently and intuitively adopts when he or she enters into communicative action with another.[22] In my opinion, these should mainly describe *characteristics of the discourse* that influence how competent shared constructions of reality can be made, as opposed to presuppositions about the abilities of the other discourse participants.[23] Typically, competence is seen as a quality of actors, rather than rules. But clearly, there are better and worse ways to resolve validity claims. If we want to know the geology underlying a landfill, we do not vote, but conduct certain engineering studies. We have developed rules for resolving validity claims in some (but not all) circumstances. Rather than talking about "competent rules" however, these could be seen as "rules for discourse that promote competent actions by the discourse participants."

Habermas's conception of competence revolves around the individual. If he were to spell out the conditions for a competent speech situation he might say: everyone must assume the others have cognitive competence to employ logic, linguistic competence to form sentences, pragmatic competence to make

---

22 Participants must assume that these rules will be realized approximately enough to allow argumentation (Habermas 1991:91-92).

23 Here there is a need to distinguish conventions from rules. Rules are unavoidable presuppositions that participants must make. These can be proven by showing how a person who challenges these rules gets caught up in a performative contradiction. Conventions are ways that make these rules operational under empirical conditions (Habermas 1991:88-92).

validity claims, and interactive competence to engage in discussion.[24] While it is clear that people participating in a discourse must make presuppositions about the competence of the other participants if communicative action is to occur, this is not useful as an evaluative yardstick. The problem with this approach, besides the obvious practical and ethical difficulties of evaluating individuals' competencies and selecting out the misfits, is that people may possess the skills, yet not apply them. Furthermore, since communicative competence can only be learned through communication, barring people from discourse on the basis of incompetence means society will not reproduce communicatively competent citizens.

A conception of competence that does not center on the individual, but on the rules that coordinate interaction is needed. I defend this claim with:

(1) an observation about the qualities and origins of the four aspects of communicative competence,

(2) a recognition that any given public participation discourse has to achieve a level of competence that exceeds that of the immediate participants (through gaining access to the stored knowledge of previous learning experiences),

(3) an observation about how rules (methods, procedures) for selection of knowledge in discourse are established, and

(4) some pragmatic points about encouraging competent discourse.

## Origins and qualities of the four kinds of communicative competence

Cognitive and lingual competence refer to rudimentary skills that people must have in order to employ pragmatic or role competence, but they are not linked to the social context. In other words, a speaker can be linguistically competent, but still not say anything meaningful or relevant. Pragmatic and role competence, on the other hand, are inherently social skills, rooted in the social context of the discourse and oriented toward meeting collective as well as personal needs.

The difference between pragmatic and role competence is that the former refers to the ability to insert validity claims in a social reality and the latter refers to the ability to redeem those claims. A validity claim establishes a *relationship* between the speaker and the four sources of validity: language, objectified world, society, and self. To take a stand on the validity of an

---

24 Alexy proposes this rule. See Habermas 1991:89.

assertion, one must take on a *role*. For example, a man who challenges an expert's calculations of a watershed must assume a role in which he holds his actions accountable to criteria for proper scientific rationality.

Distinguishing between competently tailoring one's actions to the social setting and competently employing rudimentary skills that are the same regardless of context is important. In a practical setting for public participation in a western democratic country, cognitive and lingual competence have to be taken more or less for granted. Except for the most obvious cases of mental illness and the inability to use language (and even here it is sometimes difficult to draw a clear line), excluding participation opportunities based on assessments of individuals' cognitive competence is unethical.[25] Even lingual incompetence in the dominant language is not an excuse to legitimately exclude a citizen – translators should be hired.[26] This is a common solution when the lingual incompetence refers to technical jargon and a consultant (translator) is hired.

Pragmatic and role competence cannot be taken for granted, simply because there are too many kinds of social and problem contexts, too many conflicting interests at play, and too much to know.[27] No one is ever fully competent. Instead, procedures and rules that promote competent constructions of understanding – rules that partly define pragmatic and role competence – are needed.

### Access to stored cultural knowledge

It is of course unreasonable to expect members in every public discourse to be fully competent in all aspects of the subject at hand. Indeed, public participation should offer people a chance to learn new knowledge and skills. One of the practical problems then, is to answer how discourse with imperfect participants can achieve a level of competence that is required to make good decisions.

---

25 Robyn Eckersley has challenged Habermas's claim that only humans capable of speech should be permitted to participate in a discourse. She notes that this is an arbitrary determination and anthropocentrically prejudiced. Although there are obvious practical problems with including non-humans in discourse, it remains an important question to ask what impact this "unfairness" has on the quality of our understandings, especially in cases of environmental decision making. One should also take note that many earlier cultures did have various strategies for including non-human nature in the decision making, although the competence of doing so is difficult to establish, especially from a rationalist point of view. [See R. Eckersley, "Habermas and Green Political Thought," *Theory and Society* 19:739–776 (1990).]

26 This is precisely what has been done in many cases. Two well-known examples involved Native Americans in Northern Canada: the Berger Inquiry and the Hydro-Quebec project.

27 Consider the different kinds of knowledge and skills (scientific, social, political) needed to participate in decisions about local road planning, regional education, national nuclear waste storage, and global climate change.

My answer is to take advantage of preexisting knowledge and rules for selecting among competing claims. These are part of the stored knowledge and experience of the culture (in other words the products of past learning endeavors). Participation models should do everything within reason to encourage access to the stored knowledge and experiences with which the current participants are not familiar. One example is to use outside panels to peer-review information packets.

Adopting rules to enable competent actions in discourse can facilitate social learning and individual development and sidestep the problem of having to restrict participation to educated elites. By adopting rules for interaction and selection that are tried and tested – for example, invoking the rule that normative claims cannot be validated if they contradict higher values – the competence of the discourse can exceed the combined competence of the participants.

## How rules for knowledge selection are made

Even if rules for the competent construction of understanding are not imported, they will develop and emerge during the discourse itself. Just as the validity claimed in assertions originates in the background consensus of the lifeworld (the shared experience of the participants and the stored knowledge of the culture), so do the rules and guidelines used to redeem conflicting validity claims. The motivation and authority of communicative competence does not lie completely with the individual, but in the collective experience of the culture and, to some degree, in the physical nature of the human mind. In other words, we only know that we are speaking competently when other people, references to stored cultural knowledge, and our personal experiences affirm it.

Rules for knowledge construction and selection are not adopted randomly but are products of generations of experience. They are accessible in differing degrees to people. Some are so basic and common in everyday use that people may not even notice that they are using them or conceive that they could be changed. An example is the simple logic of cause and effect: we better understand something when we can explain how it is related to other things and events. Some rules are formal, in the sense that they are spelled out explicitly. Examples include rules for parliamentary debate (calling for motions, voting) and scientific protocols (how to take a surface water sample for testing). Others are traditional (soliciting the opinion of all value leaders in the community) or conventions for coordinating interaction with which people have become comfortable (such as raising your hand when you want to speak and waiting to be called upon). Rules may seem intuitively right to some, they may be habits or established patterns of behavior that people cannot easily replace, or they may be invented and established only after complicated argument.

This observation about the origins and qualities of rules illustrates that it is possible to discuss, and to an extent predict, which rules the participants of a given discourse may find relevant. We all have access to knowledge about the various rules – which are better to use in each instance, how much reliability we can place in them, and so on. Thus, it is possible to compose a list of rules that could serve as a starting point for a discourse. Of course it would be improper to assign any permanent authority to any particular set of rules. They will change over time and they may be more or less relevant depending on the social context and the needs and the desires of the participants.

## Encouraging competent discourse by adopting constraints

While specifying protocols for resolving validity claims competently or relying on outside expert advice will constrain the discourse and implicitly bias it, the alternative – merely to ignore the issue of competent process and to rely wholly on the competence of the individuals to self-design adequate communicative procedures – is impractical, may result in more bias, and less likely to be competent. (It is taken for granted here that any pre-proposed constraints are merely suggestions and they can be removed by the consensual will of the group. Thus, even the most perfect rules cannot guarantee competence, because the participants may choose to reject them.)

Imposing established rules on discourse participants will inevitably disadvantage certain interest positions. This imposition is justified because it is not intentional, in most cases it is not even known. When a systematic disadvantage is located, there is a moral obligation to correct it. Of course, constraints should not be used to coerce understanding, merely to suggest a starting form discourse may take, but they do have the power of an existing social consensus and this power should be employed to prevent the strategic manipulation of rules.

There are several practical reasons to support the choice of constrained discourse over unconstrained. First, discourses about environmental decision making do not simply appear. They are initiated by individuals who have to deal with practical problems of getting people to work together. To some extent adopting constraints is unavoidable, but it is also desirable. People try to be efficient, they do not want to "re-invent the wheel" every time they have a problem to solve. Citizens may not trust experts or members of interest groups, but this does not mean they want to disregard all advice and make all validity determinations themselves. They have no desire to rediscover known facts and laws. People recognize that expertise is valuable, but they are suspicious that experts may purport a political agenda.

Second, there is less room for strategic action when the discourse rules are the result of a social consensus. When only a few people are involved in the determination of rules, as in unconstrained discourse when only those present can participate, the outcome is more dependent on the skills and intentions of each person. A person skilled at manipulation may be able to convince the others to adopt a less competent rule that favors his or her interests.

Third, when a discourse is standardized by use of proven rules, the outcome is no longer totally reliant upon the competence of the participants. Two separate discourses can be expected to produce similar results. This is an indication that the understandings reached have a firm basis. Participants will be more satisfied knowing that they reached lasting understandings, rather than merely creating a random outcome.

Fourth, constraints manage the problem of unlimited public demand for self disclosure.[28] Without rules that enable closure to be reached, listeners could conceivably demand the speaker give deeper and deeper justifying arguments for a validity claim. The group must have some mechanism for forcing a determination.

Finally there is the problem of time. Habermas suggests that communicative action will always produce agreement, given enough time. In practice, time is often in short supply. Participants are not necessarily willing to spend their time working for perfect agreement. The fact of the matter is that discourse has to reach closure if it is to satisfy its purpose to coordinate action and often that means terminating discussion before all disagreement or misunderstandings have been remedied.

## *A Procedural view of competence*

### Definition of competence

These four points – that pragmatic and role competence are secured in rules for interacting and selecting knowledge in discourse, that each discourse has to achieve a level of competence greater than the sum of its participants, that rules are grounded in the background consensus of the lifeworld, and that constrained discourse is more consistent with practical concerns of the participants – lead me to define competence in discourse as: *construction of the most valid understandings and agreements possible given what is reasonably knowable at the time.*[29]

---

28 Moon criticizes Habermas on this precise matter. See Moon, 1991:224.

29 This is not so far from Habermas. He noted that the purpose for ensuring discussants are communicatively competent is so that they will establish rules for discourse that produce the best possible understandings (Habermas 1970). More recently, Habermas restated the conditions of

Competence is related to the performance of the participants in the discourse, compared to what can be reasonably expected of them, given the current information and knowledge available. We cannot expect them to be perfect. We can expect them to follow procedures for the collection and verification of knowledge that have been tried and proven over time to be reliable and accurate.

Employing the best available techniques for the construction of knowledge cannot guarantee that the participation model will produce decisions with completely foreseen outcomes. When decision making extends into realms where uncertainties are high, the best available procedures may not be good enough. In any case, it is also necessary to look at outcomes over the long-term to ensure that the procedural criteria make sense. The point is that we learn to make better decisions by making mistakes. If other procedures can be shown to produce better understandings, then they should supersede less adequate ones.

## Conditions for an ideal speech situation that ensure competence

Based on these observations and this definition, I propose the following rules for the competent ideal speech situation:

1. Every potential discourse participant must meet minimal societal standards for cognitive and lingual competence.

2. Every discourse participant must have access to the knowledge needed to make validity claims and criticize others.'

3. Speakers must verify the results of any attempt to translate expressive claims.

4. Judgments about conflicting validity claims must be made using the most reliable methodological techniques available.

These rules are parallel to the rules oriented around fairness. Table 2 illustrates the links.

---

the ideal speech situation in a way that includes a concern for competence (Habermas 1992:260). See also his discussion about the two goals of communicative action (to reach understanding and the effectively deal with the problem) in Habermas 1987:120.

*Table 2.* Conditions for the fair and competent ideal speech situation.

| Fairness | Competence |
|---|---|
| Anyone may participate | Minimal standards for cognitive and lingual competence |
| Assert validity claims | Access to the knowledge |
| Challenge validity claims | Consensually-approved translation scheme |
| Influence final determinations of validity | Most reliable methodological techniques available |

## Further observations on competence

Before moving ahead to operationalize these new rules, a few further observations are relevant. First, participants in a discourse achieve their goal of coordinated action by constructing a new shared understanding. Competent agreement for action can only evolve out of a competent understanding. Through competent interaction (assertion, explanation, and understanding), shared beliefs, interests, and values are constructed.[30] Agreement does not necessarily follow from understanding, but understanding is a prerequisite of rationally-motivated agreement. Agreement may also be achieved without understanding, but this does not count as rationally-motivated agreement in Habermas's sense. Whereas understanding requires people to detect what it is about statements that gives them their validity, agreement requires people to choose which statements are the best and most desirable. Therefore, discourses should not be evaluated primarily on whether or not they produce agreement. Progress could be significant if the discourse achieves mutual understanding.

30 As a consequence, attempting to evaluate competence from personal assessments made from a single interest position will not work. To evaluate the competence of understanding, the initial expectations cannot be compared to the consequential understanding, because the initial expectations are too temporary and linked to particular or underdeveloped collective interests.

Second, the social context of the discourse will constrain the ability of the participants to employ or access the best available procedures for knowledge selection. When evaluating a discourse, such constraints should not be accepted uncritically, because they may be strategically imposed by actors in the discourse. An evaluation must consider how the potential competence of the discourse is limited by the social context. This is a matter of how the discourse is embedded in the social realty and it may have impacts on fairness and/or competence. Therefore, the participants of the discourse should have the opportunity to reconsider the constraints to which they subject themselves.

Third, even competent understandings are impermanent, because our understanding of the world is always changing. When the knowledge is not available or when it is inaccurate, competent communication may produce inaccurate understandings and decisions with unpredictable consequences. Therefore, it may be necessary to postpone the discourse until a later date, if the knowledge available is not deemed adequate by the participants to produce a competent decision.

Fourth, competence is achieved when the composite of rationalities present and actively exercised in the communicative process is complete and without gaps. That is, neither technical, political, social, or cultural information is excluded from the discourse on each issue raised. Therefore, discourse should not merely focus on including a great number of people, but also on getting all points of view represented.

## Operationalizing the Conditions of the Ideal Speech Situation: The Discursive Standard Criteria

### *Institutional constraints*

Conditions of the ideal speech situation only state the general presuppositions that actors who wish to cooperate must hold. To acquire the level of detail that is needed to evaluate public participation discourses about environmental decision making, institutional constraints need to be considered. Institutional constraints are imposed rules and guidelines intended to facilitate communication in a given social context, complete with the practical limitations of time, space, money, and so on. Constraints do not share the moral imperative power of the conditions of the ideal speech situation, because they are grounded in solving practical problems. In this section I offer a definition of the constraints associated with public participation in environmental decision making, in the form of a set of criteria and indicators for fairness and competence (the actual listing appears in the Appendix).

These comprise a normative model for public participation in this topic field in western industrialized democracies.

## *Discursive standard criteria for evaluating fairness*

### Scope of fairness

Fairness refers to the distribution among participants of opportunities to act meaningfully. When people come together to reach understanding and make a decision about an environmental issue, there are four fundamental actions that every actor must be free to assume: *attend* (be a participant in the discourse), *initiate discourse* (make speech acts), *discuss* (challenge and defend claims), and *decide* (influence the collective consensus). These are the four *needs* of a fair process and they are each grounded in a condition for the ideal speech situation. The four needs of fair discourse are relevant in each of the three activities which comprise a public participation discourse: *agenda and rule making, moderation and rule enforcement, and discussion.* (See Figure 1).

### Agenda and rule making

Agenda and rules establish the framework for the discussion. Setting the agenda is much the same as defining the problem. By participating in making the agenda, people have the chance to make sure that their concerns will be addressed as well as to shape the group definition of the problem. This does not mean that every person necessarily has to play an active role in making the agenda. For example, a preliminary agenda could be composed by a subcommittee and the final version approved at the start of the discourse. Besides the agenda content, time must be allotted to each item and the schedule ordered. The agenda can unfairly influence the ensuing discourse by not allotting enough time, by framing a topic in a particular way, or by scheduling a topic at an unfortunate time. Moreover, an agenda that omits a topic that a participant considers relevant is obviously unfair.

Rules are made to manage interruptions, resolve stalemates, define appropriate behavior, and so on. Many rules come from norms for conduct in everyday life and are not formalized, others are highly formalized, depending on the discourse. Interruptions, threats, raised voices, and derogatory gestures and statements are all signs that a speaker's access to the discourse activities are threatened. Rules provide a means to quell these manipulations, be they strategic, accidental, or merely spontaneous. Just as people must have fair access to all the needs of agenda making, they must have fair access to the process of formally establishing rules.

*Figure 1.* Evaluation framework showing the relation between sub-criteria in the Appendix and the metacriteria: fairness and competence.

| EVALUATION FRAMEWORK | | | | |
|---|---|---|---|---|
| **FAIRNESS** | **NEEDS** | | | |
| **ACTIVITIES** | Attend | Initiate | Debate | Decide |
| AGENDA AND RULE MAKING | A1, A2, A3 | A1 | A2 | A3 |
| MODERATION AND RULE ENFORCEMENT | B1 | B1 | B2 | B3 |
| DISCUSSION | C1 | C2 | C2 | C3 |

| **COMPETENCE** | **NEEDS** | |
|---|---|---|
| **ACTIVITIES** | Access to Knowledge | Best Procedures |
| EXPLICATIVE DISCOURSE | D1 | D2, D3, H1, H2 |
| THEORETICAL DISCOURSE | E1, E2, E3 | E5, E6, E7, H1, H2 |
| PRACTICAL DISCOURSE | F1, F2, F3, F4 | F5, F6, F7, F8, H1, H2 |
| THERAPEUTIC DISCOURSE | G1, G2 | G3, G4, G5, H1, H2 |

## Rule enforcement

Discourse participants also must agree on a means to enforce rules. One of the most common solutions is to appoint a facilitator or moderator. Both serve as neutral parties responsible for enforcing the rules fairly. They differ, however, in that a facilitator merely tries to catalyze the discussion without guiding, while the moderator exhibits more leadership. A facilitator may encourage quieter people to contribute to avoid domination of the discussion by a few, but largely, he or she merely keeps the group on its agenda and enforces rules for interaction. A moderator may make proposals and participate in the debate by presenting information and arguments which are missing from the discussion. In any case, the behavior of the facilitator or moderator should be subject to the scrutiny and approval of the participants.

## Discussion

During the discussion on any agenda item everyone potentially affected by the problem must have an opportunity to participate. For example, everyone should have the chance to participate in defining terms. There is power in being able to determine the definitions of terms, especially when it amounts to making a normative choice. Likewise, participants need to be able to make statements of fact, even if they are not grounded in formal scientific observations. Anecdotal knowledge about local conditions and personal experiences are also valuable. In other words, one does not have to be a scientist to participate in discourse about states of affairs in the world.

This realization has to be tempered by the reality of social organization – the division of labor. (Here fairness and competence begin to collide.) Consider the problem of discussing scientific results. Although claims made to the validity of factual statements are still ultimately grounded in the shared lifeworld, the access to verifying those claims – through scientific methodologies – demands considerable resources and expertise. Experts who have committed themselves to developing a competence in these methodologies are obviously more familiar with the requirements of redeeming factual validity claims about systematic evidence. The same is likely to be true of citizens who "specialize" in collecting anecdotal evidence.

These observations put Habermas's ideals about the background consensus of the lifeworld and democratic theorists' ideals for equality against the fact that different people have different specialties. On the one hand, individuals must be free to argue for what they believe and to participate in making final judgments. On the other hand, it is in the group's interest to detect strategic behavior and errors in judgment. Some balance must be struck between giving everyone equal rights to participate and assigning higher credibility to certain speakers on the basis of experience or specialization. This is not an unusual problem, in everyday life people make judgments about others' credibility and competence. It is only necessary to recognize that there is not a universal prescription for how to make these choices.

## *Discursive standard criteria for evaluating competence*

### Scope of competence

Competence refers to the construction of the best possible understandings and agreements given what is reasonably knowable to the participants. In accordance with the conditions of the competent ideal speech situation, there are two basic needs: *access to information and its interpretations* (the second and third conditions in Table 2), and *use of the best available procedures* for knowledge selection (conditions one and four). Information and its interpretation (knowledge) are the raw material that a discourse processes into collective understandings and agreements. Access is always a matter of time, effort, and cost. Although it is impossible to make final determinations about how accessible information and knowledge must be, unreasonable inaccessibility are grounds for criticizing the discourse.

### Access to knowledge

It is clear that simply having access to information is not enough. Experts and others who are capable of explaining the relevance of the facts need to be retained. For example, a report from a water testing laboratory may contain a lot of information about the composition of the water, but until the relevance is explained in terms of what matters – impacts of human health, corrosive effects on copper, impacts on a species of amphibians, or whatever is at issue – the information is not knowledge.

### Procedures for knowledge selection

The other ingredient necessary to produce competent understandings is the evaluation and selection of knowledge by the participants. Understandings of reality depend on the rules and procedures used to select and construct knowledge. Determining competence requires defining those procedures and characterizing their scope limitations. Time tested methods for gathering information and constructing knowledge have been developed and it is reasonable to expect that people should use these methods when selecting and employing information and knowledge.

Criteria and rules for redeeming the different types of validity claims can be explicitly discussed in the discourse. Some of these rules are *objective* in the sense that systematic observation has verified their reliability and we can rely on them to select between different factual (cognitive) validity claims. This includes some of the knowledge gathered in the sciences. Experts have a professional obligation to warrantee knowledge produced in this way and to make it available to citizens and others. Participants of the discourse have an obligation to recognize the validity of knowledge that meets these objective standards, provided that the experts can defend their results in a peer review process. For example, if two independent licensed laboratories produce the same analysis of a water sample, then the participants are obliged to accept the results as true (provided that collusion has been ruled out). The point is that participants are not supposed to rewrite the laws of physics. These so-called objective rules are subject to change, but only in a manner consistent with scientific convention.

Other rules are *socially constructed* and need to be consensually imported into the participation discourse. These are partly reflected in norms and mores of society, some of which are expressed in law, but, more importantly, they represent the collective interests of the community or society to which the discourse belongs. Since there may not be consensus among all in society about these rules, and since norms are always changing, the discourse participants need to establish which rules will be adopted. In consideration of the first goal of fairness, this decision must be made consensually – no one should be forced to compromise on their own values. For example, if the two water laboratories are in different states and one has stricter requirements for drinking water safety, then the determination of safe has to be made by the participants. This may or may not agree with their own state's legal definition.

## *Competence in each type of discourse*

### Types of discourse

In each of the four types of discourse different rules and procedures are effective for constructing consensual understandings of reality. Each of the two needs of competence are relevant to the four types of discourse. (Please refer to Figure 1 and Appendix.)

### Explicative discourse

In an explicative discourse, the comprehensibility of assertions is discussed. Comprehensibility includes: pronunciation, style, grammatical correctness, spelling, and using the proper definitions of words. The first four items are rarely problematic for a native speaker, and when they are, the matter is usually quickly dispatched. Native speakers have an immediate access to the validity source of comprehensive speech via their socialization. In the case of non-native speakers, these can present substantial problems. Access to the source of validity for these people can only be guaranteed by making interpreters available to them. As I stated above, these are considered to be trivialities that are taken for granted in the societies being discussed here.

Definitions are more problematic. In one sense, confusion over definitions is a kind of comprehensibility problem (because people are not understanding one another), but definitional disputes can also be normative disputes in disguise. In a discussion about drinking water quality, for example, the specific definition of safe is neither a linguistic problem nor a technical problem, it is a normative one. In explicative discourse, only issues of comprehension are discussed. There are commonly accepted definitions and everyone should have access to sources that define terms. The authority of references should be clarified and disputes resolved by appealing to the validity of these sources. Examples of sources include textbooks, value leaders, and experts.

### Theoretical discourse

Theoretical discourse addresses truths of the objectified world (nature and society). These facts are gathered through scientific methodologies (qualitative and quantitative) as well as through daily life experience. The content of such facts include, for example, soil properties, technologies, traffic flow patterns, insurance fees, laws, and existing public preferences and concerns. Natural and social sciences seek to reveal causal relationships by systematic observation and analysis. Rules and procedures for selecting from among validity claims (for example: the rule of experiment repeatability) have been established by the scientific community. These rules are subject to revision according to conventions developed in that domain.

Experts have a strong role to play in making systematic evidence and interpretations available to everyone in theoretical discourse. Every discourse participant must be able to access the information that he or she feels may be relevant. Effective access usually means hiring consultants who can explain laws, figures, and procedures; but it may also be as simple as copying reports for the participants. It includes not only the available data and knowledge, but also the collection and analysis of new data.

Without challenging the right of citizens in discourses to make final judgments of cognitive validity claims, it is possible to recognize that citizens have a right, and often an interest in exercising that right, to delegate some determinations of validity to groups or people whom they consider to be more expert than themselves. Delegation can only be legitimately done when the consent is unanimous. Without the protection of unanimity, individuals in the discourse may have their interests subjugated by the will of the majority, which could select an expert review panel that supports a particular interest position.

Information and knowledge about nature and society not gathered through formal scientific inquiries are also important to producing accurate understandings. These can take the form of anecdotal observations ("He knows three people who got brain cancer after using cellular telephones."), idiosyncratic observations – especially about local conditions ("That pond is completely dry in August."), or it may be traditional knowledge passed down over generations ("The full moon brings the first frost."). Local knowledge is particularly relevant in some cases, because expert knowledge is based on generalizable principles that may be incomplete or inaccurate in the given context.

Just as scientific experts are hired to enhance access and use of scientific knowledge, "local knowledge experts" can be sought out as consultants. Additionally, participants may be encouraged to develop and improve upon their own local knowledge by gaining more personal experience. For example, if the problem concerns a specific site, a visit to the location and a walk around may help to give people a feeling for the site, thereby providing valuable knowledge that is not available through blueprints or maps.

No matter how the information is gathered, depictions of existing states of affairs are evaluated according to their consistency with what is already known. People must choose among conflicting truth validity claims by deciding which provides the better description of reality. If a claim challenges fundamental understandings, then it will likely be rejected. If it challenges weak or incomplete understandings, then it may be cautiously adopted pending more information (example: "Electromagnetic fields may cause leukemia in children."). If the evidence is compelling, however, beliefs may be revised immediately (example: "Silicon breast implants regularly leak and induce breast cancer."). Decisions to reject or adopt claims should be based on the consensual opinion of the expert community (Is the data valid?) and the common-sense opinion of the lay participants (Does it seem likely?). Local knowledge can also be "peer-reviewed" by other people in the area. When this is not possible, the reputation of the source individual can be investigated as a clue to reliability.

There is always the possibility of different types of validity claims being convoluted with each other. One well known problem is the re-framing effect – for example, when normative disputes are disguised as disputes about objective fact. To discourage this strategic or accidental activity, the speaker's intent should be clarified through questioning and re-phrasing.

## Practical discourse

Practical discourse involves disputes over claims about the appropriateness of social relations (norms). More than with any other kind of discourse, practical discourse requires broad-based participation of all affected people. Deciding who is affected is always a tricky part of public participation. The safest approach to take is to employ both objective methods (which uncover possible causal pathways and inform unsuspecting people of their potential affectedness) and subjective techniques (which allow people to decide for themselves whether or not they are affected).

When every single person cannot be accommodated, then a surrogate way to restrict access that does not disadvantage any particular interest must be developed. Some interests – some people – will be disadvantaged by a selection routine, that is unavoidable in any practical setting. All that can be done is to ensure that no one person or interest group is strategically or systematically disallowed to participate. The simplest example is that the meeting should not held in a location that is extremely difficult for one person or a group to reach.

Accessing relevant knowledge and data about normative claims must include the ability to hear and question individuals and groups in the population, as well as receive information about the factual implications of normative choices. Regarding the former, discourse participants may decide to organize some mechanism for public comment (hearings, surveys, open houses, etc.). Regarding the latter, they may ask expert panels to prepare scenario analyses of what might happen were a certain decision made. Normative choices must not only be preferable, but also possible and this requires information about the objectified worlds.

Procedural rules for selecting among competing normative arguments are one of the most sensitive parts of any participation program. Indeed, one of the reasons for this is that there is no explicit consensus on how to make normative choices. But there are common sense rules that encourage an open discussion about shared preferences.

A basic requirement of normative choice is that it be consistent within itself and within the belief system in which it is embedded. Normative choices should conform to established norm orderings. Established norms may be common sense (example: "We should not spend any more time than absolutely necessary on this.") or they may be formalized in law. Laws appear as objective conditions within which norms must operate. Unlike cognitive aspects of law, which are straightforward ("Is this chemical included on the list of 189 toxic chemicals in the Clean Air Act?"), normative aspects of law often call for interpretation ("Is this equipment considered BACT [best available control technology]?"). One way to test is to have legal experts review the proposed normative choice.

## Therapeutic discourse

Therapeutic discourse does *not* refer to empty pacification of legitimate citizen concerns. Expressive claims have two aspects to their validity: authenticity (is this truly what the speaker feels, thinks, has experienced, etc.?) and sincerity. For example, suppose in a discussion about permitting logging in a new region, a man asserts that he is afraid the state will soon loosen restrictions on logging near streams, thereby increasing the potential for harm to fish. He is not making a cognitive or normative statement, but expressing a fear or apprehension. The listeners are now charged the responsibility of redeeming or not redeeming his claim. They must ask themselves if he is speaking authentically and sincerely.

Of course another person might choose to challenge the cognitive claims implicit in this concern, that logging near streams can increase silt runoff into streams, or that increased siltation will hurt fish. This shift in the discussion to theoretical discourse is an example of how translation of expressive claims might occur, but that is not the point of therapeutic discourse itself.

Therapeutic discourse is likened to a conversation between a psychotherapist and a patient in which the patient is encouraged to explore the authenticity of his or her own subjectivity. The conversation consists of the doctor *passively* assisting the patient – by asking for clarifications and suggesting specific investigations – in order to seek out authentic feelings, experiences, desires, beliefs, etc. Authentic understandings of one's own subjectivity can only be made knowable to the self through introspection. Only the patient can know when the validity has been established, the doctor may guess but cannot make accurate judgments.

There is no way for participants to directly verify the speaker's subjective experience or to guess at his or her motivations, but there are ways to promote authentic expression and to expose aspects of truthfulness. One way that people come to understand their own opinions is through discussions with others. Promoting small group discussions and allowing time for personal reflection (perhaps by stretching the discourse out over several days or weeks) gives people to opportunity and incentive to inquire into their true subjectivities. Second, a person's reputation and association may help listeners to critically judge what a speaker reveals about him- or herself. This requires information about past promises and behavior. Honesty and integrity of the speaker provide a basis for making a judgment about the likelihood that the speaker is truthful. In addition, motivations for lying such as conflict of interests must be examined.

People gain insights into the sincerity of others' subjective experiences through empathizing, by helping that person to explain the cognitive or normative basis behind the expression, or by examining his or her reputation.

To empathize, people need an open mind and a sensitivity to others. There is little that a discourse structure can do to promote empathy, except for adopting at the onset a list of commitments that state the shared interest in empathizing with another.

## Evaluating Models of Participation with the Discursive Standard Criteria

### *Qualitative assessment*

Figure 1 illustrates how the criteria listed in the Appendix relate to the metacriteria of fairness and competence. Completing an evaluation is not a quantitative exercise, but requires interpretations and judgments. Attempting to quantify scores for each indicator and add these to arrive at a score for each criterion would be to unjustifiably assume that the different items share some ordinal dimension.

The best way to employ these criteria is to qualitatively assess how well the model performed (or would be likely to perform) on each criterion, making one assessment based on the indicators comprising the criterion. This assessment is best represented in terms of plus, neutral, or minus. Combining the assessments for different criteria, for example, to arrive at an overall rating of fairness, is done in a similar way.

This is a list of standard criteria that can accommodate a wide variety of public participation applications. There will certainly be applications where some criteria matter more than others. Inevitably, one can locate examples where this or that criterion does not matter. Unlike mathematical theorems, which can be disproved with one exception, these evaluative criteria must be treated with a degree of interpretation and flexibility. Every criterion and indicator in the listing may not be essential to the success of every public participation exercise. However, each and every criterion and indicator are consistent with the goals of achieving competence and fairness in public participation.

### *Adopting a moral attitude*

Acting on “higher” motives than one’s own interests is moral action. This is the key ingredient to cooperation – the willingness of the participants to set their personal interests aside as they search for the best decision for all.

Neither the standard discursive criteria nor Habermas’s discourse ethics or ideal speech situation can force people to act morally. Nonetheless, a moral attitude is necessary for the success of communicative action. This chapter began with the assumption that citizens who participate invoke a moral attitude. Obviously, this is not always the case. People participate for a variety of reasons, some more self-centered than others.

Work needs to be done to understand how structural conditions of the discourse can promote participants to adopt a moral attitude. This will add to previous work to identify aspects of the problem setting that facilitate moral action, such as gridlock – when no one can gain from the status quo. People are also induced to act morally when they require each other's cooperation to acquire a common good or when the level of social cohesion and the sense of social responsibility is high.

## *Importance of participation as education*

Public participation is much more than conflict resolution. On an immediate level, its goal is the construction of an image of self at both the personal and community levels. This happens as people explore the questions: "What are our needs, desires, and interests?" and: "How can the wide assortment of different and often conflicting particular wills be integrated into a generalized will?" Rarely is all conflict erased. But in answering these questions citizens place themselves within a larger definition of self that expresses itself in the generalized will. Participation is a kind of "on the job training" for citizens. It is an educational experience that should leave the participants with a better idea of what it means to work out differences cooperatively and with the skills and knowledge to do so. Altogether, participation is a way to socially construct an holistic picture of reality depicting personal and shared relations to inner nature, outer nature, and society.

## *The relevance of Habermas's work for citizen participation*

Current understandings of public participation emerge out of macro-level analyses that focus either on the function of participation in maintaining social and political order or on the opportunities it provides for social change. These approaches have been useful toward revealing the strengths and limitations of the different models of participation as an institution of democratic society, as well as illustrating the ramifications of social and political changes in modern life. However, because they are macro-level approaches, they do not engender an understanding of how public participation relates to individuals at both a personal and social level. The significant feature of the approach taken here is that public participation is viewed from an interactionist perspective, one that focuses on the communication between individuals, which lies at the core of public participation.

Habermas's theory of communicative action has been used because it provides a normative theory of human interaction. It contends that a better society can only come about when people are free from all forms of arbitrary domination and when non strategic and cooperative reasoning is as highly developed and widely employed as instrumental reason. The concept of the ideal speech situation embodies these normative ideals and connects the theory of interaction to the practice of evaluation.

This does not mean that other forms of democratic decision making cannot be fair and competent. It is true that Habermas often conveys the idea that strategic forms of interaction are inferior and ought to be eliminated. Few people find that view acceptable. Indeed, strategic action does have a place in "right" citizen participation. Accurate impact assessments require that the reactions of individuals be taken into account, for instance in computing how many people might leave town, how property values may be affected, etc.

Habermas's theory of communicative action is a consensus theory of truth. The standard discursive criteria describe how one aspect of consensual democracy can be realized, but this should not be interpreted as a condemnation of adversarial decision making. Both consensual and adversarial forms of democracy have their advantages and disadvantages. My only claim is that, if we want to engage citizens early on in environmental policy making, it is best to adopt a consensual, rather than an adversarial approach. When citizens are involved in other aspects of governing, adversarial approaches may be more appropriate. Another set of criteria would be needed to evaluate those procedures.

Habermas's practical insight is that societal decisions should be made through a discursive process in which *collective* preferences, interests, and needs *are defined by the participants* in accordance with their own free will. As long as people are not forced to surrender their values in the selection process, the final decision should reflect the best possible agreement. Equally important is that non-coercive communication also serves as a vehicle for self reflection and that through reflection needs evolve in response to changing environmental and social pressures.

The part that seems to be missing from Habermas's theory is that discourse needs some structure to enable the competent rationally-motivated redemption of validity claims. Structural conditions of the discourse (rules, facilitation, agenda) provide the boundaries that keep discourse moving in the direction of competence. But Habermas does not describe these conditions. He does outline conditions for the ideal speech situation – to ensure fairness, and he does discuss what communicative competence in the individual means, but he never integrates these two concepts into a workable definition of fair and competent discourse.

## *Implications for practice*

In conclusion, merely putting people into a room and telling them to work out a non-coercive consensual agreement is not always good enough. Anyone who has served on a jury knows how frustrating and inefficient such settings can be. While participating in such a setting can be very educational and a valuable experience, as a routine tool for making decisions outside of the courtroom this approach is simply impractical. Some have even questioned its practicality in the courtroom (Bownes 1990).

People need the benefit of procedures that society has developed over time for reliably selecting knowledge and guaranteeing individual autonomy. These cannot be left up to democratic, charismatic, or autocratic mechanisms. Rules that are the result of generations of trial and error and the exploitation of expertise should bracket the liberties of the participants. However, only the participants themselves have the legitimacy to place themselves under these constraints. The justification I have offered for the specific set of criteria presented here is that these describe a mode of discourse that almost all people would find competent and fair. It does not intentionally prefer one group's interests to another group's interests, and it is oriented toward producing the most reliable understandings and agreements that are reasonably achievable.

Two normative premises form the basis for the approach taken here. First, no one should be allowed to force his or her own interpretation upon the collective construction of knowledge. Second, the group should formulate a means to reach discourse closure. Saying that public participation must be fair and competent is another way of expressing these premises and the literature on public participation supports the choice of these two goals as reasonable.

## *"Right" citizen participation*

I have chosen the word "right" to signify a universalistic grounding for the normative legitimacy of these criteria. The origin of this legitimacy lies in fundamental facts about how people use language. Adopting an attitude of cooperation, recognizing everyone has access to common standards of validity, and respecting individual autonomy are all inherent features of language use that are reconfirmed every time people speak to each other. These are realized in "right" citizen participation through constraints and liberties promoted in the structure of the discourse models.

The discursive standard criteria and indicators describe ideal discourse in environmental decision making. "Right" participation encourages multi-way communication. Unlike one way communication (education or persuasion), or two-way communication (between individuals), multi-way communication refers to a group setting where people are actively participating as speakers and listeners. The discussion is diffuse and simultaneously involving all the participants.

"Right" participation is also consensual and non-hierarchical. That means it is based on the principle that no one should be forced to surrender his or her right to contribute to determining discourse closure. There are many reasons why a group member may choose to surrender his or her veto right, this is why one may find voting or sending to committee happening in "right" participation.

"Right" participation requires respect for the autonomy of the individual and trust that the person will abide by reasonable rules for social interaction. It does not require trust in the sense that the speaker's validity claims are blindly accepted. That would put the speaker in a position of being able to proclaim a specific definition of reality. Over time, certain people may acquire a reputation of producing particularly reliable validity claims, and others may, consequently, choose to adopt that person's word on given occasions, but this is a convenience not a necessity. The illusion that trust is required for discourse seems based in a desire to instill an authoritarian definition of reality in one particular person. That is neither competent or fair.

"Right" participation relies on the reasonableness of the citizenry to produce workable decisions. Only by giving the people the power to make determinations about their own government can it be assured that their interests are protected and that the collective choice is competent. This is the only way to "repair the lifeworld." The end product will be that culture is reproduced, society is integrated, and healthy personalities are nourished.

Finally, "right" participation promotes critical self reflection among the participants. In making decisions, people are actually determining what kind of society they want to have. Individual interests, beliefs, and values become linked with those of others into an idea of self that has expanded to include other interests, values, and beliefs. In making collective decisions everyone needs to ask: "Whose interests are being served here?" This is the way to expose injustice and incompetence. Good and just decisions can only come about when people critically reflect upon the connections between themselves, their neighbors, the economic structure and system, the environment, and government, and realize their future course together in mutual interdependency.

## References

Alexy, R, "Eine Theorie des praktischen Diskurses," in: W. Oelmüller (ed.) *Normen begründen, Normendurchsetzung* (Paderborn, 1978).

Austin, J., *How to Do Things with Words* (Harvard University Press: Cambridge, MA., 1969).

Benhabib, Seyla, *Critique, Norm, and Utopia* (Columbia University Press: New York, 1986).

Bernstein, Richard (ed.), *Habermas and Modernity* (MIT Press: Cambridge, MA, 1985).

Bownes, Hugh, "Should Trail by Jury be Eliminated in Complex Cases?," *RISK:Issues in Health and Safety* 1(1):73-78 (1990).

Braaten, Jane, *Habermas's Critical Theory of Society* (SUNY Press: Albany, NY, 1991).

Brulle, Robert, "Jürgen Habermas: An Exegesis for Human Ecologists," *Human Ecology Bulletin* Spring/Summer 1992 (No. 8):29-40.

Dietz, Thomas, "Social Impact Assessment as a Tool for Rangelands Management," in: National Academy of Sciences (eds.), *Developing Strategies for Rangelands Management* (Westview: Boulder, CO, 1984), pp. 1613-1634.

Dietz, Thomas, "Theory and Method in Social Impact Assessment," *Sociological Inquiry* 57:54-69 (1987).

Dietz, Thomas, "Social Impact Assessment as Applied Human Ecology: Integrating Theory and Method," in: Richard Borden, Gerald Young, and Jamien Jacobs (editors) *Human Ecology: Research and Applications* (Society for Human Ecology: College Park, MD, 1988), pp. 220-227.

Dietz, Thomas, "The Challenges of Global Environmental Change for Human Ecology," in: Lars O. Hansson and Britta Jungen (eds.) *Human Responsibility and Global Change* (University of Göteborg: Göteborg, Sweden, 1992), pp. 30-45.

Dietz, Thomas, and Alicia Pfund "An Impact Identification Method for Development Program Evaluation," *Policy Studies Review* 8:137-145 (1988).

Fiorino, D. J., "Citizen Participation and Environmental Risk: A Survey of Institutional Mechanisms," *Science, Technology, & Human Values* 15(2): 226-243 (1990).

Forester, John (ed.), *Critical Theory and Public Life* (MIT Press: Cambridge, 1985).

Forester, John, *Critical Theory, Public Policy, and Planning Practice: Toward a Critical Pragmatism.* (SUNY Press: Albany, NY, 1993).

Habermas, Jürgen, "Toward a Theory of Communicative Competence," *Inquiry* 13(4):360-376 (1970).

Habermas, Jürgen, "Wahrheitstheorien," in: H. Fahrenbach (ed), *Wirklichkeit und Reflexion: Walter Schutz zum 60 Gebürtstag* (Pfüllingen: Neske, 1973), pp. 211-263.

Habermas, Jürgen, *Communication and the Evolution of Society* translated by Thomas McCarthy, (Beacon Press: Boston, 1979).

Habermas, Jürgen, "A Reply to my Critics," translated by Thomas McCarthy, in: John B. Thompson and David Held (eds.) *Habermas: Critical Debates* (MIT Press: Cambridge, MA, 1982), pp. 219-283.

Habermas, Jürgen, *Theory of Communicative Action. Vol. 1: Reason and the Rationalization of Society* translated by Thomas McCarthy, (Beacon Press: Boston, 1984).

Habermas, Jürgen, *Theory of Communicative Action. Vol. 2: System and Lifeworld* translated by Thomas McCarthy, (Beacon Press: Boston, 1987).

Habermas, Jürgen, "The Public Sphere: An Encyclopedia Article," translated by Sara Lennox and Frank Lennox, in: Stephen Eric Bronner and Douglas MacKay Kellner (editors), *Critical Theory and Society: A Reader* (Routledge: London, 1989).

Habermas, Jürgen, *Moral Consciousness and Communicative Action* translated by Christian Lenhardt and Shierry Weber Nicholson, second printing (MIT Press: Cambridge, MA, 1991).

Habermas, Jürgen, *Autonomy and Solidarity* Revised Edition, Edited by Peter Dews (Verso: London, 1992).

Horkheimer, Max, *Critical Theory: Selected Essays* (Herder and Herder: New York, 1972).

Johnson, James, "Habermas on Strategic and Communicative Action," *Political Theory* 19(2):181-201 (1991).

Kemp, Ray, "Planning, Political Hearings, and the Politics of Discourse," in: John Forester (ed.), *Critical Theory and Public Life* (MIT Press: Cambridge, 1985).

Laird, Frank, "Participatory Analysis: Democracy and Technological Decision Making," *Science, Technology, and Human Values* 18(3):341-361 (1993).

Moon, Donald, "Constrained Discourse and Public Life," *Political Theory* 19(2):202-229 (1991).

Rosener, Judy, "Citizen Participation: Can We Measure Its Effectiveness?" *Public Administration Review* 38:457-463 (1978).

Stern, Paul C., "Learning Through Conflict: A Realistic Strategy for Risk Communication," *Policy Sciences* 24:99-119 (1991).

Stern, Paul C., Oran R. Young, and Daniel Druckman (eds.), National Research Council *Global Environment Change: Understanding the Human Dimensions* (National Academy Press: Washington, D.C., 1992).

Stern, Paul C., and Thomas Dietz, "The Value Basis of Environmental Concern," *Journal of Social Issues* (forthcoming).

Stallen, Peter Jan, "Risk of Science of Science of Risk?" in: Harry S. Otway (ed.) *Scienctific and Cultural Approaches to Risk* (Dordrecht: D. Reidel, 1985), pp. 131-148.

White, Stephen, *The Recent Work of Jürgen Habermas* Second Printing (Cambridge University Press: Cambridge, 1989).

# *Appendix to Chapter 3*

## *The Discursive Standard Criteria*

### 1. Speech Acts and Validity Claims

Habermas uses speech act theory to explain what people do in discourse. He focuses on four types of speech acts, each of which draws attention to a specific "world." Each speech act also makes an implicit *validity claim, i.e. the appeal implicit in a statement that makes the message meaningful. It is the essential Truth-oriented component of the statement.*

Communicative speech acts are "trivial," because they merely claim to be understandable utterances. Their validity claim is to *comprehensibility*, and is grounded in the "world" of *language*.

Ex: *During a discourse, people speak to one another.* – Grammatically correct sentence.

Constantive speech acts claim to represent something in the *objectifiable world of nature or society*. The claim they make to validity is *truth or correctness*.

Ex: *Average annual rainfall here is 42 inches.* – Verifiable fact.

Regulative speech acts propose an appropriate set of interpersonal relations or norms. Their validity claim is to *rightness* and they appeal to the world of *culture or society*.

Ex: *We should not trade the loss of an endangered species for jobs.* – Normative contention.

Representative speech acts reveal a selective element of the speaker's own *subjectivity*. In so doing, they make validity claims to *truthfulness and sincerity*.

Ex: *I am pleased to hear that you agree.* – Expression of personal subjectivity.

All rationally-motivated communication, then, centers around these four types of speech acts. Speech act theory contains other types of speech acts, but these are the ones associated with non-strategic action.

Habermas believes that the different validity claims require fundamentally different means for *redemption*, that is, the process by which the discourse participants collectively decide whether or not the speaker's validity claim is valid (and therefore, acceptable or redeemable). Thus, he distinguishes between four *types of discourse* in which the four validity claims are discussed. These are called: explicative, theoretical, practical, and therapeutic discourse.

## 2. The Discursive Standard Criteria for Fairness

Fairness is an assessment from the point of view of an individual of the sufficiency of opportunities that each and every person will have to protect and express their legitimate individual interests and to contribute to the development of the collective will. Individuals in a group must be able to initiate actions that enable them to accomplish their legitimate goal. In a fair discourse, each participant should feel safe and comfortable about advancing any type of statement and participating in judging the validity of others' claims. Whether or not the individual has the ability to make maximum use of these opportunities is something that will be taken up in the discussion about competency.

There are four major elements to the fairness criterion:

1) Attendance at the event.
2) Initiation of different types of speech acts.
3) Participation in debate for and against validity claim redemption.
4) Participation in the group resolution of disputes over claims.

From an interest in fairness, three activities in public participation are important to distinguish: agenda and rulemaking, moderation and rule enforcement, and the discussion. Below is listed the criteria and accompanying indicators that evaluate fairness in that activity.

### A. Making of the Agenda and the Process Rules

**A1. The model should provide everyone with an equal chance to put their concerns on the agenda and to approve or propose rules for discourse.**

A1-1 Does the model provide an opportunity for everyone to suggest items for the agenda?

A1-2 Does the model provide an opportunity for everyone to suggest items for the rules?

**A2. The model should provide everyone with an equal chance to debate and critique proposals for the agenda and the rules.**

A2-1 Does the model provide everyone an equal opportunity to debate proposals for the agenda?

A2-2 Does the model provide everyone an equal opportunity to debate proposals for the rules?

A2-3 Does the model provide enough time to accommodate all agenda items that the group wants to discuss?

A2-4 Does the model provide an opportunity for everyone to suggest changes to the agenda or the rules?

**A3. The model should make certain that everyone has an equal chance to influence the final decision about the agenda and the discourse rules.**

A3-1 Does the model provide a consensually-approved means to resolve conflicts about the agenda?

A3-2 Does the model provide a consensually-approved means to resolve conflicts the rules for discourse?

### B. MODERATOR AND RULE ENFORCEMENT

**B1. The model should provide everyone with an equal chance to suggest a moderator and a method for facilitation.**

B1-1 Does the model provide an opportunity for everyone to suggest a moderator?

B1-2 Does the model provide an opportunity for everyone to comment on the facilitation style?

**B2. The model should provide everyone with an equal chance to challenge and support suggestions by others for a moderator and a method for facilitation.**

B2-1 Is there a setting for discourse among all who wish to debate proposals for moderator?

B2-2 Is there a setting for discourse among all who wish to debate proposals for how moderation should be carried out?

**B3. The model should provide everyone with an equal chance to influence the final selection of moderator and moderation method.**

B3-1 Does the model provide a consensually-approved means to resolve conflicts about the choice of moderator, either through selection or verification?

B3-2 Does the model provide a consensually-approved means to resolve conflicts about the style of facilitation?

### C. DISCUSSION

**C1. The model should provide everyone who is potentially affected by the decision proposal (positively or negatively) an equal chance to be present or represented at the discourse.**

C1-1 Does the model attempt to identify the individuals or groups that are potentially affected by the problem?

C1-2 Does the model provide all people in the greater affected population an equal chance to participate?

C1-3 Does the model provide all people who feel they are affected an equal chance to participate?

**C2. The model should make certain that everyone has an equal chance to put forth and criticize validity claims about language, facts, norms, and expressions.**

C2-1 Does the model provide all an equal chance to make communicative validity claims?

C2-2 Does the model provide all an equal chance to make cognitive validity claims?

C2-3 Does the model provide all an equal chance to make normative validity claims?
C2-4 Does the model provide all an equal chance to make expressive validity claims?

**C3. The model should make certain that the method chosen to resolve validity claim redemption dispute be consensually chosen before the discourse began.**

C3-1 Does the model make certain that disputes over communicative validity claims will be resolved using a procedure that was consensually approved before the discourse began?
C3-2 Does the model make certain that disputes over cognitive validity claims will be resolved using a procedure that was consensually approved before the discourse began?
C3-3 Does the model make certain that disputes over normative validity claims will be resolved using a procedure that was consensually approved before the discourse began?
C3-4 Does the model make certain that disputes over expressive validity claims will be resolved using a procedure that was consensually approved before the discourse began?

## 3. The Discursive Standard Criteria for Competence

Competence refers to the ability of the participation decision making process to provide the participants with the procedural tools and knowledge needed to make the best possible decisions. This evaluation is not necessarily tied to the character of the decisions produced and is not determined by a preset valuation of decision outcome (i.e., a decision not to site a facility may be competent).

There are two major elements associated with this criterion:

1) Assess to knowledge and interpretations.
2) Implementation of the best procedures for resolving disputes about knowledge and interpretations.

There should be competence in all four types of discourse that occur:

Explicative discourse – references are made to language, terms, definitions, grammar.
Theoretical discourse – reference are made to the objectified world (nature or society).
Practical discourse – references are made to social needs and the appropriate forms of social interaction.
Therapeutic discourse – references are made to the subjectivity of the speaker.

**D. RULES FOR REDEEMING COMPREHENSIBILITY VALIDITY CLAIMS**

**D1. The model should provide everyone equal access to the sources for commonly-agreed-upon standards and definitions.**

D1-1 Does the model provide every participant equal access to the commonly-agreed-upon sources for definitions of terms that are relevant to the discourse?

D1-2 Does the model provide the flexibility in time that is needed to resolve comprehensibility problems?

**D2. The model should confirm that everyone has an understanding of each others' terms, definitions, and concepts.**

D2-1 Does the model make certain that all terms, definitions, and concepts are made explicit?

D2-2 Does the model make certain that all participants acknowledge that they understand the agreed-upon definitions?

**D3. The model should make certain that disputes about definitions, terms, and concepts take advantage of preestablished references standards.**

D3-1 Does the model encourage the resolution of disputes through appealing to commonly-agreed-upon standards (such as a dictionary, or a textbook)?

**E. RULES FOR REDEEMING TRUTH VALIDITY CLAIMS**

**E1. The model should provide everyone equal access to the available and relevant systematic knowledge about the objective world.**

E1-1 If expert advice is to be brought into the group, does the model assure that the agreement to do so is consensual?

E1-2 If consensus on how to bring expertise into the group cannot be achieved, does the model provide the financial means for every participant to hire their own expert help?

E1-3 Is the model flexible enough to allocate time to consult with experts and to have experts collect data?

E1-4 If there is an educational component, is the material reviewed by independent experts and/or stakeholder groups?

**E2. The model should provide everyone equal access to the available and relevant anecdotal and intuitive knowledge about the objective world.**

E2-1 Does the model promote the consideration of anecdotal and intuitive knowledge?

E2-2 Does the model promote ways for the people to improve their own anecdotal and intuitive knowledge by being exposed to relevant experiences (field trips, lectures from other similarly impacted people, site visits, etc.)?

**E3. The model should make certain that the uncertainty of factual information is considered along with content.**

E3-1 Does the model provide a means for the uncertainty of factual information to be considered?

**E4. The model should include a mechanism to check if factual claims are consistent with the prevailing opinion in the expert community or consistent with the anecdotal knowledge of other people not involved in the discourse.**

E4-1 Does the model promote peer review and independent verification of scientific data and knowledge?

E4-2 Does the model promote "peer-review" and independent verification of anecdotal knowledge?

E4-3 Does the model provide enough time for participants to collect the scientific data and anecdotal experience they feel is relevant and to discuss it thoroughly?

**E5. The model should provide a means to separate cognitive claims from normative claims.**

E5-1 Does the model provide a means to translate claims into their cognitive and normative constituent parts?

E5-2 Does the translation require verification by the speaker?

**E6. The model should provide the participants with the option to delegate determinations of factual truth to an outside expert panel.**

E6-1 Does the model permit the participants to select an expert panel consensually and ask for its recommendations?

E6-2 Does the model ensure that the decision to rely on expert advice is consensual?

E6-3 Does the model provide information about the range of expert opinions and positions in that particular subject?

**E7. The model should make sure that cognitive legal claims are examined by legal experts.**

E7-1 Does the model ensure that legal experts will verify how well the decision outcome conforms to the technical definitions in the law?

### F. RULES FOR REDEEMING NORMATIVE VALIDITY CLAIMS

**F1. The model should not contain any implicit barriers that will bias the distribution of interests that participate.**

F1-1 Does the model provide adequate notice of all activities?

F1-2 Does the model have a purpose that is made clear to all beforehand?

F1-3 Are the physical, social, economic, and symbolic barriers to participating in the model removed?

F1-4 Does the model make a connection between purpose, process, and outcome?

F1-5 Does the model include an effort to achieve representation of formal interest group organizations in the discourse?

F1-6 Does the model include an effort to achieve representation of ad hoc interest group organizations in the discourse?

F1-7 Does the model include an effort to randomly select participants for the discourse?

**F2. The model should determine the affected population using objective criteria but also allow the people in the general region to make subjective determinations.**

F2-1 Does the model employ an objective method to determine who makes up the potentially affected population?

F2-2 Does the model permit citizens to make their own personal determination of whether or not they are a member of the affected population (and so, should have an equal opportunity to participate)?

F2-3 Does the model attempt to inform the greater population about the potential impacts so that they can make informed judgments of whether or not they feel affected?

**F3. The model should promote both the discovery and the development of mutual understandings of values among all the participants.**

F3-1 Does the model promote the elicitation of values from the community, its government, and the stakeholder groups?

F3-2 Does the model inform everyone of each others' values and interests?

F3-3 Does the model promote introspective reflection among individuals or groups into the currently existing values and interests of the community through techniques such as small group discussions?

F3-4 Does the model provide a mechanism by which the impacts of the proposed decision options on the generalized will can be characterized relative to the definition of the generalized will?

**F4. The model should make certain that the factual implications of normative choices are considered in practical discourse.**

F4-1 Does the model provide a mechanism to evaluate the cognitive implications of proposed normative choices?

F4-2 Does the model make sure that all participants know the anticipated physical and social consequences of their normative preferences before making a decision?

**F5. The model should promote, through rational and formal discourse procedures that build compromises, the discovery and development of a mutual understanding of values in order to formulate a generalized will.**

F5-1 Does the model provide flexibility in terms of the time available?

F5-2 Does the model provide information or training to the participants on how to build compromise and resolve disagreements?

F5-3 Does the model promote the use of small group discussions?

F5-4 Does the model discourage people from prejudging the moral beliefs of others?

**F6. The model should make certain that normative choices are not inconsistent with themselves or with the general will.**

F6-1 Does the model provide a systematic structuring of values?

F6-2 Does the model encourage the participants to pay attention to the consistency and contradictions among norms and to use these standards in judging others' claims?

**F7. The model should make certain that normative choices are not incompatible with laws.**

F7-1 Does the model provide a means to check that the decision choice is consistent with the intent of legal provisions?

**F8. The model should make certain that normative choices are compatible with present expectations.**

F8-1 Does the model provide the means to check that the decision choice does not violate a higher norm in pursuit of a lower one?

F8-2 Does the model promote reciprocal validation of values and their interpretations between those who promote them and those who have to live with the consequences?

## G. Rules for Redeeming Truthfulness Validity Claims

**G1. The model should promote discussion about the authenticity of the speaker's expressive claims.**

G1-1 Does the model promote personal reflection?

G1-2 Does the model provide participants with the opportunity to informally discuss their feelings with their friends and colleagues?

G1-3 Does the model encourage the participants to try and empathize with the speaker?

**G2. The model should promote an examination into the speakers' sincerity.**

G2-1 Does the model promote a discussion about the commitment of the participants to cooperation?

G2-2 Does the model promote a discussion about the promises, past behavior, and future performance of the participants?

**G3. The model should promote an examination into the qualities of the situation.**

G3-1 Does the model promote a discussion about the organizational limitations that may impact on the project?

G3-2 Does the model promote a discussion about the capability of the actors?

G3-3 Does the model promote a discussion about or provide information about the availability and uncertainty of factual information when discussing expressive claims?

**G4. The model should provide individuals time enough to accurately state and defend their expressive claims.**

G4-1 Does the model provide speakers with the time they need to discuss expressive claims?

**G5. The model should use a translation scheme that is acceptable to everyone.**

G5-1 Does the model promote the use and development of a method to translate expressive claims into cognitive or normative bases?

G5-2 Is translation verified by the person expressing the claim?

**H. ALL CATEGORIES**

**H1. The model should reduce the misunderstanding before reaching for agreement.**

H1-1 Does the model encourage the participants to reach compromise on redeeming validity claims only after they have been clarified?

H1-2 Does the model attempt to clearly state the existing consensus of the group?

H1-3 Does the model feedback the final statement for verification?

**H2. The decision as to which validity claims are redeemed by the group should be made using a technique that was consensually pre-approved.**

H2-1 Does the model use a technique to resolve disagreement about validity claims that was pre-approved consensually?

Chapter 4

# *The Redemption of Citizen Advisory Committees: A Perspective from Critical Theory*

*Frances M. Lynn*
*Jack D. Kartez*

## Introduction

Without doubt, the citizen advisory committee (CAC) has been the longest-used participatory technique among those covered in this volume. This durable and, in the United States, ubiquitous approach, even with its serious deficiencies in terms of fair and competent discourse, offers important lessons. This is because the validity of the CAC approach is periodically redeemed through changes in the way the bodies are organized and used.

We chose the word redemption to signal an analogy to Habermas's concept of communicative ethics in which he considers the necessary conditions for non coercive discourse aimed at achieving consensual agreements and understandings. Habermas contends that in the dynamics of political life "fundamental differences in beliefs and values block the initiation or continuation of (the) communicative relations that are necessary to consensus or accord" (McCarthy 1985). Habermas's conception of competent and legitimate communication is built on a philosophical presupposition about the "possibility of instituting or reinstating a consensual basis for interaction without resort to force in any of its forms from open violence to latent manipulation" (McCarthy 1985:291). Latent manipulation, such as asking citizens to present their views about a pending action when the basic decisions have already been made, has often characterized government consultation with the public. The "truth, freedom and justice" which, Habermas argues, potentially exist in human communication are destroyed by such distortions (Habermas 1970).

In this chapter we first briefly review the historical context and present an overview of the current debate about the strengths and weaknesses of citizen advisory committees. We then, using the broad criteria which guided the evaluation of all the techniques in this book, analyze the legitimacy and utility of the CACs and efforts to redeem them. We also examine important changes currently taking place in the venue and structural context in which CACs are being used (for example, the removal of government as the intermediary in citizen-industry discourse).

## History and Overview of Citizen Advisory Committees

In the United States, experimentation and reform of the citizen advisory committee for environmental issues spans more than twenty years, though the precedent of citizen advisory committees in government goes back almost to the nation's founding. Most use of CACs in American environmental affairs has revolved around their role in local, statewide, or regional issues. That is the level closest to the affected parties, where issues of legitimate and competent participation are most acute. A major historical precursor for CACs is the city planning and zoning commissions. The development of planning commissions illustrates an important contradiction in the use of citizen advisory bodies in the American context: when such bodies become formalized, they often become out of touch with the citizens they represent. Unfulfilled demands for legitimate citizen representation eventually re-emerge.

City planning commissions, for example, began as voluntary, non-governmental groups of leading citizens concerned about the waste, pollution and corruption of pork barrel public works spending and land development under late 19th century city machine politics. Such "civic associations" were formed to prepare downtown land plans in Cleveland (1902) and San Francisco (1904). Their members represented liberal reform interests who both controlled capital and wanted rational "good government" through the sponsorship of professionally-prepared city plans. Another motivation was fear of the instability and crises that could be brought on by the increasing degradation of the working class living conditions in major cities (Hall 1988).

Formalization of these citizen bodies soon followed. The Connecticut legislature sanctioned the first official city planning commission for Hartford in 1907, and Milwaukee, Chicago, Baltimore and Detroit had institutionally-endorsed commissions within two years. In 1928, the federal Department of Commerce developed the model Standard City Planning Enabling Act for state legislatures. The commissions, as quasi-independent, appointed citizen bodies, were to represent the community's true interests by virtue of their supposed insulation from politics.

Nonetheless, despite their reformist tilt, planning commissions often tended to represent major organized interests, such as real estate and business. Moreover, the increasing loss of community consensus regarding the assumed benefits of rapid local growth in the 1960's and 1970's often created conflicts between appointed planning commissions and the community at large. In a new wave of land use reform, questions were raised about how well commissions represented citizen interests (cf. Walker and Heiman 1981;

Rosenbaum 1976). That and other pressures led to the frequent use of citizen planning advisory committees (CPACs) to augment the official commission's deliberations in order to build broader legitimacy (cf. Solnit 1982, 85).

In the late 1970's and the early 1980's increasing conflict and stalemate over the siting of radioactive, hazardous, and solid waste disposal facilities and the discovery of contaminated properties spurred a new wave of efforts to involve the public in forums supplementary to the obligatory and increasingly contentious public hearing. States, localities, and federal agencies convened citizen advisory committees to provide input for siting and even clean-up situations. A few industries with particularly severe contamination problems established citizen advisory committees to review facility operations annually as a means of regaining and maintaining trust and credibility. (Smith, Lynn, Andrews, 1986). Increasingly, local communities are establishing environmental affairs boards (EAB's), patterned after planning commissions but without the decision making power of the latter.

The debate over the role of citizen advisory committees in government decisions was well-defined over two decades ago by Arnstein (1969) and Burke (1968). On the one hand, Burke argued that citizen participation, through boards, commissions, and committees, serves important legitimating and efficiency functions for public institutions by educating the public. Within limitations, Burke suggested, governmental institutions also benefit from the public's knowledge.

On the other hand, Arnstein, in her oft-cited article, "A Ladder of Citizen Participation" stood on its head the assumption about the connection between citizen participation and institutional legitimacy. She argued that token participation through placation or socialization and therapy for the public are illegitimate functions of participation. The role of citizen participation advanced by Arnstein was to create citizen power through a partnership with authorities, if not outright citizen control. In Arnstein's ladder of citizen participation, citizen advisory committees rank in the middle in terms of decision making authority and oversight.

The chief benefits claimed for CACs are: they inform public agencies about broad community attitudes; they educate citizens about proposed institutional actions; they increase ultimate acceptance of those decisions; and they allow government and industry to deal with one relatively small body of citizens rather than the entire community. As well, it is possible for a CAC to influence institutional decisions by signaling errors made by technical experts or political leaders, and/or demonstrating that substantial public opposition exists (Howell, Olsen and Olsen 1987; Creighton 1980; Pierce and Doerksen 1976). Nonetheless, failed use and misuse of CACs has elicited almost as much criticism of the legitimacy and usefulness of citizen involvement efforts in environmental conflicts as has the "decide-announce defend" approach often used by government and industry (Klapp 1988; Anderson 1986; Ducsik 1984).

The term CAC is generic and simply describes any of "several techniques in which relatively small group of citizens who are called together to represent ideas and attitudes of various groups and/or communities (Rosener, 1978:118.)" In short, no single model of CACs can possibly capture the wide variations in their formal charges, organization, discourse procedures, or outcomes in specific situated political contexts. As we evaluate CAC's in terms of the common criteria used in this book, we will illustrate how these factors have been addressed by different reforms.

## Embeddedness, Legitimacy, and Accountability

One of the chief issues in citizen involvement efforts in the United States, and one of the key factors which define this book's concept of embeddedness, has been whether the parties involved are representative of the body politic. Similar to boards and commissions, the institutionalization of citizen advisory committees has traditionally been found to be associated with less representativeness, at least in terms of how committee members compare to the general public on attributes like income and education (Hutcheson 1984). Typically, the members have been leading citizens appointed by governmental institutions to be representative of public opinion, or at least to produce recommendations that the general public will accept as legitimate. On the other hand, there are instances where citizen advisory committees have been initially organized without government involvement, such as the original planning commissions. There are also a growing number of instances in which committee members, although appointed, are representative of a wide spectrum of community organizations.

Another issue which falls under the rubric of embeddedness is the power and authority under which CAC's operate. CAC's are typically given a charge by a dominant political-legal institution, such as a state or city government, to address a defined problem or agenda. This creates an inherent bias toward upholding the institution's goals. Still, there is wide variation in how much latitude CAC's have to review the issue and define the problem.

CACs are rarely formed to fundamentally redistribute political power. More often, they are used to rationalize established power through some degree of shared governance. Unlike appointed citizen policy boards, CAC's do not endow citizens with direct power over final decisions. It is possible, however, for a CAC to influence institutional decisions by flagging errors made by technical experts or political leaders, and/or demonstrating that substantial public opposition exists.

Ultimately, CAC recommendations can be ignored by governing bodies or regulators. This sets this approach apart from mediated negotiation, where the disputants define the agenda (usually outside established legal-institutional procedures) with the express purpose of solving a problem and resolving a clash of power (Susskind and Cruikshank 1987).

In terms of legitimacy and accountability, there are great variations in the extent to which a CAC is held or holds itself accountable to those it represents or the public at large. A CAC may work in isolation from the general public or it may involve others through its own open hearings or educational public workshops. A CAC's public education roles often depends on the level of monetary and resource support provided by the institutions that convene the group.

## Efforts Toward Redemption: The Oregon CAC Reforms and Canada's Advisory Group Guidelines

The approach of Oregon's Statewide Planning Program illustrates one approach to redeem CACs vis-a-vis issues of embeddedness, accountability and legitimacy. Oregon's landmark planning law, adopted in 1975 after a year of hearings and workshops involving 10,000 Oregon citizens, was passed in part because of concern among state elites that parochial decisions by local planning commissions no longer represented the public interest – or citizens concerned about uncontrolled growth.

Goal 1, with the legal force of an administrative rule, required that citizens "be involved in all phases of the planning process" that every city and county undertook. Prior to passage of the law, citizens' sole opportunity to participate in land use decisions was typically a contentious and inadequate public hearing.

Goal 1 requires that every city or county governing body include as a component of their citizen involvement program an "officially recognized citizen advisory committee (CAC) ... broadly representative of geographic areas and interests." (Land Conservation and Development Commission 1975). To ensure enforcement, the state's planning program created a Statewide Citizen Involvement Advisory Committee (SCIAC) to watchdog local governments, particularly regarding the right of citizens to ask for membership on a CAC rather than merely allowing local officials to appoint a narrow membership of their choice.

In addition to setting up the CAC's, Goal 1 had five other components designed to support their valid functioning. They included:

- Communication: citizens must have effective communication with elected and appointed officials.
- Citizen influence: citizens must be involved in all phases of the planning process, such as technical plan preparation, implementation and revision.

- Technical information: technical information necessary to reach policy decisions shall be available in simplified, understandable form.
- Feedback mechanisms: citizens who participated and made recommendations shall receive from policy makers a written response and "rationale" for final decisions.
- Financial support: governing bodies shall allocate adequate human, financial and informational resources to the citizen involvement program (Land Conservation and Development Commission, 1975).

The need for those additional requirements may seem self-evident now, but the formal adoption of these basic standards was well in advance of the academic and bureaucratic interest in public participation related to technological development and the public risk anxiety and protest that would emerge in the 1980's (cf. Edelstein 1988; Hadden 1989: 144-148; Matheny and Williams 1988).

Oregon's requirements address several weaknesses of CAC's. For example, merely to require that citizens have effective communication with public officials, with no further conditions, is to invite false dialogue that is (from a communicative ethics view) insincere, unintelligible, and ultimately illegitimate. The requirements that citizens be involved before final decisions are made (citizen influence), and have aid in understanding the content of the issues at hand (technical information and financial support) are, in principle, aimed at recovering the intelligibility and sincerity of the advisory communication process. The requirement that citizens hear from public officials regarding the rationale for final decisions (feedback mechanisms) is a form of accountability guarantee that adds legitimacy.

Almost 15 years after the adoption of Oregon's Statewide Planning Program, the Canadian Federal Environment Review Office released a manual, Planning and Implementing Public Involvement Programs, which included guidelines for establishing, selecting, and operating advisory groups (Praxis, 1988: 48-56). The Canadian process, which resulted in the siting of one of the few new comprehensive hazardous waste management facilities, contained seven general principles for advisory groups:

1. Representation by the full range of interests and values of the interested public;
2. clear definition of the group's role in decision making;
3. limitations on the group's or member's tenure;
4. regular communication between advisory group members and the constituents they represent;
5. participation by agency line managers;
6. agency use of the public's language when communicating technical information, and;

7. time and logistic support by the agency.

The manual also acknowledged the need for external consultants when issues of agency credibility arise. One of the unique contributions of the Canadian model is its emphasis on the presence of responsible line managers so that "the group feels that it is being heard by the people with genuine authority, and so that managers hear public concerns first hand (Praxis: 50). The model is silent, however, on feedback mechanisms for the rationality of final agency decisions.

## Criteria for Fair and Competent Discourse

Evaluating the fairness and competence of interactions *within* a CAC is difficult because discourse can run the range from one-way communication and manipulation of citizens by authorities to sophisticated efforts to promote meaningful and legitimate discussions and decisions as seen in the Oregon example, above. Whether or not a CAC meets the criteria used in this volume to define internal fairness (e.g., a role in agenda setting, moderator determination and agreement on discussion and decision making procedures) often depends upon how the sponsoring agent organizes it. In addition, CAC's are almost always, in principle, deficient in this volume's fairness criterion of "ability to attend" (Table 1) because institutional sponsors often select participants (with some exceptions as we noted). In theory, the CAC format provides opportunities to initiate discussion in all modes of discourse discussed in this volume (i.e., definitional, factual, values, and feelings).

Nonetheless, the ability to meaningfully debate and decide on claims is not always present. That is especially true for technical/factual information if the CAC has been organized with no charge or too few resources to collect its own evidence or query its own experts or intuitive informants as an aid.

This is perhaps the most crucial mode of discourse within any CAC because CAC's are usually convened to review and respond to the technical proposals or imperatives of an agency or institution.

If a CAC is not provided with resource support and encouragement to analyze technical issues, there will be a tendency for normative and expressive claims to become divorced from the focal decision about a project or policy. In other words, convening agencies can disable the CACs recommendations by claiming, "that's their story" implying that it is opinion based on values and emotions, not fact. In practice today, however, most innovative CAC efforts, such as those currently underway in the United States' comparative state based comparative risk projects, emphasize citizen analysis and debate about theoretical validity of factual claims as well as the factual (technical and legal) implications of normative assertions (Patterson, 1992).

*Table 1.* Evaluation of how well the Citizen Advisory Committee model for participation performs on the goals of fairness and competence. (Note: "+" means satisfies most criteria, "S" means satisfies some criteria, and "-" means satisfies few criteria.)

| **FAIRNESS** | | | | |
|---|---|---|---|---|
| **Activities** | Attend | Initiate | Debate | Decide |
| Agenda and Rulemaking | - | + | S | S |
| Moderation, Rule Enforcement | - | + | S | S |
| Discussion | - | + | S | S |

| **COMPETENCE** | | |
|---|---|---|
| **Discourse Type** | Access to Knowledge | Best Procedures |
| Explicative Discourse | + | S |
| Theoretical Discourse | S | S |
| Practical Discourse | + | S |
| Therapeutic Discourse | + | S |

The competence of the CAC participants for engaging in the different forms of discourse varies greatly. Participants can arrive and begin the CAC process with little or no knowledge or expectations about the procedures for discourse and validation of each others' claims, other than to strategically press their own positions. Nothing in the structure of CAC's ensures that procedures for competent communication will be present. As a result, it is usually easier for CAC members to discuss basic definitions, norms, and values, respectively, because that is inherent to the process of "getting to know one another," which is promoted by this approach to public participation. Institutions that convene CACs implicitly count on development of a group identity and mutual empathy as a means to promote consensual decision making. In practice, access to knowledge about each others' basic understanding of the focal problem and underlying value positions can be happenstance, and agreed procedures to resolve conflicts and judge the validity of fears and feelings nonexistent.

Efforts to redeem the fairness and competence of a CAC's discourse grew in the 1980's. One good illustration is the Citizen Involvement Program (CIP) model designed by community development specialists for the U.S. Department of Energy's nuclear repository siting program (Howell, Olsen, and Olsen 1987). The model, tested in Washington state counties experiencing proposals for new hydroelectric power facilities, is built around the use of a Citizen Task Force (i.e., a CAC) that is convened by the project proposer and/or governmental regulator. The Task Force's purpose is to inform citizens (and themselves) of the proposal, identify citizen values and preferences, and report a citizen's recommendation to the government and the project proposer.

A key feature of this model is the joint adoption of a "Memorandum of Agreement" between the project proposer, Citizen Task Force members, and the professional facilitators for the public involvement program. The model memorandum includes goals that specify that citizens should be involved before the final decisions are made, that "two-way dialogue" between citizens and government must be facilitated so that "all parties will be fully aware of different needs and concerns," and that the proponent must respond formally to the final citizen recommendations. Such preliminary agreements are vital to conducting CACs that deal adequately with the underlying embeddedness, accountability, and legitimacy problems to which CACs can be very vulnerable.

It is in the operational details of the CIP model that a number of fairness and competence issues are resolved, typical of efforts to redeem CACs in the 1980's. The task force operates in a planned context with the aid of a neutral facilitation team which assists from the development of the citizen involvement program agreement and identification of community representatives through the conclusion. The initial meetings of the task force are expressly intended to allow citizen members to identify specific issues and information needs relating to the proposed facility, allowing for substantive influence of the agenda. Theoretical discourse is particularly well supported in this model by the planned formation of working groups to analyze and report back on specific issues. Furthermore, each working group is supported by the assignment of a technical specialist to aid in data collection and interpretation. An equally important aid to the competence of this approach is the planned provision of training in group process, conflict resolution, and analysis-decision skills.

The CIP model also incorporates several linked steps to overcome the major weakness of CACs: their lack of broad representation of public values and interests. First, a major purpose of the task force is to inform the public about proposed actions. Second, each working group holds one or more public workshops on their specific problem area, with the intended purpose of collecting anecdotal and experiential information from the larger population. Those meetings are followed by public forums on the actual conclusions of each working group. Third, the facilitation team has the responsibility to conduct, with task force involvement, a methodologically sound survey of community attitudes towards the proposed action as a check on any bias that results from narrow attendance at the forums.

Individual features of the CIP model are often found in CAC efforts, but taken together they resolve many competence issues in using select citizen bodies (particularly in theoretical discourse). The major limitations to fairness are also mitigated, especially in providing quite broad opportunities to involve the public at large throughout the procedures. The sophisticated evaluation of the model in the state of Washington, funded by the U.S. Department of Energy (USDOE), incorporated pre- and post-tests of task force-member attitudes, as well as comparisons of general public workshop/forum participants to non-participants, and comparison of the project county's populations to a control county with a similar project proposal.

Interestingly enough, as many task force members' attitudes towards the project changed from positive to negative as vice-versa. Nonetheless, general public participants tended to be more favorable towards the project after the CIP process than were those citizens who experienced no contact with the CIP Program's activities (Howell, Olsen and Olsen 1987). This illustrates the influence that CACs can have on general public attitudes even when the internal discourse among members makes efforts to be fair and competent.

## New Venues for CACs: Transferability and Evolution

In the mid-eighties environmental groups, in particular the National Toxics Campaign (NTC) began calling for more direct interactions with local industry. Impatience with government as an intermediary in solving the underlying citizen-industry conflicts has spurred a search for new approaches. The first requests were for plant inspections by the groups' own experts and for the signing of "Good Neighbor Agreements" to ratify changes that industry and citizens agreed were needed for safer plant operation. Citizen groups expect that the company will abide by the settlement and that an on-going working relationship will develop.

One effort to formalize the direct participation of citizens in industrial affairs was the introduction into the New Jersey legislature of the Hazard Elimination Through Local Participation (HELP) Act, which was endorsed by over 145 groups. The rationale behind the Act, which did not pass in the 1991 legislative session, was to "encourage a dialogue between companies and those potentially affected by pollution...(to)...increase trust and reduce environmental and workplace risks" (Chess: 1991b). Local Emergency Planning Committees (LEPC's), established under the 1986 Federal

Emergency Planning and Community Right-To-Act, would have been given expanded inspection powers. Local "Qualified Community Organizations" would have similar rights as would workers on site who would form Hazard Prevention Committees. But in the past five years, an even broader phenomenon of direct citizen-industry participation emerged in which U.S. industry has begun to establish its own citizen advisory committees (Electric Power Research Institute, 1991; Chess, 1991a). This effort was given a major boost in late 1989 when the Chemical Manufacturers Association recommended to its membership that they establish community advisory panels (CAPs) as part their Responsible Care Program. To date close to 200 such committees "predicated on the concept of open and honest exchange of ideas and views" are in place (CMA, 1991:3).

The CMA recommends selecting an independent facilitator, although it leaves open the possibility of third parties (such as the League of Women Voters) establishing independent panels. It is also recommended that a company provide the preliminary agenda, a sample mission statement, and preliminary guidelines for the panel's review, but that the panel then be encouraged to develop its own mission statement. Nonetheless, a CMA company reserves the right to establish parameters such as topics which they would prefer not be discussed (i.e. product costs, profits or personnel matters) and to "determine issues which need public input." The CMA is clear that "the panel is not a decision making body for the local plant" (CMA: 16). Model industry approaches include BASF Corporation's "Community Advisory Panel Handbook" (1989:23). The handbook recommends balance in selection of members, with people coming from three groups: likely advocates, likely opponents, and prominent leaders whose opinions of the environment and industry are not known. The charge for the BASF CAPs is to review and discuss the facility's operations, environmental performance, safety and emergency preparedness, the company's social commitment to the area and other issues the CAP may decide are important. BASF managers, however, do not ask their CAPs to formally approve or disapprove any company actions.

These developments illustrate the durability of the CAC approach to seeking positive consensus about an institution's activities from a broad constituency. The newness of the Citizen Advisory Panels means that we do not have a valid measure of how actions match up with intentions. One of the few evaluations of such industry-initiated community advisory committees was conducted by Rutgers University's Environmental Communication Research Program of Sybron Chemical. It illustrates how many of the fairness and competence issues that have dogged governmentally-sponsored CACs also emerge when industry is the sponsor.

Sybron, a small specialty chemical manufacturer, created a Neighborhood Involvement Council (NIC) in the wake of a chemical release and fire. The Rutgers study concluded that "the extent of the NIC's independence is unclear." While members "frequently raise topics not on the agenda," in other respects "the agenda seemed to be set by the company." It was unclear as to who actually ran the meeting and whether members had enough information to judge company proposals (Chess, 1991a: 22-26). Another CAP set up by Jenner Nuclear for the purpose of "two way communication" received mixed evaluations, with a sense that while citizens are allowed to voice their opinions and concerns, sometimes it appeared that the company was "more concerned with panel members spreading the 'good word' than with responding to the public's health and environmental apprehensions" (EPRI 1991: 2-17).

The industry literature still conveys the impression that the capacity of citizens to exercise different rationalities (e.g., about rightness of values or normative claims) and to contribute to informed decision making (through discourse on the factual and instrumental implications of values) is still unacknowledged. Moreover, there is no mention in the CMA manual of funding for independent experts to aid in citizens' competence, with the exception of a professional facilitator. As in our evaluation of twenty years of redemption of government-sponsored CACs, the competence of theoretical discourse about facts, and the relationship between normative claims about values made by citizens and instrumental technical issues is still often unclear.

Nonetheless, the industry sponsored advisory panels offer the potential for citizens to become informed about plant operations, while also allowing the opportunity for a powerful entity to be influenced by the knowledge of citizen preferences. That process is an improvement in industry-citizen relations where the intermediary imposition of governmental processes can obscure as much as clarify how much influence citizens have and how their input is used. On the other hand, if industry-sponsored CACs are perceived as illegitimate, more "regulatory" community advisory groups like those envisioned by New Jersey's HELP Act will emerge.

## Conclusions

Ultimately, the practical impact of CACs is contingent on the intentions of the institutions being advised. It makes little difference how well-orchestrated the process of citizen deliberation is with respect to facilitation of all views and values if the underlying purpose of those discussions is insincere and illegitimate – such as when CACs are being used deceptively to claim "representation" of the public. Clearly, even an apparently meaningful discourse within a CAC is invalid if the institution(s) involved has no intention of following or even responding argumentatively to citizen recommendations.

Likewise, discourse is distorted or impossible when citizen abilities to understand the issues of truth validation (e.g. about pollution from a waste facility) are limited due to lack of expertise or other resources. Nonetheless, from the perspective of communicative ethics, there are practical steps that have been taken to improve the validity of CACs as a means of broad citizen involvement in environmental issues. Practical experience and reflective analysis from a critical perspective both suggest that initial agreements about the ground rules and procedures under which CACs will operate can be the vehicle for such improvements.

Citizens, institutions, and helpers (such as facilitators) should explicitly agree on how citizen disadvantages in expertise or access to information will be addressed. This is essential to the intelligibility and sincerity of dialogue. Likewise, agreements about how the CAC recommendations will influence the decision-making authorities is a major step toward redeeming the sincerity and ultimate legitimacy of the CAC model.

Such agreements have been used (i.e., the WSU model for DOE and Oregon's Citizen Involvement requirements). The effect has been to make CAC's a little more like voluntary negotiation bodies, such as those citizens and industries have formed to settle acute site- specific environmental conflicts (Bacow and Wheeler 1984; Baughman, this volume). Successful voluntary negotiations such as these almost always involve preliminary agreements about resources, purpose of arguments, and legitimacy (who can promise what). As with successful environmental negotiations, competent CAC processes demand adequate support for citizens to collect and analyze information necessary to redeem theoretical (e.g., scientific, technical) claims made by government or industry. Nevertheless, CACs are primarily aimed at finding citizens who are "representatives" of a dispersed public, rather than acutely affected parties to a specific dispute or problem.

As legitimation crises developed, CACs historically appeared as a tool of local government reform politics. They have now begun to appear in the area of industry public relations. Nonetheless, one advantage CACs have over direct negotiations is that they can, in principle, take into account the broader publics' interests and values and address multiple ongoing issues when conducted with the features of the successful models that have evolved.

## References

Anderson, Richard, "Public Participation in Hazardous Waste Facility Location Decisions," *Journal of Planning Literature* 1(2):145-161 (1986).

Arnstein, Sherry, "A Ladder of Citizen Participation," *Journal of the American Institute of Planners* 35:216:224 (1969).

Bacow, Lawrence, and Wheeler, Michael, *Environmental Dispute Resolution*. (Plenum Press: New York,1984).

BASF Corporation, *Community Advisory Panel Handbook*. (BASF Public Affairs: Parsippany, NJ, 1989).

Burke, Edmund, "Citizen Participation Strategies," *Journal of the American Institute of Planners* 35:287-294 (1968).

Chess, Caron, Saville, Alex, Greenberg, Michael, and Tamuz, Michel, *From Crisis to Credibility: Behind the Scenes of the Risk Communication Program of Sybron Chemicals, Inc.* (Environmental Risk Communication Research Program, Rutgers University: New Brunswick, NJ, 1991a).

Chess, Caron, "Opening Factory Gates Measure Worth Taking," *Sunday Star-Ledger* (Newark, N.J., Section 3, page 4, June 16, 1991b).

Chemical Manufacturers Association, *Listening and Responding Through Community Advisory Panels* (Chemical Manufacturers Association: Washington, D.C., 1991).

Creighton, James, *Public Involvement Manual: Involving the Public in Water and Power Resources Decisions* (Government Printing Office: Washington, D.C., 1980).

Ducsik, Dennis, "Power Plants and People: A Profile of Electric Utility Initiatives in Cooperative Planning," *Journal of the American Planning Association* 50(2):162-174 (1984).

Edelstein, Michael, *Contaminated Communities: The Social and Psychological Impacts of Residential Toxic Exposure,* (Westview Press: Boulder, CO, 1988).

Electric Power Research Institute, *Risk Communication Manual for Electric Utilities* 3 vols. (EPRI EN-7314), (Electric Power Research Institute: Palo Alto, CA, 1991).

Forester, John, "Critical Theory and Planning Practice," *Journal of the American Planning Association* 46:275-286 (1980).

Forester, John, "Planning in the Face of Power," *Journal of the American Planning Association* 48:67-80 (1982).

Habermas, Jürgen, "Towards a Theory of Communicative Competence," *Inquiry* 13:363-372 (1970).

Habermas, Jürgen, *Communication and the Evolution of Society* (Beacon Press: Boston 1979).

Hadden, Susan, *A Citizen's Right-to-Know: Risk Communication and Public Policy* (Westview Press: Boulder, Co., 1989).

Hall, Peter, *Cities of Tomorrow* ( Basil Blackwell, Inc.: Cambridge, MA, 1988).

Howell, Robert, Marvin Olsen, and Darryl Olsen, *Designing a Citizen Involvment Program: A Guidebook for Involving Citizens in the Resolution of Environmental Disputes* (Western Rural Development Center, Oregon State University: Corvallis, Or, 1987).

Howell, Robert, and Darryl Olsen, *Citizen Participation in the Socio-Economic Analysis of Nuclear Waste Repository Siting. Report to the Western Rural Development Center*, (USDA-DOE Interagency Agreement DE-IA-97 80/ET46623, Washington State University: Pullman, WA, 1980).

Hutcheson, John, "Citizen Representation in Neighborhood Planning," *Journal of the American Planning Association* 50(2):183-193 (1984).

Kemp, Ray, "Planning, Public Hearings and the Politics of Discourse," in: John Forrester (ed.), *Critical Theory and Public Life* (MIT Press: Cambridge, MA, 1985), pp. 177-201.

Klapp, Merrie, "Challenges from Smart Publics," *The Environmental Professional 1989*, 189-197 (1988).

Land Conservation and Development Commission, *State Wide Planning Goals and Guidelines, adopted by the Land Conservation and Development Commission,* (Department of Land Conservation and Development, State of Oregon: Salem, OR, 1975).

Lewis, Sanford, *The Good Neighbor Handbook: A Grassroots Strategy for Clean Technologies and Job Security for America's Neighborhoods* Draft Document (The Good Neighbor Project: Acton, MA, 1992).

Lynn, Frances, "Citizen Involvement in Hazardous Waste Sites: Two North Carolina Success Stories," *Environmental Impact Assessment* 7:347-361 (1987).

Matheny, Albert, and Bruce Williams, "Rethinking participation: assessing Florida's strategy for siting hazardous waste disposal facilities." In: C. Davis and J. Lester (eds.), *Dimensions of Hazardous Waste Politics and Policy,* (Greenwood Press: New York 1988).

McCarthy, Thomas, *The Critical Theory of Jürgen Habermas* (MIT Press: Cambridge, MA, 1985).

Paterson, Christopher J., *Comparative Risk Assessment and the States: Science, Values and Public Participation in Siting Regulatory Priorities* (Unpublished dissertation proposal, Department of Environmental Sciences and Engineering, University of North Carolina at Chapel Hill: Chapel Hill, NC, 1992).

Pierce, John, and H. Doerksen, "Citizen Advisory Committees: the Impact of Recruitment on Representation and Responsiveness," in: Pierce and Doerksen (eds.), *Water Politics and Public Involvement* (Science Publishers: Ann Arbor, MI, 1976).

Praxis, *Public Involvement,* (Praxis: Calgary, Alberta, 1988).

Renn, O., T. Webler, and Brandon Johnson, "Public Participation in Hazard Management: The Use of Citizen Panels in the U.S.," *Risk-Issues in Health and Safety* 2:197-226 (1991).

Renn, O., and T. Webler, "Anticipating Conflicts: Public Participation in Managing the Solid Waste Crisis," *GAIA* 1(2): 84-94 (1992).

Rosenbaum, Nelson, *Citizen Involvement in Land Use Governance* (The Urban Institute: Washington, D.C., 1976).

Rosener, Judith, "Matching Method to Purpose: The Challenge of Planning Citizen Participation Activites." in Stuart Langton (ed.), *Citizen Participation in America* (Lexington Books: Lexington, MA, 1978).

Smith, Martin, Frances Lynn, and Richard Andrews, "Economic Impacts of Hazardous Waste Facilities" *Hazardous Waste and Hazardous Materials* 3 (2) 195-204 (1986).

Solnitt, Albert, *The Job of the Planning Commissioner, 3rd ed.* (Wadsworth: Belmont, CA, 1982).

Susskind, Lawrence, and Alan Weinstein, "Toward a Theory of Environmental Dispute Resolutions," *Boston College Environmental Affairs Law Review* 9(2):311:357 (1981).

Susskind, Lawrence, and Jeffrey Cruishank, *Breaking the Impasse: Consensual Approaches to Resolving Public Disputes* (Basic Books: New York 1987).

Walker, Richard, and Michael Heiman, "Quiet Revolution for Whom?" *Annals of the Association of American Geographers* 71(1): 67-83 (1981).

*Chapter 5*

# *Citizens' Advisory Committee as a Model for Public Participation: A Multiple-Criteria Evaluation*

*Anna Vari*

## Introduction

Public involvement in environmental decisions in Western democracies has continuously increased in recent decades. There has been a growing demand for evaluating and comparing public participation processes. Webler (this volume) developed a set of evaluation criteria for fair and competent discourse where fairness is defined as the sufficiency of opportunities that affected citizens have to protect their interests, and competence is related to the adequacy of knowledge selection tools.

Public involvement, however, is only one aspect of environmental decision processes. There are various other expectations about these processes, including for example accountability, efficiency, and adaptability, which may contradict public involvement. A conceptual framework was developed to investigate the relationships among these characteristics of decision making processes based on organizational analysis (Quinn 1988; Quinn and Rohrbaugh 1981, 1983). The paper presents a multi-criteria evaluation of the Citizens' Advisory Committee (CAC) model of public participation taking into consideration both the criteria for fair and competent discourse and additional criteria derived from organizational analysis.

## The Embeddedness of Citizens' Advisory Committees in the Social-Political Context

During the last decade, in the United States, Canada, and Western Europe a number of CACs were created to represent affected interests in a variety of environmental decisions, including the development of legislation and regulatory standards, issuance of permits, land use decisions, and the planning of industrial and infrastructure projects. In most cases, CACs were established to provide advice to federal, state governments, and local governments, and, less frequently, to private companies.

With regard to governmental policy making, CACs were usually established to provide general guidance on implementing environmental law, promulgating regulations, and issuing permits for and planning of potentially polluting facilities. In several cases, CACs were also formed to participate in drafting or revising legislation.

- In the early 1980's, the U.S. EPA appointed a RCRA (Research Conservation and Recovery Act) Permit Advisory Committee to obtain advice on permitting hazardous waste facilities. More recently, the EPA established an advisory committee to provide guidance on how it should implement recent amendments to the Clean Air Act (Bloom 1991).
- In 1983, New York State appointed an Advisory Committee to provide assistance to the State Energy Office to study options to low-level radioactive waste management in the state. In 1987, another Advisory Committee was established to provide advice on selecting a site and a technology for a low-level radioactive waste disposal facility (Vari, Mumpower and Reagan-Cirincione 1993/a).
- In 1985, the Governor of Pennsylvania appointed a CAC to draft and revise legislation on the siting procedure of a low-level radioactive waste disposal facility. Later, two separate CACs were created to serve the regulator and the developer, respectively (Vari, Mumpower and Reagan-Cirincione 1993/b).

CACs were also created to provide guidance to local governments on environmental policy issues.

- In 1989, the Canadian government established Community Liaison Groups in 14 communities to advise municipal councils regarding interest to be considered as a potential site for a low-level radioactive waste management facility (Armour 1991).
- Finally, private firms established CACs to provide advice on planning projects which may have environmental impacts.
- In 1986, US Ecology, a private company planning to build and operate a low-level radioactive waste disposal facility in California, created a statewide CAC to provide advise on the identification of preferred sites. After preferred sites had been selected, the company established local CACs in the affected communities to communicate local concerns on the disposal technology and environmental monitoring (Anderson 1988).

The charge given to a CAC is usually defined by government agencies or private companies responsible for resolving the problem at hand (problem owners). It is also the problem owners who identify the problem to be resolved and the model for involving the public in certain phases of the decision making process. Members of CACs may be nominated by citizen and environmental groups, trade and industry groups, and governments, but their appointment typically has to be approved by the problem owners.

In general, members of CACs include representatives of interested citizens' organizations and the general public. Lynn and Kartez (this volume) indicate that there is a growing number of CACs that also involve representatives of the problem owners and other institutions affected by the decision. CAC organizers typically seem to use two criteria to select participants:

- They represent established interest groups or interest positions that the problem owner deems to be relevant,
- they are seen as legitimate representatives of a constituency.

The primary function of the CAC model is value reconciliation among the participants. There are only few cases where CACs are instrumental in solving a problem or educating citizens.

## Fairness and Competence of the Discourse Occurring in Citizens Advisory Committees

### *Participation in setting the agenda and the rules*

The agenda for CAC activities is usually defined by the problem owners. As a consequence, many important issues may be excluded from the discourse, including the questions of alternative solutions or the social need for the project. The problem owners may select the agenda strategically, so as to support their interests.

CACs were established in several states, including California, Illinois, Nebraska, Pennsylvania, and Texas to provide advice on the siting of low-level radioactive waste disposal facilities. None of these CACs questioned the necessity of building a permanent disposal facility, whereas most citizen and environmental groups in the above states expressed their preference for above-ground, monitored, temporary storage facilities (Vari, Mumpower and Reagan-Cirincione 1993/b). In most cases, the rules for discourse are defined by the problem owners before CACs are established. In some cases, however, CACs have the opportunity to change these rules.

In Canada, the low-level radioactive waste management facility siting process was planned to include public opinion surveys to determine whether a community should remain within the process. Community Liaison Groups proposed to change further steps of the process, for example to replace surveys with public referenda. These proposals were accepted and implemented by the organizations responsible for the siting (Armour 1991).

## *Participation in selecting the moderator and the method of facilitation*

CACs are expected to formulate a more-or-less consensual position with regard to certain issues with the help of a moderator. In some cases, the moderator is selected from the members of the CAC, in other cases, he/she is a professional facilitator specialized in group decision or negotiation support. Typically, moderators are selected by the problem owners, but the methods of facilitation are determined by the moderators. In the majority of cases, CAC members have a very limited opportunity to influence the decisions regarding the moderator and the facilitation method. This situation has started to change recently, particularly with regard to public participation programs initiated by industry (Lynn and Kartez, this volume).

## *Participation in the discussions*

Legitimacy of the CAC model is based on the argument that citizens and interest positions affected by a certain problem are represented by CAC members. Group efficiency considerations, however, limit the size of CACs to 10-20 members. Therefore, in complex cases, (e.g., if citizens of a state or a nation are affected), representation by a CAC is rather problematic.

The CAC model provides the general public with the opportunity to attend committee meetings. The chance for the general public to influence the discourse is, however, limited. In addition, it is questionable whether the members of the public who are active at CAC meetings, do adequately represent the affected population.

CACs may diverge in terms of their scope. Some provide guidance on a number of policy questions, whereas others are established to address particular problems. Similarly, some CACs are involved in all phases of the decision making process, including goal setting and the generation and evaluation of alternatives, whereas other CACs are involved only in certain stages of the process.

The U.S. EPA appointed a general advisory committee to provide guidance on implementing recent amendments to the Clean Air Act. It also appointed a specialized advisory committee to advise it on specific issues related to the acid rain provisions of the act (Environmental Law Institute 1991).

The Low-Level Radioactive Waste Site Selection CAC of the State of California was involved in most steps of site selection. They participated in defining site selection criteria and their relative importance, in evaluating alternative sites in terms of each criteria, and rank-ordering the sites. On the other hand, in New York, site selection criteria were defined and alternative sites were evaluated by technical experts hired by state agencies, whereas members of the Advisory Committee participated only in the weighting of the site selection criteria (Vari, Mumpower, and Reagan-Cirincione 1993/b).

To a large extent, the scope of CAC participation determines the chance of CAC members to put forth claims and arguments and to criticize those of others. CACs are most frequently invited to participate in stages of the decision making process where public interest and values have to be made explicit, for example, in defining evaluation criteria and their relative weights. The CAC context also facilitates the translation of fears, feelings, and concerns into values and norms. On the other hand, CAC members usually have little chance to make factual judgments, for example, to evaluate the feasibility of alternative solutions, since these activities are usually assumed to be the tasks of technical experts. Similarly, CACs possess little opportunity to define the terms to be applied, because the problems are traditionally conceptualized by the problem owners before public participation gets started.

Typically, CACs do have the opportunity to criticize claims and arguments put forth by other parties. However, since CACs have merely advisory power, problem owners may or may not take their criticisms and recommendations into consideration. Under these circumstances, it is very rare that a mutually acceptable method for resolving disputes is consensually adopted by the various parties.

## *Access to knowledge*

Since the primary function of CACs is value reconciliation, discussions are expected to be highly competent with regard to the values, norms, fears, feelings, and concerns of the groups and organizations represented. There is usually less emphasis on competence pertaining to the definitions of the terms or the factual aspects of the problems at hand.

There can be significant differences in expertise among the participants. Therefore, in many cases, substantial time and resources are devoted to educating CAC members. This helps to decrease the differences in expertise not only among the CAC members, but also between CAC members and other actors in the decision making process. Education, however, is usually provided by the problem owners and no additional funding is provided for CACs to select and invite additional "independent" experts.

The CAC model itself does not include any education program for or communication with the larger public. In an increasing number of cases, however, problem owners combine the CAC model with public education and risk communication programs as well as public opinion research (Lynn and Kartez, this volume). These activities, however, are often conducted by various agencies in parallel, and coordination becomes problematic.

In 1987, in New York State, an Advisory Committee was created to represent the public in selecting a site and a disposal technology for a low-level radioactive waste disposal facility. At the same time, the New York State Department of Health was required to conduct a statewide public information program on the health and safety implications of the facility. In order to investigate public knowledge and attitude towards radioactive waste management, the Department of Health also conducted a statewide opinion poll. There was, however, little coordination between the activities of the Advisory Committee and the Department of Health (Vari, Mumpower, and Reagan-Cirincione 1993/a).

## *Conflict resolution*

In the following I analyze how the CAC model promotes the application of effective conflict resolution methods to explore, discuss, and decrease divergence in opinions.

There are three basic approaches to conflict resolution: (i) the statistical approach – aimed at finding a solution on the basis of the statistical analysis of opinions, (ii) the reconciliatory approach – aimed at finding, in an interactive group setting, a mutually acceptable solution without trying to address all disagreements, and (iii) the confrontational approach – aimed at finding a creative solution via direct confrontation of the different opinions (Vari 1989).

Of the various public participation models, public surveys and referendums are examples of the statistical approach; negotiated rule making, mediation, and Citizens Juries are examples of the confrontational approach, whereas the CAC model is a typical context for the reconciliatory approach. The essence of this approach is that it attempts to reconcile the different problem representations of the participants by group decision support. Phillips (1982, 1984) calls this requisite modeling, where the model is requisite in the sense that everything required to solve the problem is either included in the model or can be simulated in it. In CAC sessions, decision analytical models, including multi-attribute utility models or other multi-criteria structures, are frequently used to provide a framework for the iterative development of a coherent representation of the problem.

In 1983, the Texas Low-Level Radioactive Waste Authority and its contractors defined a MAU model with 17 site selection issues and 46 subsidiary factors to rank-order potential sites for a disposal facility. Technical experts rated each site on each factor and assigned importance weights to factors within each issue. Importance weights for issues were defined in a two-day workshop attended by technical experts and members of the Citizen Advisory Panel. Individual voting of group members was followed by a discussion of the rationale behind the extremely high and low votes. The voting was repeated and discussed until the range of opinions narrowed, or the average between successive votes did not change substantially. Ratings were then combined with the weights for factors and issues to produce overall site ratings which provided the basis for designating a site (Texas Low-Level Radioactive Waste Disposal Authority 1990).

The California Low-Level Radioactive Waste CAC applied a multi-criteria approach to rank-ordering candidate sites. First, evaluation criteria were jointly defined by the developer and the CAC. Then the CAC was asked to assign relative importance to the criteria. Based on the evaluation of each site in terms of each criterion, CAC members were asked individually to divide the sites into three (good, medium, and poor) categories and to give reasons for their ratings. The following discussions led to a consensus by CAC with regard to the top sites (Anderson 1988).

The CAC model does not provide for dispute resolution between CAC members and others not involved in developing the requisite model of the problem. As a consequence, in many cases, the recommendations of the CAC are rejected by the decision makers or the general public. In Texas, local communities opposed to the selected site questioned the input data of the MAU model and challenged the integrity of the site selection process (Vari, Mumpower, and Reagan-Cirincione 1993/b). In California, the legitimacy of the low-level radioactive waste disposal facility siting process, including the activity of the CAC, was questioned by local protest groups and later by the broader political community as well (Vari, Mumpower, and Reagan-Cirincione 1993/b).

## The Performance of the CAC Model in Terms of Decision Criteria

### *Conceptual framework*

Public participation is always embedded in broader social decision processes. There are a number of various expectations expressed by the various parties about the desired attributes of environmental decision processes. Some urge that such processes should be participatory and flexible, and that the decisions should be accountable and efficient. Such expectations often compete with one another, however. As more emphasis is placed on public participation, for example, less emphasis can be placed on the efficiency of the decision process. Likewise, as concern for the accountability of the decision increases, the adaptability of the process cannot also increase.

One conceptual framework that can analyze the contradictions inherent in environmental decision processes is based on the Competing Values Approach (CVA) to organizational analysis (Quinn 1988; Quinn and Rohrbaugh 1981, 1983; Lewin and Minton 1986). In the CVA framework, two basic value dimensions combine to define perspectives on effective decision making processes. The first value dimension is related to process structure; an emphasis on flexibility competes with an emphasis on control. The second value dimension is related to the focus of the process; an emphasis on the desires of those directly affected (internal focus) competes with an emphasis on the needs of the broader public (external focus). The relative emphasis on these competing values defines four distinct perspectives of effective decision making: the rational perspective (high control, external focus), the empirical perspective (high control, internal focus), the consensual perspective (high flexibility, internal focus), and the political perspective (high flexibility, external focus).

Combining the four perspectives with the distinction between means (processes) and ends (decisions), results in eight criteria describing decision processes (Rohrbaugh 1987; Reagan & Rohrbaugh 1990). These criteria include: goal-centered process, efficient decision; data-based process, accountable decision; participatory process, supportable decision; adaptable process, and legitimate decision. A more detailed description of the CVA framework and the above criteria is given by Mumpower (this volume). In the following, the CAC model is briefly characterized in terms of the above mentioned criteria.

## *Goal-centered process – efficient decision*

A process is goal-centered to the degree it focuses resolutely on the primary problem. Goal-centered processes help to make efficient decisions which are achieved with relatively few resources. CACs are usually focusing on the problems set by problem owners, and except for recent approaches, they rarely change the formulation of the problem or the agenda. Therefore, implementation of CACs as primary models of public participation usually results in rather goal-centered processes and efficient decisions, especially when compared with direct models of public participation (see: Crosby; Dienel; Midden; this volume). Compared to negotiation or mediation models, however, CAC processes are usually less goal-centered and result in less efficient decisions.

## *Data-based process – accountable decision*

A process is data-based to the degree that decisions are driven by data. Data-based processes help to make accountable decisions which are transparent, well-documented, and can be easily justified. The CAC model – due to its conciliatory approach to conflict resolution – does not encourage the confrontation of contradictory opinions. Such confrontation would stimulate the conflicting parties to collect technical evidence in support of their positions. Moreover, there is usually no or very limited funding provided for CACs to gain input from independent experts into group discussions. Therefore, implementation of CACs usually results in less data-based processes and less accountable decisions than most other models of public participation.

### *Participatory process – supportable decision*

In a participatory process, opinions of all affected parties are considered in each phase of the decision. Steps are taken to assure that these opinions not only are made known to the responsible decision makers, but also exert considerable influence on the outcome of the process. Participatory processes help to make decisions which are supported by the affected parties. Due to major differences in the degree to which CACs represent the interests of the public and other affected groups, and the degree of their influence on the final decisions, processes implementing CACs widely differ in terms of participation and supportability. Implementation of the CAC model usually results in more participatory processes and more supportable decisions than negotiation or mediation, but less participatory processes and less supportable decisions than direct methods, especially with regard to local and regional problems.

### *Adaptable process – legitimate decision*

A decision process is adaptable if, in response to unexpected events and interventions, it can be easily changed. Adaptable processes help to reach legitimate decisions that inspire general confidence in their fairness. Recent developments of the CAC models, which include informing the larger public about the proposed actions and eliciting their input, as well as allowing the public to change the process (Lynn and Kartez, this volume), significantly enhanced the adaptability of the processes and the legitimacy of their outcomes. In general, the adaptability of CAC processes and the legitimacy of decisions is higher than those of negotiation and mediation models but lower that those of direct input.

## The Performance of the CAC Model in Term of Implementation Criteria

### *Implementability of CAC outcomes*

CACs usually have review and advisory power. This model is suited for making non-binding recommendations rather than binding decisions. As mentioned earlier, the CAC model does not provide for communication between committee members and the general public. Therefore, recommendations of CACs are often rejected by ad hoc groups emerging during the decision making process, or other parties not represented by the CAC. In general, if not combined with other participation models, the implementability of CAC outcomes is rather poor.

## *Applicability and transferability of the CAC model*

The technical feasibility of the CAC model is relatively high, since it does not require any special mechanism nor any substantial funding. The effective use of the model, however, is highly contingent upon the wider legal and political context. The most important prerequisite for the effective application of the CAC model – as well as any other model of public participation – is the transparency of the decision making processes and the disclosure of all relevant information. Free access to information is a crucial factor of applicability of the CAC model in any country.

CACs are created and financed by private firms or governments. There has to exist a legal framework or some other incentive which makes industry or government recognize that public participation is ultimately also in their interests. This, however, assumes a relatively high degree of environmental awareness of the affected population and a society where the success of private business and the legitimacy of government is dependent on good public reputation.

CACs are established to represent a variety of public interests in environmental decision processes. The model can be functional only in a society where these interests are well articulated and the public is relatively well organized. Therefore, the pre-existence of pre-established organized groups is an important prerequisite for the applicability of the model. It is rather difficult to use the model in a turbulent political environment characterized by the frequent emergence and restructuring of interest groups and coalitions. On the other hand, the CAC model can effectively function only in a political context where the mutual acceptance of multiple perspectives and the cooperation between the various parties is more rewarding than direct political confrontation.

There have been several attempts to promote the transfer of various public participation methods developed in Western European and North American countries to the new democracies of Central and Eastern Europe (Petkova 1992, Vasarhelyi and McIlwane 1992). However, the lack of adequate legal framework and the decreasing concern for the environment due to pressing economic problems do not establish a favorable background for public participation in general. Short-term perspectives for the effective use of reconciliatory approaches, including the CAC model, are especially uncertain in the turbulent and adversarial political environment of Central and Eastern Europe.

## Summary

Both the major strengths and weaknesses of the CAC model originate from the relatively small number of participants and the reconciliatory approach taken to conflict resolution. The small size of CACs makes the use of effective group facilitation and decision support technologies feasible. This can assure that participants extensively discuss and clarify the terms, definitions, concepts, and data that they use. Various tools (e.g., decision analytical models) can help to separate factual and value judgments, elicit judgments from CAC members, and translate expressive claims into cognitive or normative ones.

Another advantage of the CAC model is that it allows public participation in a procedural stage where no preliminary decisions have already been taken. Participation does not have to be restricted to the final decision, it can also include the definition of goals and constraints, as well as the generation and evaluation of alternative solutions. The CAC model also allows – at least in principle – that public participation does not end as soon as the decision is made, it can include participation in the implementation phase (e.g., monitoring) as well.

The major weakness of the CAC model is the lack of representativeness of the participants in cases where a large number of affected groups are involved. On the other hand, the reconciling approach taken by most CACs assumes that there are no essential differences between the interests and values of the participants, and those involved are ready to provide each other with the information relevant to joint decisions. For the above reasons, the CAC model can most effectively be used with regard to local problems, or problems where only a few, well-defined groups are affected. In order to be effectively applied to larger-scale problems, CACs should be combined with models, providing for more representation of the general public, for example public surveys or referendums.

Another important shortcoming of the CAC model is that in most cases it does not provide for the hiring of independent experts by the public. In addition, the conciliatory nature of CAC processes and the lack of exploring the sources of conflicts does not encourage the collection of verifiable data and comprehensive expert input. For the above reasons, the CAC approach should be combined with more confrontational models, including Citizens Juries or Planning Cells.

## References

Anderson, Gloria, *Final Report to US Ecology on the Low-Level Radioactive Waste Disposal Site Selection Citizens Advisory Committee* (Sacramento, California, 1988).

Armour, Audrey, *Opting for Cooperation: An Innovative Approach to Facility Siting. Manuscript* (York University: North York, Ontario, Canada, 1991).

Bloom, Jane L., *Opportunities for Public Involvement in Environmental Policy Making Unpublished Manuscript* (Regional Environmental Center for Central and Eastern Europe: Budapest, Hungary, 1991).

Crosby, Ned, "American Citizen Juries: One Solution for Difficult Environmental Questions," (this volume).

Dienel, Peter and Renn, Ortwin, "Planning Cells: A Gate to Fractal Mediation," (this volume).

*Environmental Law Institute, Public Participation in Environmental Regulation,* Draft Working Paper, (Washington, D.C., 1991).

Lewin, Arie Y., and John W. Minton, "Determining Organizational Effectiveness: Another Look, and an Agenda for Research," *Management Science* 32(5): 514-538 (1986).

Lynn, Frances M., and Jack D. Kartez, "The Redemption of Citizen Advisory Committees: A Perspective from Critical Theory" (this volume).

Midden, Caes, "Direct Participation in Macro-Issues: A Multiple Group Approach: An Analysis and Critique of the Dutch National Debate on Energy Policy, Fairness, Competence, and Beyond" (this volume).

Mumpower, Jeryl L., "The Dutch Study Groups Revisited," (this volume).

Petkova, Elena, "Public Participation: Opportunities and Constraints: The Case of the Troyan Environmental Action Project," in: *Proceedings of the International Workshop on Public Participation in Environmental Decisions: A Challenge for Central and Eastern Europe* (Regional Environmental Center for Central and Eastern Europe: Budapest, Hungary, 1992).

Phillips, Lawrence D., "Requisite Decision Modeling: A Case Study." *Journal of the Operational Research Society,* 33: 303-311 (1982).

Phillips, Lawrence D., "A Theory of Requisite Decision Models." *Acta Psychologica,* 56: 29-48 (1984).

Quinn, Robert E., *Beyond Rational Management* (Jossey-Bass: San Francisco, CA,1988).

Quinn, Robert E., and John W. Rohrbaugh, "A Competing Values Approach to Organizational Analysis." *Public Productivity Review,* 5: 141-159 (1981).

Quinn, Robert E., and John W. Rohrbaugh, "A Spatial Model of Effectiveness Criteria: Towards a Competing Values Approach to Organizational Effectiveness." *Management Science,* 29: 363-377 (1983).

Reagan, Patricia, and John W. Rohrbaugh, "Group Decision Process Effectiveness: A Competing Values Approach." *Group and Organizational Studies,* 15 (1): 20-43 (1990).

Rohrbaugh, John W., "Assessing the Effectiveness of Expert Teams," in: Jeryl L. Mumpower, Lawrence D. Phillips, Ortwin Renn, and V.R.R. Uppuluri (eds.), *Expert Judgment and Expert Systems.* (Springer Verlag: Berlin, Germany, 1987).

*Texas Low-Level Radioactive Waste Disposal Authority, Report to the Public on the Site Selection Process and Other Analyses* (Austin, Texas, 1990).

Vari, Anna, "Approaches Towards Conflict Resolution in Decision Processes," in: Charles Vlek and George Cvetkovich (eds), *Social Decision Methodology for Technological Projects* (Kluwer Academic Publishers, Dordrecht, 1989), pp. 79-94.

Vari, Anna, Jeryl L. Mumpower, and Patricia Reagan-Cirincione, *Planning for the Management of Low-Level Radioactive Waste in New York State: A Case Study. Research Report* (Center for Policy Research, State University of New York at Albany: Albany, NY, 1993/a).

Vari, Anna, Jeryl L. Mumpower, and Patricia Reagan-Cirincione, *Low-Level Radioactive Waste Disposal Facility Siting Processes in the United States, Western Europe, and Canada. Research Report* (Center for Policy Research, State University of New York at Albany: Albany, NY, 1993/b).

Vasarhelyi, Judit, and Susanna McIlwane, "Collaborative Environmental Planning on the Community Level: An Adaptation of the Comparative Risk Model in Two Hungarian Communities," in: *Proceedings of the International Workshop on Public Participation in Environmental Decisions: A Challenge for Central and Eastern Europe* (Regional Environmental Center for Central and Eastern Europe: Budapest, Hungary, 1992).

Webler, Thomas, "'Right' Discourse in Citizen Participation: An Evaluative Yardstick", (this volume).

*Chapter 6*

# *Planning Cells: A Gate to "Fractal" Mediation*

*Peter C. Dienel*
*Ortwin Renn*

## Introduction

Ecology suffers. Humanity is far from having peace with Nature. The world population explodes. Ethnic conflicts are ubiquitous. Wars and civil wars are haunting all continents and many nations. The individual in society has lost functions and lacks cultural identity. Millions of people die from hunger. Millions are driven away from home and live in refugee camps. Problems abound!

The question is where to start when we attempt to address or even solve these problems. If we define them as purely technical, we miss the point. Most have evolved partly because of technologies, partly in spite of technologies. Changing technologies will not provide the answer, at least not in any decisive way. If we define the problems as questions of morality, we start to act in an elitist fashion assigning us superior moral responsibility compared to all the others. If we regard the problems as scientific challenges, we narrow our focus to the role of knowledge in problem solving.

Without doubt, knowledge is a key variable in coping with many problems, but the paradox of our present situation is that most problems exist in spite of better knowledge. The difficulty with these problems is that they defy any mono-causal schema of explanation. They constitute a complex web of interrelations. All these problems are caused by many factors, but they do have one characteristic in common: they demonstrate the inability of the present governing and administrative systems to cope with pressing challenges. The clock of problem generation is running faster than the clock of problem solving. We have become used to the fact that administrative systems can only react to a crisis, but not anticipate, let alone prevent crisis from occurring. Modern administrative systems are deficient in dealing with today's problems for a variety of reasons. Among them are:

- Short-term interests seem to dominate the political decision making process because election cycles run for a short period of time and voters seem to reward fast return of their trust capital invested in politicians or parties (von Kielmannsegg 1979: 31).

- Administrative systems tend to focus attention on minor problems because these problems appear manageable within a reasonable time frame (Bachrach 1967). Debating and working on minor problems leaves the impression of successful political activity, whereas addressing major problems would cost too much time and effort. In addition, it requires the willingness to take risks and increases the potential for failure or unpleasant surprises. The probability of success is much higher when attempting to resolve minor or peripheral problems.

- Societal subsystems have developed increased autonomy and self-governing structures. They often refuse to coordinate their actions with other subsystems or to be guided by integrative policies. To facilitate communication and to legitimize collectively binding decisions, political systems need to restrain themselves to those actions that find approval by all major actors (Schwarz and Thompson 1990). Such actions tend to be trivial.

- In conjunction with the autonomy of functional systems within society, individuals experience a hard time developing a sense of national or cultural identity. Facing increased disorientation within their reference society or group, they take refuge in fundamentalist movements or nationalistic-ethnic ideologies. Modern administrations have lost their ability to cope with restorative movements and often act helplessly in view of the anti-modernist protest.

Due to these and other reasons, the political-administrative system is deficient in its functions and tasks (Grottian and Murswieck 1974). This is true for the local, regional, national, and international level. Necessary decisions are delayed because of fear public support will be lost. Effort is invested in cosmetic changes intended to signal activity and responsiveness. Competent actions are not rewarded. Effective administration and problem solving have been sacrificed for shiny illusions. The continuous fight for public office and power does not leave sufficient space for accountable actions, so concluded Joseph Schumpeter when he diagnosed one of the major weaknesses of the competitive democratic system (Schumpeter 1947).

## The Role of the Citizen: Towards a New Model of Political Involvement

The inability of the administrative-political system to cope with today's problems has not gone unnoticed in the social sciences nor in public perception. Distrust in political performance (not necessarily the political system) coupled with a growing disenchantment with any type of technocratic decision making have grown over the last two decades. Public officials have responded to this crisis in two different ways. First, they perceive the need for internal reform of the administrative system. Along the line of increased efficiency, reforms aim at improving functionality of each segment, enhancing specialization and differentiation of the public institutions, extending internal and external communication efforts, and investing in training and professionalization. The second strategy is to install feed-back communication mechanisms that help administrators respond faster when the public expresses dissatisfaction with the official policies. Such mechanisms range from opinion polls to round tables. The goal of such activity is to streamline administrative action and to improve the basis for legitimation.

Both strategies are likely to fail (Burns and Überhorst 1988; Rosenbaum 1978). Expanding the functions of public administration and making it more efficient may indeed improve the cost-benefit-ratio of public actions, but it does not provide any remedies for the underlying dilemma. Public dissatisfaction is not primarily about inefficiencies but priorities and selection rules. The same is true for the second strategy. Listening to public concerns is certainly beneficial, but not enough. Nobody wants public officials to act as puppets of public opinion. More and more people demand effective leadership, but at the same time feel entitled to determine their own livelihood. The challenge for modern society is to provide political means for visionary long-term policies together with increased participation by those who are affected by the political decisions.

This is a serious challenge for political systems at the end of the 20th century. If serious reforms are not taken, the distance between the administrators and those who are or feel administrated will grow. Perfect feed-back mechanisms may keep people pacified, but on the expense of substance and reason. Public communication will function as a gigantic therapeutic discourse. In the end, people may feel like they are living in an ant colony. Being deprived of their political functions other than casting votes for interchangeable elites, they may resort to private surrogates such as excessive consumption, computer hacking, sexism, or violence. In response to these "irrational" actions of citizens, the administrative system would need more police, more prisons, more places for therapy, and more sedatives such as drugs and TV-programs. Is that the future we would like to have?

We admit that this picture of contemporary society is exaggerated. Society is never black and white. Many administrators and politicians are highly motivated to face the challenge and many steps have been taken to enrich the role of the citizen in governing the common good. The present structures of democratic institutions, however, do not support these efforts and may counteract well-intended actions by individuals or groups. A kind of structural reform is necessary that brings political decision making back to the place where it belongs: to the informed and motivated citizens.

According to all democratic constitutions, the citizen is the sovereign of the political system. The reality is, however, that political parties assume the role of the citizen and claim to be the new sovereigns of democratic systems (Leipholz 1968:245). Originally established as manifestations of the collective will, parties have evolved into organizational machines that tend to develop and sustain structures for serving the interests of their members. Parties create their own political identity that differs from the identity of those that they proclaim to represent (Scheer 1979:14). Other contestants for assuming the citizen role are organized interest groups. Again we face the problem of particularism. Interest groups are what they pretend to be: organized groups who share similar economic or political interests and values. As many analysts have shown, the mere fact that many interest groups exist or compete with each other, does not guarantee the realization of a civic society (Etzioni 1968; Olson 1982). Organizational pressure depends on the exclusiveness of interests that are represented in each group (Olson 1965). This is the opposite of what the common good approach would suggest. So we need a new definition of the politically active citizen. Who can take the role of a decision maker and commit him- or herself to long-term goals and perspectives?

Let us first start with the negative selection. There are plenty of self-proclaimed citizens whom we do not have in mind. First, there is the aggressive type of citizen who claims to speak for the rest of the world. These advocates of common causes often deny their own self-interests and know better than everybody else which policies are needed for the common good. Second, there is the apathetic citizen who would like to be on the winning team without raising his or her hand. Such citizens are hard to mobilize for any cause, but join the bandwagon as soon as all the work is done and benefits are to be distributed. Third, there is the moralist who wants to impose his or her moral standards to the rest of the world. Fourth, there is the hobby politician who wants to be celebrated for all his or her involvement and does not miss a chance of being portrayed in the media. We could add to this list other typical roles of citizens to which we have become accustomed over the years. In short, when we talk about empowering citizens, we do not think of the professional citizen, who is basically interested in a political career, charismatic leadership, personal benefits, and public image.

But who then? Is there anybody left not motivated by these mundane goals? Do we want to invent the altruistic, modest, caring, universally knowledgeable and self-effacing individual? Not at all! We do not believe in educating the new human being who combines all the virtues that one might demand from the ideal citizen. Our understanding of democratic processes is that virtually everybody (including all the individuals listed above, but not exclusively) can and should play the function of a decision maker if his or her life is affected by the decisions. Our main proposition is that those who enjoy or suffer the consequences of a decision should be the ones who ought to make them. If people feel that their representatives do not represent their interests and values, then the people should be asked to represent themselves.

How can they do this? Who can speak for the affected citizens? Who guarantees that those who are affected know enough that they will not regret their decision? How can we avoid the tendency for some to perpetuate their call and become professional citizens starting, the cycle of professionalism and distrust anew? Our model of participation provides four structural elements that relate to these crucial questions.

- Since all affected people are experiencing the positive or negative consequences of a decision, they should have an equal share in the decision making process. Since we cannot organize a session with thousands of people and since a voluntary selection process would undoubtedly attract mostly professional citizens, our solution is random selection. We want to give everybody the same opportunity to serve as a citizen. Each person drawn by this selection procedure has the right to refuse the role, but once s/he has accepted s/he will be given the opportunity to serve as citizen within a specific decision making context.
- Considering the complexity and advanced knowledge necessary to understand the context of political decisions, it would be impossible for any single individual to play the role of the universal citizen. In analogy to the division of labor, we propose that citizens share the responsibility for making informed decisions (cf. Luhmann 1968). Sharing means that each citizen who participates in the process of decision making constraints his or her involvement to a single issue at a given time. Should the same person be selected again, the topic, of course, will change. During the time period that the specific decision will be made, all relevant information and expertise is given to the groups of selected citizens. Our experience has been that most of the people selected were able and capable of digesting the information necessary to evaluate the likely consequences of different decision options (Dienel 1992).
- Considering the time constraints under which decisions have to be made we propose that many randomly selected groups of decision makers be organized in parallel. The reservoir of eligible citizens is endless. We have experienced that participants are highly motivated and committed to solve a commonly accepted problem (Renn at al. 1993). Unlike the normal political process where political problems are dealt with sequentially, groups of citizens can be organized in multiple and parallel processes. The only prerequisite is that the problem is defined and that decision options are available.
- Participants serve their role as active political citizens for a limited time only. If they decide to become full-time politicians, this is certainly admirable and welcome. However, our proposal is to commit citizens to a decision

making role briefly so that tendencies towards professionalization and special interest representation are blocked. Again it has been our experience that "normal" citizens identify themselves more easily with the role of representatives of the public good than professional spokespeople of interest groups (for obvious reasons). They also are perceived by others as one of us rather than one of them.

## Planning Cells: Organizational Structure and the New Role of the Citizen

The features of random selection, temporary assignments, parallel problem solving, and division of labor have been combined in the Planning Cell model. Planning cells are groups of (about 25) people who are released from their everyday work obligations (for a week or at least three days) and are asked officially to prepare recommendations on problems of assessment, planning, or control (Dienel 1992). The objective is to provide these citizens with the opportunity to learn about the technical and political facets of the decision options and to enable them to discuss and evaluate these options and their likely consequences according to their own set of values and preferences.

The participants are informed about the potential options and the corresponding consequences before they are asked to evaluate options. Since the process requires time for the educational program and the evaluation, the panels are conducted in seminar form over three to five consecutive days. All participants are exposed to a standardized program of information, including hearings, lectures, panel discussions, videotapes, and field tours. The idea is to conduct a process similar to a jury trial with experts and stakeholders as witnesses and advisors on procedure as "professional" judges. Most of the task-oriented work is performed in small groups of five people. Such an environment encourages all participants to play an active role and to contribute to the solution.

The idea of Planning Cells rests on a set of conditions specified in Table 1. These conditions are essential for success and create a working climate that can produce innovative results and viable compromises. Participants are entitled to receive an "honorarium" as value consultant and/or are freed from their work obligation for several days. This working contract serves as an indicator for the seriousness of the participation effort expected and creates also an atmosphere of discipline and dedication to the task.

*Table 1:* Conditional Characteristics of Planning Cells

| STRUCTURE | CONDITION |
|---|---|
| Composition | Random selection of directly and indirectly affected citizens<br>Involvement of stakeholders and public officials as witnesses, not as participants |
| Tasks | Evaluation of different decision options in accordance with personal values and preferences<br>Clear political mandate to draft recommendations for legal decision maker |
| Operation | Continuous meeting over several days<br>Receiving information about likely consequences of each option<br>Incorporation of uncertainty and dissent through public hearings and videotapes |
| Roles of participants | Identification of participants as "value consultants"<br>Need for external, neutral, and unbiased facilitator<br>Low involvement of sponsor (confined to witness role similar to stakeholders) |
| Organization | Payment of an honorarium to each participant for working a value consultant<br>Local organization committee for facilitating the invitational process |

Planning cells are organized into three major components: (i) reception of information through lectures, field tours, videos, written material, and others; (ii) processing of information through small group discussions, plenary sessions, and hearings; and (iii) evaluation of impacts of options through small group discussions, personal judgments, and consensus-building exercises in the plenary. Informing the participants about the planning options and their likely consequences is the most vital part of the whole procedure. The major requirement is that all expert camps be equally represented in the information package and that they be allowed to present their own case (Dienel 1989). A typical sequence of a Planning Cell is described in Table 2.

The second major component of the Planning Cell procedure is the elicitation of values, criteria, and attributes and the assignment of relative weights to the different value dimensions. The procedures used for this purpose differ from one application to another. In several cases we used methods derived from Multiattribute Utility Theory (Humphreys 1977; Watson 1980; von Winterfeldt and Edwards 1986). The respondents are first asked to rate each decision option on each criterion that they deem important. Each criterion is weighted against each other criterion resulting in a matrix of relative weights and utility measures for each option and each criterion. Both tasks, the transformation in utilities and the assignment of trade-offs are performed individually and in small groups.

*Table 2:* Typical Sequence of a Planning Cell

| STEPS | ACTIVITY IN PLANNING CELLS |
|---|---|
| 1 | Introduction to issue through lecture(s) and field tour. |
| 2 | Background knowledge through lectures, written material, self-educating group sessions, audio-visual information, field tours, and others. |
| 3 | Introduction of conflicting interpretations of information through written material, videos, hearings, panel discussions. |
| 4 | Introduction of options through lectures (non-controversial) or hearings (controversial). |
| 5 | Problem-structuring with respect to each option through group sessions and plenary discussions. |
| 6 | Evaluation of options elicited through individual questionnaires and group discussions (captured in group response forms). |
| 7 | Drafting of rough recommendations through work-groups and plenary sessions. |
| 8 | Articulation of recommendations in a citizens' report done by the facilitator after the completion of the Planning Cells. |
| 9 | Feedback of citizens' report to participants (usually in an evening meeting two months after the Planning Cells). |
| 10 | Presentation of the citizens' report to the sponsor, the media, and interested groups |

The discussion in the Planning Cells leads to proposals and evaluation of options. These are collected and compared on the individual, group, and plenary level. All results are then compiled into an official "citizen report." The first draft of the report is usually written by the project team and validated by the participants. The citizen report is the official document that is handed over to the legitimate decision maker for further consideration.

Planning cells need organizers not affiliated with the problem or any party involved with the issue. Academic institutions or special mediation offices may serve as facilitators. Although the selection of the facilitator is done prior to the first session of the participants, all members of the Planning Cells can reject a facilitator that they regard as biased or demand changes with respect to style and structure of moderation.

## Effects of a Dialogue-Based Participation: Opportunities for Political Renewal

Planning cells have impacts on three levels: the participating individual, the process of policy formation, and society as a whole. With respect to the individual, participants experience empowerment and shared responsibility because they function as public jurors in real decision making arenas rather than as guinea pigs of public experimentation or as objects of public opinion polls. This new experience has a taste of adventure. It establishes self-confidence and helps to build personal as well as social identity (Dienel 1992: 222). In order to be exposed to these enriching experiences, one needs to be selected for a Planning Cell. But if this method of public decision making becomes more popular, many if not all individuals may serve on one of the many needed Planning Cells in the course their life.

With respect to the process of policy formation, the output of Planning Cells differ in several areas from alternative means of decision making. The results of citizen panels can be characterized by:

- *novelty:* citizens often think of innovative ideas not seen or deliberately excluded by established decision makers or experts. Participants of Planning Cells do not lose "authority" or make themselves ridiculous in the eyes of their peers if they consider allegedly exotic or unusual solutions.
- *competence:* since citizens are given sufficient time to study a single issue and to consult all the relevant expertise, they demonstrate a high degree of competence and knowledge. They will not be experts in the field (and this should not be the goal, anyway), but they are able to anticipate the best available knowledge about the likely consequences of each decision option. The intensity of learning and debate within a Planning Cell corresponds to a general diffusion process of several years (Schenk 1993: 254-269).
- *fairness:* the random selection process ensures that everyone has equal opportunity to become a participant of a Planning Cell. In reality, not all people will accept this role, but rejecting the call should be seen as a legitimate personal choice to which each individual is clearly entitled. Previous studies on Planning Cells show, however, that there are only few systematic biases in the composition of those who reject or accept the invitation. In many cases, students and retired people are often over-represented and self-employed people under-represented (Dienel and Garbe 1985). This in itself is interesting as most other decision making bodies are characterized by an opposite bias. Otherwise the composition of people in Planning Cells is roughly representative of the population, at least with respect to gender, education, and class (Garbe 1992; Renn et al. 1985; Dienel 1992). The ability of providing a fair cross-section of the population is a major advantage of the Planning Cell method and contrasts with other more grass-root oriented participation

methods which are usually biased with respect to basic social categories (such as education, class, or gender).

- *common good orientation:* most participatory bodies, in particular those that involve interest groups, focus on bargaining procedures to find some middle ground between competing interests. The resulting solutions may be marginally acceptable to all parties, but do not maximize public utility (Renn and Webler 1994). Participants of Planning Cells have no defined constituents to whom they are obliged. They are selected to embody and represent the interests of all citizens rather than a specific group. It is interesting to note that citizens occupy the role of advocates of the common good almost from the beginning of the sessions (Dienel 1980; Garbe and Hoffmann 1992: 59-62); this may serve as an indicator that they miss the common good orientation in pluralistic societies.

- *mediating service:* the involvement of non-committed citizens provides an opportunity to find a "fair" solution in highly polarized debates. Similar to an arbitration process, Planning Cells can conduct hearings with each advocate group, collect arguments from experts and policy makers, and try to identify options that combine non-partiality with competence. The same task could be performed by a judge or a policy agency; but the involvement of citizens offers an additional advantage of making the claim of a fair judgment based on a fair procedure för choosing the judges. The U.S. jury system is based on this legitimation argument (Bownes 1990).

- *early availability:* most participatory processes create results at a stage where the decision has already been made or where corrections are very costly. Citizen initiatives are usually formed after the announcement of a decision and it takes some time before they can actively challenge the decision and make counter-proposals. In contrast, Planning Cells, can be formed at the earliest stage in the decision making process. The earlier they are organized, the higher is the probability that new options will come into play. Planning is not necessarily delayed by participation processes, although the hope of many planners that early participation will speed up licensing or planning processes will remain an illusion unless participation is taken seriously.

With respect to the social system as a whole, citizen panels can be important catalysts to create or refresh public trust in the democratic system. Trust is always a two-way-street. The common complaint of politicians and agency representatives that the public questions their credibility is only a mirror image of the fact that the people in power have no trust and confidence in the public (Renn and Levine 1990). One-sided trust is always unstable. Granting citizens the opportunity to share power sends out a different message, i.e. that policy makers have confidence in the wisdom and competence of the ordinary citizen to make prudent recommendations. Such a message will have its effects on public perceptions of policy makers. Furthermore, in their role as decision makers, citizens experience first hand the pain and challenge of resolving conflicting values and coping with scarce resources. It is always easy to ask the impossible if one does not have to implement the decision. Finally, non-participating citizens may be more willing to accept unpopular decisions if they know that a group of non-committed and non-privileged citizens were given the opportunity to review the consequences of such a decision (Eidgenössisches Justiz- und Polizeidepartement 1992: 79, 105). This argument leads to the crucial question of legitimation.

## Legitimizing Planning Cells in the Political Arena

A major problem of all citizen participation models is the legitimation of the citizen recommendations vis-a-vis competing claims by interest groups and elected or appointed decision makers. In any democratic system legitimation of power is linked with due process and ultimately public approval through elections (Almond and Verba 1963). Citizen participation is not the vehicle to replace the democratic process of decision making but should be an integral part of this process.

The delicate balance between decision making bodies, affected stakeholders, and Planning Cells is the main challenge. Many agencies in Germany endorse this approach because it provides a conflict resolution method that produces results. Administrators often do not favor the substantive qualities of one option over another, because they are not directly affected. Their judgments are guided by the concern that they may become entangled in a social conflict that might jeopardize their public image. They are attracted to participation models that help them meet their mandate or mission and at the same time produce popular support.

Stakeholders are less likely to hand over responsibility or influence to a group of randomly selected citizens. In the United States, stakeholders have a long tradition of being included in decision making. The role and function of stakeholder participation may differ from one policy arena to another, but the common understanding among stakeholders is that they have a right to be involved and that this right is not to be given up lightly. Nor is that necessarily desirable. Planning cells do not propose to replace the existing mode of decision making where it performs well, but are intended as a supplement when traditional solutions fail to resolve a conflict. Stakeholders also play a major role in Planning Cells. They have the opportunity to deliver their points of view through personal contacts, round tables, or panel discussions within or outside of the Planning Cell's agenda. All stakeholders get a fair chance to testify in front of the Planning Cell participants and clarify their position. In addition, we have given stakeholders the opportunity to be members of an oversight committee which is authorized to control the citizen participation process and the inputs from the organizers. They are free to review all information material, control the facilitating team, and suggest external experts. This way they can be sure that the organizers refrain from manipulating information or biasing the decision making process.

As powerful as the role of stakeholders may be in the political arena, they are not invited to be members of the Planning Cells and do not take part in the formulation of the recommendations. For many stakeholders such a limited influence may not be enough. They often demand more direct participation. It particular, they are often unwilling to commit themselves to a an open-ended process on which they have only limited influence. There are some arguments, however, to convince stakeholders that the Planning Cell arrangement can be of use to their interests and needs.

- When none of the stakeholders are able to resolve the conflict in their favor and a political paralysis exists, all groups hurt by the stalemate may want to involve an uncommitted mediator or jury to resolve the issues. Planning cells reckon with a stalemated system by bringing previously uninvolved citizens onto the scene.
- The stakeholders may agree to a Planning Cell if they are convinced that their arguments and evidence will sway the panels. This strategy may give them the final edge to convince the decision maker. However, if the panel decides otherwise, legitimation problems may arise.
- Planning cells may also be accepted by stakeholders if the decision making agency is powerful enough to force the model on the stakeholders. This situation is rather typical for many European countries in which stakeholder participation is not legally mandated or ensured through tradition. In the eyes of stakeholders, Planning Cells are still better than having no influence at all.

The model of Planning Cells may be attractive for all stakeholders as an alternative to the prevailing methods of mediation and arbitration. Finding a single trustworthy person or an institution to serve as mediators is often cumbersome since mediators themselves have preferences and political alliances (Brooks 1984). A jury of citizens may be more attractive because it assures more variety in viewpoints and political commitments and may carry more legitimacy as a consequence.

## Problems of Planning Cells

The model of Planning Cells has certain drawbacks and limitations. It is not well suited for issues that pose major inequities between different regions or social groups. In these cases, randomly selected citizens are not perceived as legitimate negotiators for the groups that face these inequities. In addition, decisions involving only a yes-no alternative (such as many siting issues) are inappropriate for Planning Cells because participants tend to select the "easy" solution of objecting to any new development, especially if the benefits are not equally shared by the affected community.

Another problem that is associated with Planning Cells is accountability and long-term planning. Since citizens are not responsible for implementing the final decision, they may make choices that are not financially or physically feasible in the long run. Although Planning Cells could be reconvened several times or different panels could be organized for the same subject over a longer period of time, it does not constitute the same public control as having elected officials who face elections and may be legally accountable for their actions. The question of how much authority these panels should be given was also a major point of criticism in a recent review of participation models in the United States (Fiorino 1990).

For thess reasons, we suggest that Planning Cells make recommendations to the legitimate decision maker as an input to the decision process and not as an binding vote. The political bodies entitled to make the decision should retain their responsibility and accountability. Recommendations by citizens may help make public officials more aware of public preferences and help legitimize their own decisions. In the end, publicly accountable institutions must make the final decision and face the consequences. Planning cells can help to make public decisions more rational, for they require officials to defend their decisions using arguments and evidence rather than rhetoric. Participating citizens are not easily persuaded by political jargon or unsubstantiated claims (Renn 1991). Decision makers and citizens can learn from each other during this process, recognizing that both have crucial contributions to make if they pursue the goal of reconciling rational decision making with democratic practices.

A third problem of Planning Cells is that the information and education process is always biased in one or the other direction regardless of the effort to provide comprehensive, complete, and objective information (von Alemann 1986). To avoid biased responses, experts with different attitudes review the educational material and help design the curriculum. Informational material is divided into three categories: basic factual knowledge that all experts agree on; interpretation of facts where all significant viewpoints are represented, and expert beliefs, which are sampled in short essays or videotaped statements (Renn 1991). In addition to the written material and lecture outlines, experts and stakeholders with different viewpoints are invited to testify before the Planning Cells. The same people are also asked to serve on an oversight committee.

There are, of course, additional limitations to this approach (Renn at al. 1993). Based on our experiences with Planning Cells, the following criteria should be used to evaluate the suitability of the procedure:

• *variability of options:* Do the participants have the choice of selecting one option out of a variety of options that are all feasible in the specific situation?

• *equity of exposure*: Are all groups of the community or the respective constituency exposed in some way to the potential disadvantages of the proposed options (to avoid a distinction between affected abutters and indifferent other citizens)?

• *personal experience*: Do participants have some experience with the problem and do they feel competent about giving recommendations after they are further educated about the problem and the remedial options?

• *personal relevance*: Do participants judge the problem as serious enough to sacrifice several days of their time to work on solutions?

• *seriousness and openness of sponsor:* Is the sponsor willing to accept or at least carefully consider the recommendations of the Planning Cells or does it pursue hidden agendas?

It also helps the process if the issue is not too controversial and has not already polarized the attitudes of the affected population. Our own experience in Germany shows, however, that even these issues can be handled by Planning Cells if the majority of participants are selected by random process (Dienel 1989; Renn et al. 1985). The Planning Cells work best in a situation in which an urgent problem has to be resolved in a short time period and different options, each posing different benefits and risks are available.

## Experiences with Planning Cells

### *German experiences*

The early applications of Planning Cells in Germany were devoted to urban planning. Community governments wanted to give citizens the opportunity to take part in community development. As long as the recommendations were technically feasible and economically viable, the legitimate decision maker (city or community council) had no reason to reject them. After initial test runs in the years 1972-1973 at the town of Schwelm (considering a waste disposal facility), Planning Cells were established in many communities in Germany. From the 1970's to today approximately 26 cities or communities used Planning Cells as a method of local planning. More than 2,600 adults were involved in these panels for an average of 3-5 days each. In one case a relocated community decided to entrust Planning Cells to do energy planning for the community (Friedrich and Garbe 1983). In another case, the city of Cologne asked Planning Cells to review plans for urban renewal of one of the central districts (Bongardt et al. 1985). These two cases were not characterized by major social conflicts or pre-structured ideological positions of stakeholders. The community government welcomed the citizen input and adopted the recommendations to the degree that they were technically feasible.

This situation is different from a national or regional issue in which the decision stakes are high and stakeholder groups have formed strong positions on the issue. Four national cases should be mentioned here:

- The Federal Ministry of Postal Service and Telecommunication sponsored a Planning Cell process with 22 panels. The goal was to debate and evaluate the "future telephone." Issues ranged from desired telecommunication services to governmental regulations of private utilization of public networks. During three four-day sessions the participants invested more than 2,000 workdays and issued 85,000 individual or collective statements (Garbe 1992: 651). The citizen report (Bürgergutachten ISDN 1991) included 66 substantial recommendations. The German ministry adopted several of these recommendations, in particular those concerning data protection in telecommunication (Garbe 1992: 656).
- The Federal Ministry of Research and Technology commissioned a citizen report on "Handling societal consequences of new information technologies." A total of 14 Planning Cells investigated and reviewed policies in the field of industrial production, tele-homework, hospital application, leisure activities, and adult education. In September 1986, a citizen report was published which offered 78 proposals and listed 43 areas for further research and investigation (Bürgergutachten, Regelung... 1986).
- The German Consumer Association commissioned a citizen report on its methods and procedures of testing new products (Garbe and Grothe 1985). The goal was to measure public preferences of consumer policies and of testing procedures practiced by the Consumer Association. In this case, a stakeholder group used the method to revise its own policies and to be more in line with its clients.
- The most comprehensive study dealt with the evaluation of national energy policies. In August 1982, the German Ministry of Research and Technology initiated a large research project to investigate the preferences of the German population with respect to four energy policy options developed by a parliamentary commission in 1979 (Renn et al. 1985; Renn et al. 1984; Dienel and Garbe 1985). The Government was interested in eliciting reliable information on which energy scenario was most appealing to the population and on what basis citizens would evaluate the policy options laid out in each scenario. A research team in which both authors served as senior investigators conducted a three-year study to collect data on public preferences and to analyze the motivations and underlying reasons for the judgment process of evaluating the predefined energy scenarios. The study operated with 24 Planning Cells (each including approximately 25 participants) drawn from seven communities in different parts of West Germany. The panels

unanimously rejected a high energy supply scenario and opted for an energy policy that emphasized energy conservation and efficient use of energy. Nuclear energy was perceived as non-desirable but – at least for an intermediate time period – as a necessary energy source. The panelists recommended stricter environmental regulation for fossil fuels even if this meant higher energy prices. They developed a priority list for policies and drafted recommendations for implementing high priority policies (Dienel and Garbe 1985).

In summary, the German applications of Planning Cells provides some evidence and reconfirmation that the theoretical expectations linked to the Planning Cell method can be met on the local as well as on the national level. It is a valid instrument to elicit preferences and educated responses of citizens in a rather short time period. Far from being an established planning tool, it has proven its viability and feasibility in different contexts and constitutes at least a serious alternative to other forms of public involvement.

## *Swiss experiences*

In 1992, The Building Department *(Baudepartement)* of the canton Aargau (Northern part of Switzerland) asked a research team at the Swiss Federal Institute of Technology (including two of this book's editors O. Renn and T. Webler) to organize Planning Cells for siting one or several landfills in the eastern part of the canton. The Building Department proposed, and the Cantonal Government approved the plan to construct a one million cubic meter landfill, occupying a site of ten to twenty hectares and remaining in operation for about forty years. Before our involvement in the project, the Building Department characterized the need for new disposal facilities and chose 13 potential sites through a mapping-elimination process. The mandate of the Planning Cells was first to develop criteria for comparing the different sites; second, to evaluate the geological data that were collected during that period; third, to eliminate the sites that should not be further considered; and fourth, to prioritize the remaining sites with respect to suitability to host a landfill. We managed to meet these objectives during the time from November 1992 to September 1993 (Webler 1994).

In late October of 1992, we asked the representatives (Gemeinderäte) of the thirteen communities in which the potential sites were located to send one member of the town council to serve on an oversight committee (Behördendelegation). The oversight committee consisted of one member of each town council and the director of the building department. The oversight committee had the legitimate right to make the final recommendation to the Building Department. In addition, they were asked to inform the public about the site selection process, to review and critique the participation process, and to select the representatives from each of their communities for the citizen panels.

The selection of representatives for the citizen panels differed from our theoretical approach. Rather than use random selection, we gave the oversight committee the task to recruit and select citizen participants. The sponsoring agency was concerned about the legitimacy of the recommendations issued by the panels and felt that random selection would not be seen as a legitimate way of choosing representatives. Using lotteries as a political means of achieving equity is alien to the Swiss political culture. In substitution we proposed that either a town meeting or the community government nominate the representatives, with some assistance by the research team to encourage consideration of all relevant social and political viewpoints. We asked each community to select eight representatives.

Once the representatives were chosen, four panels were formed, each consisting of two representatives from each potential site community. With the exception of one community, every town sent eight people to the panels. Not a single one of these people dropped out during the process. Between January and June 1993 the panels met 7-9 times before they attended a workshop of two days to come up with the final decision. All participants rated each site on the basis of their self-selected evaluative criteria, their personal impressions, the written and oral information, and the results of consultations with experts.

All four panels composed a list of prioritized sites for the landfill. The most remarkable outcome was that each panel reached a unanimous decision. Even those participants whose towns were selected for the short list of recommended sites agreed with the panel's recommendations. Furthermore, the outcomes of the four groups were rather similar. The first priority site was the same for all panels. There were some minor differences in the order of the remaining priorities. To resolve this conflict, each panel appointed five representatives to a superpanel. The superpanel met in September 1993 and issued a consensual list of five sites ordered in a priority list. This list was later approved by the oversight committee and forwarded to the Building Department. In December of 1993, the result of the participation process was made public and the canton government entered the next phase of the licensing procedure.

Looking over the results, it seems as if the selection process was so obvious that no major controversy was to be expected. All panels agreed that one location (Schinznach) was the best choice. This site, however, was not the most favored choice of the building department. We had asked the technical director for the siting process to conduct his own analysis and to make a priority list before the panels would make their final decision. The director performed this analysis in conjunction with other staff members. The first choice on the department's list was another town. The most favored location of the four panels was Number 3 on the department's list. The main reason for this difference was the high importance that the officials from the building department assigned to geological stability. whereas the citizens included social and aesthetic criteria in their preference judgment.

## *Experiences in the Basque region of Spain*

The Basque region within Spain is an area characterized by public distrust in official policies and exposed to violent and intense ethnic conflicts. After one of the authors, Peter C. Dienel, introduced the Planning Cell method at a regional meeting, several communities facing serious opposition in planning public buildings and new municipal infrastructure seized the opportunity and implemented the first Planning Cells (nucleos de intervencion participativa) in their communities (Dienel, in press). During the early 1990's two printed citizen reports (dictamen ciudadano) were completed and handed over to the local authorities (Laboratorio 1992). In one case a long standing conflict about the construction of a gymnasium was resolved peacefully with the aid of Planning Cells.

Encouraged by these local experiences, a larger project has been initiated The regional department of transportation commissioned a study for siting a highly disputed highway through the Basque region (Autovia Urbina-Malzaga). All previous siting attempts had failed and resulted even in violent clashes between the police and protesters. The new study, which is presently underway, involves a total number of 14 Planning Cells and 350 randomly selected participants. These panels are equally distributed throughout the region and are roughly proportional to the population. The mandate of the panels is to review the existing plans for siting the highway, consider alternative routes, and assess the social and political effects of each option.

It appears unusual that the Basque region, which is so sensitive to outside interventions and has suffered so much from ethnic violence (some 80 policemen have been killed during the last seven years) has opted to experiment with a new and foreign model of participation. One reason may be despair: all models of conflict resolution appear to have failed during the last years. Another reason may be the presence of an institutional citizen advocate or ombudsman (Arateko) who is in charge of promoting direct citizen participation. Such people are always grateful for new suggestions. We hypothesize, however, that the main reason is the appeal of fairness and legitimacy linked to the Planning Cell method. In situations where all selection processes look biased and interest-driven, random selection and openness to public demands and preferences seem to convey the message to the majority of skeptical people that this is a serious and honest effort to let the affected population determine its own fate. It is too early to make any more general conclusions, but so far the expectations of the organizers as well as the participants appear to be met.

## *American experiences*

Using randomly selected citizens for policy making and evaluation is not alien to the United States. The Jefferson Center in Minneapolis has conducted fourteen projects with Citizen Juries similar to the Planning Cells (Crosby et al. 1986; Crosby, in this volume). Several community planners have experimented with citizen panels which were composed to reflect a representative sample of the population (cf. Kathlene and Martin 1991). There has been one attempt to implement the original version of the Planning Cell method in the United States. In July 1988 the Department of Environmental Protection of New Jersey asked a research team of Clark University directed by one of the authors, Ortwin Renn, to apply the model to sewage sludge management problems. The project started in August 1988 and was completed in September 1989. The objective of the project was to give citizens of Hunterdon County, New Jersey, the opportunity to design the regulatory provisions for an experimental sludge application project on a Rutgers University research farm located in Franklin Township (New Jersey).

Although much smaller in scale, the project provided many new insights and experiences that partially confirmed our German observations and partially documented the need for adjustments to the U.S. political culture. The project was organized in a fashion similar to the German energy study. We conducted the Planning Cells on two consecutive weekends. The desired goal was to elicit recommendations for regulatory provisions that should be included in the permit for the land application of sewage sludge on the site in question.

The envisioned program for the citizens panel was radically altered after the participants, in particular the land owners abutting the site, made it clear that they rejected the project of land application and that they felt more comfortable conducting their own meetings without assistance of a third party. The citizens met several times without the assistance of a facilitator and formulated recommendations that were forwarded to the sponsor (New Jersey Department of Environmental Protection). The proposed sludge management project at the Rutgers Experimental Farm was rejected by the Planning Cell. As a result of this recommendation, Rutgers University withdrew its proposal.

In addition to the policy recommendation to reject the proposal of land application, the process provided us valuable information about citizen concerns and values. Whereas most of our consulted experts were convinced that citizen concerns focused on issues such as odor, traffic, and contamination of ground water, the value tree analysis of the citizens revealed that their major concerns were the expected change of community image from an agricultural community to a "waste dump" and the long-term effects of pollutants on farmland (Renn et al. 1989). In addition, the questions of equity and fairness played a major role in the citizen deliberations.

The unexpected change of the panel's structure to exclude us from further meetings was clear evidence that the U.S audience is more sensitive to due process and methods of participation. Whereas in West Germany participants were almost grateful and pleasantly surprised that someone made the effort to pre-plan and structure a procedure for their participation, U.S. citizens distrust pre-fabricated participation models and suspect hidden agendas with such an approach (Stewart et al. 1984). In response to the desire of the participants to have control over the process we think that it is advisable to have a meeting with the participants two weeks before the actual Planning Cells to discuss the agenda and the information material. During that preliminary meeting, the participants can be informed about the process and the importance of the given time schedule. They can also add points to the agenda or change the allocated time frame. This prevents surprise discussions or rebellions during the actual Planning Cell procedure.

## Conclusion

Our experiences from previous projects and the implementation of our model indicate clearly that the public has something to contribute to the planning process. Experts and regulators are often restricted in their assessment of a project and confine their analysis to the typical risk factors. Local specifics or other dimensions of concerns are often neglected. Public participation helps to include these concerns in the decision making process and to avoid potential consequences that the experts missed.

The model of Planning Cells is only one of many possible ways to involve the public in decision making and policy designing. It is characterized by several features usually not found in other proposals for citizen involvement (cf. Fiorino 1990). In contrast to joint commissions of experts and citizens, in this model each participating group is assigned a specific function. In contrast to negotiations with stakeholder groups, Planning Cells are inspired by the normative goal of a fair and impartial representation of all citizens' values and preferences, be they organized or not. The objective of Planning Cells is not to include the already organized stakeholders or local officials, but to provide citizens with the opportunity to take part in a policy making process. In contrast to elected bodies, the Planning Cells have a single issue mandate working on a specific problem in a pre-defined time period. They do not depend on special constituencies, such as voters or interest groups.

At the present date, more than 90,000 hours of work have been invested by ordinary citizens to provide their input to political decisions. In a time in which paid work is becoming a scarce resource, the original idea of voluntary public service may experience a revitalization (Dienel 1971:152). Dividing the population into those with access to paid work and those without it invites conflict and inequities. Sharing productive work and regulative work among the people who are willing to provide both services may be one way to overcome or at least mitigate this growing schism. Planning cells are an ideal instrument to provide responsible working opportunities to those who have not been able to compete in the private sector as well as for those who do so successfully. Sharing responsibility and trust in a realistic setting (and not in a therapeutic game), provides opportunities for mutual understanding with respect to people's needs, interests, values, and fears. The common complaint of many social analysts that modern society has been fallen apart into almost autonomous subsystems with their own reasoning, morale, and rationale may be counteracted by the situation of randomly collected citizens who are obliged and paid to cooperate and develop consensual policies. In this respect, Planning Cells do not only contribute to tackling social problems, but also to establishing a common ground for communication and cooperation among and between different social groups. Planning cells may serve as "fractal" mediators in a socially and culturally fragmented world.

We envision Planning Cells as a learning experience in problem solving and public service. Participants have confirmed that the process helped them become better communicators and made them more confident in creating cooperative working styles. At the same time, they recognized their role as consultant or advocate of the public good rather than their egoistic interests. Many analysts have articulated a social need for new communitarian values (most recently Etzioni 1993). Our experiences show that development and effectiveness of these values depend on the opportunity of ordinary people to take responsibility in the public sphere. There is no doubt that professional and specialized politicians and administrators are needed in modern societies, but the role and function of the non-professional citizen in a democratic society needs to be revised. Voting and protest are not sufficient. We need opportunities for political involvement for each person, at least for a limited amount of time after which people may return to their "normal" life. If the public is to have a role to play in the decision making process, the model of Planning Cells is certainly one that provides a rational and structured approach toward this goal.

## References

Almond, G. A., and S. Verba, *The Civic Culture: Political Attitudes and Democracy in Five Nations* (Princeton University Press: Princeton 1963).

Bachrach, P., *The Theory of Democratic Elitism: A Critique* ( Little Brown: Boston 1967).

Bongardt, H., P. C. Dienel, and H. Henning, *Bürger planen das Rathausviertel. Planungszellen erarbeiten Ausgangsdaten für den städtebaulichen Wettbewerb "Rathaus/Gürzenich Köln"* (Campus: Frankfurt/Main 1985).

Bürgergutachten, *ISDN,* compiled by P. C. Dienel, B. Moog-Kopp, and A. Reinert. (Participation Research Unit: University of Wuppertal 1991).

Bürgergutachten, *Regelung sozialer Folgen neuer Informationstechnologien,* compiled by P. C. Dienel, M. Huppertz and H. Poschmann. Fourth Edition (Participation Research Unit: University of Wuppertal 1986).

Brooks, H., "The Resolution of Technically Intensive Public Policy Disputes," *Science, Technology, and Human Values*, 9 (Winter 1984), 39-50.

Burns, T. R., and R. Ueberhorst, *Creative Democracy: Systematic Conflict Resolution and Policymaking in a World of High Science and Technology* (Praeger: New York 1988).

Crosby, N., J. M. Kelly, and P. Schaefer, "Citizen panels: A New Approach to Citizen Participation," *Public Administration Review*, 46 (1986), 170-178.

Dienel, P. C., "Partizipation an Planungsprozessen als Aufgabe der Verwaltung," in: *Die Verwaltung,* 4 (1971), 151-176.

Dienel, P. C., *New Options for Participatory Democracy.* Werkstattpapier No.1 of the University of Wuppertal (Participation Research Unit: University of Wuppertal 1980).

Dienel, P. C., "Contributing to Social Decision Merthodology: Citizen Reports on Technological Projects," in: C. Vlek and G. Cvetkovich (eds.), *Social Decision Methodology for Technological Projects* (Kluwer: Dordrecht 1989), pp. 133-152.

Dienel, P. C., *Die Planungszelle. Eine Alternative zur Establishment-Demokratie* Third Edition (Westdeutscher Verlag: Opladen 1992).

Dienel, P. C., *NIP – Nucleos de Intervencion Participativa: Un paso hacia la democracia* (Zarautz: in press).

Dienel, P. C., and D. Garbe, *Zukünftige Energiepolitik. Ein Bürgergutachten* (HTV Edition "Technik und Sozialer Wandel": Munich 1985).

Eidgenössisches Justiz- und Polizeidepartement (ed.), *Mitwirkungsverfahren und -modelle* (Bundesamt für Raumplanung: Bern 1992).

Etzioni, A., *The Active Society. A Theory of Societal and Political Processes* (The Free Press: New York 1968).

Etzioni, A., *The Spirit of Community. Rights, Responsibilities and the Communitarian Agenda* (Crown: New York 1993).

Fiorino, D. J., "Citizen Participation and Environmental Risk: A Survey of Institutional Mechanisms," *Science, Technology, and Human Values,* 15, 2 (Spring 1990), 226-243.

Friedrich, T., and D. Garbe, *Bürgergutachten Energieversorgung Jüchen-Nord* (Participation Research Unit: University of Wuppertal 1983).

Garbe, D., "Social Compatibility of Telecommunication Technologies. Results of the Citizen Report on ISDN," *Telecommunications Policy,* 11 (1992), 646-656.

Garbe, D., and A. Grothe, *Bürgergutachten Testkriterien und Testplanung* (Participation Research Unit: University of Wuppertal 1985).

Garbe, D., and M. Hoffmann, *Soziale Urteilsbildung und Einstellungsänderungen in Planungszellen* (Participation Research Unit: University of Wuppertal 1992).

Grottian, P., and A. Murswieck, "Zur theoretischen und empirischen Bestimmung von politisch-administrativen Handlungsspielräumen," in: P. Grottian and A. Murswieck (eds.), *Handlungsspielräume der Staatsadministration* (Hoffmann und Campe: Hamburg 1974), pp. 15-34.

Humphreys, P., "Application of Multi-Attribute Utility Theory," in: H. Jungermann and D. de Zeeuw (eds.), *Decision Making and Change in Human Affairs* (Reidel: Dordrecht 1977), pp. 165-205.

Kathlene, L., and J. A. Martin, "Enhancing Citizen Participation: Panel Designs, Perspectives, and Policy Formation," *Policy Analysis and Management,* 10 (1991), 46-63.

Laboratorio de Sociologia Juridica (ed.), *Dictamen Ciudadano Astigaaraga* (San Sebastian, Spain: December 1992).

Leibholz, G., "Parteienstaat und repräsentative Demokratie," in: H. Rausch (ed.), *Zur Theorie und Geschichte der Repräsentation* (Wissenschaftliche Buchgesellschaft: Darmstandt 1968), 222-259.

Luhmann, N., "Die Knappheit der Zeit und die Vordringlichkeit des Befristeten," *Die Verwaltung,* 1 (1968), 3-30.

Olson, M. E., *Participatory Pluralism* (Nelson Hall: Chicago 1982).

Olson, M. E., *The Logic of Collective Action. Public Goods and the Theory of Groups* (Harvard University Press: Cambridge, MA. 1965).

Renn, O., "Premises of Risk Communication: Results of Two Participatory Experiments," in: R.E. Kasperson and P.J. Stallen,(eds.), *Communicating Risks to the Public: International Perspectives* (Kluwer Academic: Dordrecht 1991), pp. 457-481.

Renn, O., H. U. Stegelmann, G. Albrecht, U. Kotte, and H. P. Peters, "An Empirical Investigation of Citizens' Preferences Among Four Energy Scenarios," *Technological Forecasting and Social Change*, 26, 1 (1984), 11-46.

Renn, O., G. Albrecht, U. Kotte, H.P. Peters, and H.U. Stegelmann, *Sozialverträgliche Energiepolitik. Ein Gutachten für die Bundesregierung* (HTV Edition "Technik und Sozialer Wandel": Munich 1985).

Renn, O., R. Goble, D. Levine, H. Rakel, and T. Webler, *Citizen Participation for Sludge Management*, Final Report to the New Jersey Department of Environmental Protection (CENTED, Clark University: Worcester, MA, 1989).

Renn, O., and D. Levine, "Credibility and Trust in Risk Communication," in: R. Kasperson and P.J. Stallen (eds.), *Communicating Risk to the Public* (Kluwer: Dordrecht 1990), pp. 175-218.

Renn, O., T. Webler, H. Rakel, P. C. Dienel, and B. Johnson, "Public Participation in Decision Making: A Three-Step-Procedure," *Policy Sciences,* 26 (1993), 189-214.

Renn, O., and T. Webler, "Konfliktbewältigung durch Kooperation in der Umweltpolitik – Theoretische Grundlagen und Handlungsvorschläge," in: Umweltökonomische Studenteninitiative OIKOS an der Hochschule St. Gallen (ed.), *Kooperationen für die Umwelt. Im Dialog zum Handeln* (Ruegger Verlag: Zürich 1994), pp. 11-52.

Rosenbaum, N., "Citizen Participation and Democratic Theory," in: S. Langton (ed.), *Citizen Participation in America* (Lexington Books: Lexington 1978), pp. 43-54.

Scheer, H., *Parteien kontra Bürger? Die Zukunft der Paretiendemokratie* (Econ: München und Zürich 1979).

Schenk, M., "Die ego-zentrierten Netzwerke von Meinungsbildern," *Kölner Zeitschrift für Soziologie und Sozialpsychologie,* 2 (1993), 254-269.

Stewart, T. R., R. L. Dennis., and D. W. Ely, "Citizen Participation and Judgment in Policy Analysis: A Case Study of Urban Air Quality Policy," *Policy Science*, 17 (1984), 67-87.

Schumpeter, J. A., *Capitalism, Socialism, and Democracy.* Second Edition (Harper and Brothers: New York 1947).

Schwarz, M., and M. Thompson, *Divided We Stand: Redefining Politics, Technology, and Social Choice* (University of Pennsylvania Press: Philadelphia 1990).

von Alemann, U., "Partizipation oder Akzeptanz. Bemerkungen zur Verträglichkeit von Demokratie und Technologie," in: H. Jungermann, W. Pfaffenberger, G.F. Schäfer, and W. Wild (eds.), *Die Analyse der Sozialverträglichkeit für Technologiepolitik – Perspektiven und Interpretationen* (HTV Edition "Technik und Sozialer Wandel": Munich 1986), pp. 28-35.

von Kielmannsegg, P. Graf, "Politik in der Sackgasse?", in: Ministerium for Forschung und Technologie (ed.), *Energiediskussion,* Vol. 3 (BMFT: Bonn, June 1979), pp. 31-44.

von Winterfeldt, D., and W. Edwards, *Decision Analysis and Behavioral Research* (Cambridge University Press: Cambridge, MA, 1986).

Watson, S. R., "Multiattribute Utility Theory for Measuring Safety," *European Journal of Operational Research,* 10 (1982), 77-81.

Webler, T., *Experimenting with a New Democratic Instrument in Switzerland: Siting a Landfill in the Eastern Part of Canton Aargau* (Polyproject: Risk and Safety of Technical Systems, Swiss Federal Institute of Technology: Zürich 1994).

*Chapter 7*

# *Review of "Planning Cells:" Problems of Legitimation*

*Hans-Jörg Seiler*

## General Remarks

The Planning Cell seems to be a procedure which, under certain circumstances, could produce good results. I do not want to give the impression that Planning Cells constitute an unsatisfactory procedure, but instead try – from the viewpoint of a Swiss lawyer – to indicate some weak points of the model. In Switzerland there is a very strong tradition of citizen participation on all political levels. This review is based on personal and general experiences with Swiss procedures of citizen participation rather than on theoretical reflections.

## Embeddedness of the Planning Cell

The embedding of the procedure in the political context seems to be more important than the internal procedure of the model. For this reason it warrants careful consideration. The Planning Cell-procedure is intended to be purely consultative (Dienel 1991:131); it should not replace the existing decision making procedures and authorities, but improve the procedure. Traditional organizational law of the states mainly concentrates on the question of competence as a basis to decide. But over the past few decades there has been a general tendency—influenced by the English and American legal tradition of procedural due process (a point further raised by Habermas)—to give more importance to the procedure of decision ("legitimation through procedure"). For this reason it is important to consider both the procedure and the competence when deciding legal issues. The problem is further complicated by interaction between procedure and competence. I will focus on this topic below. At the same time the Planning Cell is intended to have a integrative function and to avoid increasing distrust in the state. While this is a very important point we must carefully think about whether this can be achieved by the means of Planning Cells.

The role of the model in the political context depends on its embeddedness. Planning Cells could be consultative bodies of a parliament, a government or an administrative authority on different levels (national, regional, local). The applicability of the model could be very different depending on the respective level of governance.

## Criteria for Fair and Competent Discourse

In this section I concentrate on the internal procedure of Planning Cells and evaluate it using the proposed criteria. Before I begin I would like to make a personal remark concerning the relevance of these criteria. Habermas-based ideas for procedure (and the discursive standard criteria as well) are a normative ideal which could perhaps apply to a little group of people who aspire to produce political actions without any hierarchy. But Habermas himself admits that in reality such an egalitarian discourse is nearly impossible to achieve. In most political decisions time pressure is present which prevents even the most democratically minded decision maker from discussing all the relevant topics with all the relevant and affected people. For this reason the main question is perhaps not whether these ideal, unrealistic criteria are met, but rather how much deviation is possible without denying the elementary democratic principles.

*Criterion A: Equal chance for everyone to put concerns on the agenda.*

This criterion is not met. The Planning Cells are concerned with given problems, defined by the authority willing to make use of a Planning Cell. In this concern the position of the Planning Cell is the same as the position of an expert, who is asked to contribute expert advice to a political body. If some members of the Planning Cell are not willing to accept this limitation because they think – perhaps justly – that other problems are strongly connected, the success of the whole procedure could be jeopardized. This could be a problem, especially in highly politicized and very controversial matters, where many different aspects have to be taken into account. On this basis, one might conclude that the procedure is mainly applicable for questions that can clearly be isolated from others.

*Criterion B: Everyone should take part in selecting a moderator and moderation style.*

This criterion is not met, but it is perhaps less important. An inordinate amount of time would have to be expended if Planning Cells would have to first discuss the moderation method. Based on my experience in Swiss citizen commissions, it seems to me that people are generally quite willing to accept a pre-structured moderation method provided they feel the method for selecting the moderator and the moderator's style is fair, even if they have not taken part in the selection. Nevertheless I would stress the importance of some kind of control over the moderator. The moderator has a very strong position and could easily manipulate the whole procedure. In the legal system we assign a great importance to the integrity and independence of the judges, and there are many procedural rules to make certain that the procedure is fair and both parties have the same opportunities and obligations. Other rules provide some control over administrative bodies. Such a system of preestablished legal rules in order to guarantee the integrity of the moderator and the fairness of the procedure would be indispensable.

*Criterion C: Chance of participation/representation for everyone who is potentially affected and equal chance to put forth and criticize claims.*

In the internal procedure of the Planning Cell this criterion is met. But the main problem is the position of those who are not participants of the Planning Cell: the existence of the panel prevents all these who are not members of the panel to bring in their concerns. This is a crucial point of external legitimation of the procedure.

*Criterion D: Equal access to comprehensibility claims*

This criterion is well met. Every member of the Planning Cell has the possibility to ask questions in order to understand the meaning of comprehensibility claims.

*Criterion E: Equal access to relevant knowledge about the objective world.*

The Planning Cell is based on the idea that at the beginning every member of the panel has the same information and that the information needed is prepared and packaged by the staff/moderator (Dienel 1991: 96f, 113). This provokes some criticisms.

The preparation of information by the staff could easily be misused for manipulation. Even if the staff is sincere about producing fair information, true objectivity is unachievable. In the legal procedure this problem is avoided insofar as both parties can produce their own information and supporting argumentation. Nobody would accept a legal procedure where the jury decides only on the basis of information given by the staff of the court. In the legal procedure each party can propose witnesses and the court is – up to a certain point – obliged to listen to these witnesses. It would not be acceptable if only the staff of the court would have the right to propose witnesses. Second, the idea that all the members of a Planning Cell have the same knowledge in the beginning is highly unrealistic. Every citizen has some previous knowledge and preconceptions about the topic to be discussed.

Third, all the members of the Planning Cell will try to acquire more knowledge during the procedure and/or get further information from people outside the Planning Cell. This cannot be avoided unless the members of the Planning Cell were sequestered during the procedure. This point is very important for comparing the Planning Cell procedure with other methods of political decision-making. The Planning Cell is intended to be less influenced by lobbies than, for example, a parliamentary body; but as soon as the Planning Cell becomes an important body (because their recommendations gain political legitimacy), lobbies will concentrate on the participants. Many experiences with extra-parliamentary commissions suggest that there is a correlation between the importance of a commission and the efforts of the lobbies to influence the members of these commissions.

For all these reasons it is neither desirable nor realistic to conceive the procedure as based only on objective information given by the staff. Rules must be developed in order to guarantee that information can be acquired and not simply delivered by one-sided lobbies. The legal rules for providing information to a court (every party has the right to produce information and propose witnesses) could be taken as an example.

*Criterion F: Rules for redeeming normative validity claims*

For the internal procedure of the group this criterion is more or less met, as well as in all the small discussion groups which are formed. But the real problem cannot be solved. Defining normative values is always a question of power. Therefore it is of little use to consider just the internal procedure of the Planning Cell. The question must be seen in the context of the social and legal distribution of power; otherwise it remains a theoretical discussion with pure academic interest. This concerns the question of external legitimacy.

*Criteria G und H: Rules for redeeming truthfulness validity claims.*

Planning Cells meet these criteria for all interactions within the group. The main problem relates to the question of whether the results of the group discussion will be accepted outside the group.

## External Legitimation

Several times we have seen that the main problem is not so much the procedure inside a given group (Planning Cell or other form of citizen panel), but the question of external legitimacy of the procedure. If independent panels discussed the same questions but reached different solutions, the fundamental problem remains unresolved. That is the question of who should decide. Planning Cells try to circumvent this question by being purely consultative. In reality, however, it is impossible to make a clear distinction between the preparation of a decision (including consultation) and the decision itself. Any precursory activity always has some influence on the decision. Therefore we also have to ask about the legitimacy of consultation.

External legitimacy of a body depends on three components.

1. The procedure of the body. Fair procedural rules should be followed in order to make sure that the relevant aspects are discussed.

2. The people who are involved in the decision-making process. They may be members of the deciding body or participants in the procedure (e.g. as witnesses, parties).

3. The formal competencies of the body, as defined by the legal system.

The first point has been discussed in the previous section and the third will be discussed in the next. I will discuss the second point in this section.

Traditionally in the legal system of Switzerland and other democracies – especially in European legal systems – a clear distinction was made between the procedure for *legislation* and the procedure for *administrative decisions. Legislation* was seen as the matter of the whole nation. Those decisions, therefore, should be made by the parliament – as the representative body of the whole. Ideally every citizen takes part indirectly in the legislative procedure. In *administrative decisions* there is, on the one side, the administrative authority, and on the other side, the people who are affected by the decision. These people have a legal right to put forward their arguments and opinions in the administrative procedure (Right to be heard; *Anspruch auf rechtliches Gehör)*. All other people are rightly excluded. In administrative decisions public interests are supposedly taken into account by the authority that applies the law, as written by the people's representatives. In this manner the authority represents the public interest.

In practice this distinction is not so clear. First, many legislative matters do not affect the whole nation, but merely a part of it. Yet everyone has an (indirect) say. Likewise many administrative decisions affect a great part of the population, making it difficult to identify an affected population other than the whole country. Therefore the distinction between legislation and administrative decisions is less clear than it might seem. Second, laws are not made by the representatives of the whole nation, they are influenced greatly by lobbies that push their partial opinions. Third, many laws state only general intentions, leaving considerable room for interpretation by administrative authorities that apply the law. Fourth, many decisions do not affect people in the same way, but affect some people more than others. Fifth, the administrative authorities are not always the impartial representatives of the public good and defenders of the affected population, but are also influenced by one-sided views and partial lobbies. They may try to implement projects which are rejected by the concerned population.

For all these reasons citizen participation was postulated not only (and sometimes even not mainly) for the procedure of legislation, but also for administrative decisions. In most of the participatory models the significance of involving the *affected* population is stressed: the people who are especially affected by this decision should be able to take part in the procedure in some way. Indeed this is the reason for most of the newly promoted participatory models.

One of the dangers is that the participation procedure could become dominated by private or partial interests. To counter this danger the Planning Cell model seeks to empower general interests by mkaing sure that partial interests are not allowed to dominate (Dienel, 1991:94f, 104). In other words the aim is not participation of the affected population, but the participation of the general public. This is seen as a consequence of the principle of political equality (Dienel, 1991:95) and is an important distinction to many of the other models presented in this book. It warrants further reflection about the relationship between general and partial interests and about legitimacy of representation of the whole population or, rather, of affected parts of the population.

This juxtaposition highlights the fundamental distinction between American and European conception of democracy. In America democracy is based on the idea that the population consists of different factions with distinct interests (Federalist Papers No. 10). The political procedure is therefore a "legislative judging" in order to make compromises between these factions' interests. The European conception of democracy is based on the idea of a common interest (*volonté générale*). The political procedure is seen as one which tries to implement this common interest against the particular interests. This is the reason for the wide-spread skepticism against the so called *Betroffenheitsdemokratie* ("democracy based on affectedness," a notion which is often used in a pejorative manner) in the European countries.

To give further relevance to this matter, consider the fundamental idea on which the classical European model of democratic legislation is based: that the laws concern everybody. Liberty, Kant says, consists of not being forced to obey any law unless one had the opportunity to vote on it (Kant, 1797: § 46). As a consequence, voting is exchanged for obedience. There is no reason to be allowed to vote on a law for which one's obedience is not expected. In other words there is no reason for one to take part at a decision which does not affect the one who takes part. The idea of participation is connected to the fact that the participant is affected. I do not agree with these who criticize *Betroffenheitsdemokratie*. *Betroffenheitsdemokratie* is not the perversion, but the fulfillment of the democratic ideal. Nevertheless I can agree with the criticism that, in many cases, special and partial interests are too strong in our political life, compared with the so-called general or public interests. This is not a contradiction to what I said above. The problem is, that the so-called general public is, in reality, also affected by (legislative or administrative) decisions, which seem at first to affect only a small population. Critics of *Betroffenheitsdemokratie* often think of it as a system where *only* the *especially* affected people could decide. Of course this is a wrong conception. *Betroffenheitsdemokratie* should not be conceived as a system where *only* especially affected people are allowed to participate. On the contrary, the public, which is affected in a more general manner, must also be allowed to participate. However, the ones who are affected in a special way must be allowed to participate in a special way.

The contradiction between public interests and special interests refers to contradictions between the interests of all the people or of a great majority of the people and some particular interests of some of the people. Therefore it is impossible to make a clear distinction between being affected and not being affected. On the contrary there are a lot of different methods and degrees of being affected (Uebersax, 1991: 91ff, develops a typology with about twelve different modes of being affected). Theoretically there should be a correlation between the degree of being affected and the degree of being allowed to participate at the decision. It would be the democratic ideal to allow anybody to take part at the decision corresponding to his or her degree of being affected by this decision (Rhinow 1984: 179f; Seiler 1986: 80ff; Seiler 1991: 5-17; Seiler 1991a: 125ff; Uebersax 1991: 97ff). The problem is that it is very difficult or even impossible to define the degree of being affected for all the decisions which have to be made. Consider, for instance, the decision to build or not to build a road between points A and B. Who is affected? Some people would like to have this road in order to travel faster between A and B. Others are landowners who will sell their land for the building of the road. Others living on the road are expected to tolerate the emissions. Others are owners of restaurants and businesses along the planned road who will profit from increased business. So many people are affected in completely different ways and to different degrees. In addition, we have to include the general effects for the entire social system. While a new road might improve the general economic position of the whole country, thereby affecting the well-being of the whole population; it may also cost a lot of money, taxes paid by the whole population. Furthermore, the construction of the road may increase traffic and so contribute to the consumption of fuel and to global warming. Seen from this perspective, everyone is affected, not only the whole nation, but also the global community: not only the present generation, but also future generations and the human race in its entirety.

This makes it very difficult to realize *Betroffenheitsdemokratie* in a fair way The legal system of competencies tries to avoid these practical difficulties by defining some typical levels of standardized competencies to decide (e.g. local, regional, national or international level). Nevertheless it is necessary that particularly affected people have the opportunity of special participatory rights. The main difficulty is to find a fair balance between the interests and opinions of the generally affected public (or majority) and the particular interests and opinions of the particularly affected people (or minority).

Even if it were accepted that political systems should reduce the importance of particular interests and even if Planning Cells were able to promote some interesting propositions, it is not realistic to believe that

Planning Cells could substantially improve upon existing procedures for balancing interests. This difficulty must be taken into account when evaluating Planning Cells. It may help to consider two opposing emphases that Planning Cells could adopt.

1. The particular impacts of the decision on the minority are integrated into the procedure of the Planning Cell. The panel is expected to take due care of the special position of particularly affected people. This means that these people must have a legally guaranteed possibility to promote their arguments before the panel. The Planning Cell would therefore become some kind of Court which would have to take into consideration all the aspects of a problem and to make a fair decision.

2. The panel is expected to focus on the general interest and to represent the "public good." In that case the deciding authority would have to strike a balance between the public good promoted by the Planning Cell and the special position of particularly affected people. It is doubtful whether the Planning Cell would be very useful, because it will not have solved the most difficult problem.

Consequently I would prefer the first emphasis, but this would mean that especially affected people would have the same legal guarantee to participate as they would have before the deciding authority.

Another problem regarding legitimacy should be considered. If we talk about legitimacy in a factual and not a legel sense, then it depends mainly on the *confidence* people have in an institution. A procedure – even if it is theoretically perfect – which is not accepted by the public, will never be able to create trust in the state and persuade people to accept the decision. Whether people trust in an institution depends on many factors (e.g. traditions, experiences, the historical situation). That is why it is difficult to predict whether a model would be successful in a given society. At a minimum we can say that in a political culture like Switzerland public trust in an institution depends on the democratic legitimation (elected representatives or public vote) or on the special integrity and knowledge of the persons invested with power (members of a government or of a court, experts). The Planning Cell model tries to combine both aspects. Therefore, it is possible that people consider it as an ideal method. On the contrary it is also possible that, lacking democratic legitimation nor the necessary special knowledge, the method is not accepted at all.

One important factor in creating trust and confidence seems to be the possibility of some kind of control. An important point when talking about control is the *publicity* of a procedure. Ever since the French Revolution an essential postulate has been that the parliaments and courts are public. For the courts it is even considered as a human right (Art. 10 General Declaration of

Human Rights, 1948; Art. 6 European Convention on Human Rights, 1950). The principle of publicity was introduced also for governmental and administrative bodies many countries in the last decades (Seiler 1992: 433ff). Therefore it would have to be discussed whether the procedure in the Planning Cell should be public or not. On the one hand, a general publicity would perhaps reduce the group dynamic effect of the panel. On the other hand keeping the procedure closed could create a climate of distrust against the entire institution.

## Implementation

The Planning Cell model is intended to be merely consultative. This means that the authority with the legal competence to decide is completely free to accept or reject the recommendations. Therefore we have to consider the lieklibood that the authority will adopt the proposed solution.

I have already mentioned the importance of procedural questions in establishing the usefulness of the Planning Cell model. Procedural due process is usually specificed for a certain domain of responsibilities. For example, an agency charged to make decisions about new drugs may not be compentent to decide about industrial emissions. Procedural due process for drug approval may be different than that for industrial pollution.

The Planning Cell model is intended to improve any number of decision making processes. Competence, however, is weakened when the decision making is not in the hands of the competent authority. The Planning Cell procedure attempts to overcome this limitation by ensuring that the panel has time to acquire information and to evaluate it. The value of the procedure, however, seems to lie in the possibility that out of the time-consuming procedure will come a "better" solution (or at least different than it would be without this procedure). Otherwise the entire procedure would not be very useful (unless it increases the likelihood that the decision is implementable). In the end the panel will present its recommendations to the deciding authority, which did not have the benefit of such a learning process. Now the authority has five possibilities.

1. It accepts the result without completely reviewing the procedure. It is not very probable that this will be the case. Usually authorities accept propositions more or less without question only when the proponent is an expert. This is not the case here.

2. It accepts the recommendation because the result is identical to the opinion the authority had already before. In that case the Planning Cell has the function of a supplementary legitimation enhancing argument for the decision, but it did not really influence the decision.

3. It tries to undertake the same learning process as the Planning Cell did (and comes finally perhaps to the same result, but perhaps not). This is time-consuming for the authority and will be done only in a few cases, if at all. But if the authority has that time to go through the process itself, we can question what sense it makes to conduct the Planning Cell. In cases where the deciding authority itself is able to make a decision making procedure which is fair and competent, there is no need for parallel procedures.

4. It does not try to undertake the learning process but decides in the same way as it would decide without the Planning Cell. If the recommendations of the Planning Cell are not convenient, they can easily be rejected with the argument that the Planning Cell is neither a representative democratic body nor a expert committee. The recommendations of the Planning Cell have more or less the same importance as all the other expertise, recommendations, and opinions the authority has received. The result is probably not much different than it would be without Planning Cell.

5. It accepts the recommendations even without having reviewed the entire procedure, because it trusts in the fairness and competence of the Planning Cell, unless there are some strong arguments in favor of an other solution, which were obviously not taken into account by the Planning Cell. Of course this would be the way in which most would benefit.

Which of these possibilities will come true seems to depend on the type of authority. A parliamentary body or a people as a whole (in a referendum) will probably follow the fourth way. I do not see any chance that the results of the Planning Cell could be treated differently than all the other kinds of scientific advice to politicians. All these recommendations are used as arguments in the political discussions by those who already have an opinion, but are opposed by others with different opinions. Of course the possibility exists that some of the members of parliament or some of the voters will change their opinion given the recommendations of the Planning Cell, but this case is unlikely. Altogether, Planning Cells are of a reduced value when conceived as consultative for parliamentary or direct democratic bodies, at least when they deal with controversial opinions about value-laden problems. It could be different if Planning Cells were used to prepare decisions that have not been controversial and that demand concrete solutions to an accepted problem. This would be similar to the commissions which are well known in Swiss communities.

Administrative bodies could reach different conclusions. But even here we have to distinguish between cases in which the administration has to implement a general policy defined by the law and decisions on governmental or parliamentary level. It is not probable and not desirable that the administration follow a Planning Cell recommendation which would be contrary to that policy. The Planning Cell model could be useful in cases where the administration has legally and politically different options available.

Another problem of implementation exists in cases where the aim is not to consult a deciding authority, but rather to improve the awareness of the problem among the population (e.g. people should become more conscious about environmentally-compatible behavior). In these cases Planning Cells would have a more *educational function* for external audiences. These Planning Cells could be useful, but we should not overestimate its effect. People acquire much of their knowledge by the media. Without being published by the media, the results of a Planning Cell will have an educational effect only for those who took part in it. In addition, there are other ways to educate people.

Perhaps the strongest promise of the Planning Cell model refers to cases in which one can expect a bridging of the gap of distrust between the authorities and the population. In many cases this gap is the consequence of a lack of information by citizens (e.g. the government tries to implement a policy in order to reduce the $CO_2$-emissions; many citizens may not be aware of the need to do this because they do not know enough about the problem, therefore they may not be ready to act, especially if they distrust government). But here again it is more important to provide information to the general public than to a Planning Cell, because, in the end the policy has to be accepted by *all* the people.

## Applicability

In summary, Planning Cells do not constitute the key to the solution of the most urgent political problems, but they could be useful in some cases and under certain circumstances. They seem most useful as a consultative body for administrative authorities in cases where different decisions are legally and politically possible. They could have some importance for preparing solutions for rather technical, not very controversial problems. They seem to be less useful as consultative bodies for parliamentary or direct democratic bodies. Planning Cells are most problematic when applied to very controversial matters or matters in whith strong equity concerns.

Some further restrictions must be considered. One of the aims of the Planning Cell model is to avoid building an establishment which has self-interests. The Planning Cell does not offer possibilities of a career, therefore it is deemed to be free from external considerations of their members (Dienel 1991:85f). As a consequence, one single Planning Cell must be limited in time (Dienel proposes Planning Cells of maximum duration of three weeks). This means that the Planning Cell can only deal with problems which can be solved by a more or less unique decision. Many of the political problems do nit fit this mold. They need a permanent process of revisitation rather than a single decision. That is why the Swiss citizens' commissions are usually established for four years, some members take part for eight or even more years. This allows the continuity that is not possible in Planning Cells, which last days or weeks and are then dissolved.

The limitation of each Planning Cell to one single question implies not only isolation in time, but also isolation from the larger issue. Planning Cells address a single question. However, the main political problem is to see each problem in the greater context and to maintain coherence in making public choices. The tendency towards fragmentation is strengthened when problems are delegated to Planning Cells. Participants of a Planning Cell will perceive the problem they are asked to consider far more important than the connected problems that they were allowed to ignore.This has two consequences.

1. The selection of questions establishes the political significance of the problem. Selection should, therefore be made by a legitimated authority following certain selection rules.
2. The results and recommendations of the Planning Cell need political evaluation and control taking into consideration other political aspects and inter-linkages.

As stated above Planning Cells seem unfit to deal with questions where fundamental disputes exist between a more generally affected majority and a particularly affected minority.

## Transferability

In this section I focus on the question of whether the Planning Cell model would be transferable to the Swiss political system. It is well known that in Switzerland there are a lot of democratic procedures which allow citizens to take part in the formation of political decisions, not only in consultative bodies, but also in decision roles. One of the aims of the Planning Cell model – namely the emancipatory goal to make sure that the role of citizen is not reduced to merely a consumer and subordinate status, but that he or she feels as being the sovereign and the owner of the State – is strongly internalized in the Swiss political culture. (Which does not necessarily mean that it corresponds to the reality).

The direct democratic institutions of Switzerland on a national or cantonal level are generally well known. In addition to the right to elect members of parliament, they are:

- Every amendment of the constitution and some international treaties have to acquire majority vote by people and by cantons (mandatory referendums).

- Every new law or amendment of an existing law and some international treaties are subject to the vote of the people if 50,000 citizens demand a vote (optional referendums).
- 100,000 citizens can propose an amendment of the constitution. The proposition has to pass the vote by the people and by cantons (initiative).

Most cantons offer identical opportunities. Furthermore, the people can not only propose amendments of the constitution, but also new laws or amendments to existing laws.

Less known are perhaps the many possibilities citizens have to collaborate on the community level. Most Swiss communities are quite small. Switzerland has seven million inhabitants and about 3,000 communities, making an average community size of less than 2,500 inhabitants. Of course things are different in the cities, where one community may have more than 100,000 inhabitants.

In all communities there is a Community Council (*Gemeinderat*, "executive body" of the community). It is elected by the people and consisting of five to fifteen members. With the exception of the big towns, these are not professional politicians, but citizens who perform their political duties in their free time. They may have political ambitions (e.g. to become a member of the cantonal or national parliament or government), but most of them do not. That is why they usually do not have overwhelming self interests, but rather consider their political function a "burden" they are glad to give up after some time.

In small communities (up to 10,000 inhabitants), the decision body (town meeting) is the communal assembly. Here every citizen can participate. Although usually the participation rate is between five and twenty percent. There are between three and six assemblies per year (depending on pressing matters), usually in the evening (about three hours long). The assembly decides on propositions made by the *Gemeinderat* or on initiatives made by citizens. Every participant can take part in the discussion. Votes in the assembly are usually not secret, but by raising the hands. Larger communities have a community parliament in place of the communal assembly. Such a parliament has between 20 and 60 members (nonprofessional-politicians, as well), elected by the people. For more important matters communities organize referendums. Matters to be decided by the people on communal level are: local planning, the building of communal infrastructure (communal roads, schools and so on), and all kinds of communal expenses above a certain amount.

In addition, there are commissions for most community affairs (e.g. school, local planning, social assistance). Commissions meet weekly or biweekly and consist of five to fifteen citizens elected for a period of usually four years by the *Gemeinderat*. Sometimes the slate of candidates is nominated by the local sections of the national political parties. Members quickly become competent in the committee's subject matter. In smaller communities this allows a great portion of the citizens to be involved with political authorities. For all these reasons, small Swiss communities are characterized by a high degree of trust between the population and the communal authorities. The authorities are seen as trustworthy representatives of the population's interest.

This system seems better suited to achieving fairness and competence than do the Planning Cells proposed by Dienel. I would emphasize four points.

- As opposed to Planning Cells, selection of members for Swiss commissions is not at random. People have to show some interest in becoming a member. This seems likely to produce more competent membership. Commissions do not only receive information from the administration, they also make use of private knowledge.
- Everyone has a real opportunity to become a member of such a commission.
- The commissions have not only an informal, but also a formal legitimation, because they are elected by a political authority.
- The commissions are suited not only for isolated questions, but also for continuing and permanent policies.

As a consequence, the introduction of Planning Cells in smaller Swiss communities is simply not necessary. Whatever we could reasonably expect from Planning Cells is already achieved by the actual system. At the same time, I do not believe that Planning Cells would be very useful for questions which lie in the competence of parliament or people on cantonal or federal levels. Planning Cells could promote some interesting ideas, but would present only one opinion among many, because the Planning Cell would not be accepted by the public as a legitimate body. If a matter is important enough that it must be decided by the government, a Planning Cell would probably neither be seen as legitimate in the public opinion nor have any chance of its recommendation being implemented.

Planning Cells might serve a valuable function between these two levels, i.e. for matters which are in the competence of federal or cantonal administration and affect larger communities. These institutions are removed from citizens and often face public distrust or opposition. Planning Cells could help to find feasible solutions for problems which lie in the competence of a governmental body, but have some local importance. A typical case in which a Planning Cell might work well is the project managed by Ortwin Renn and Tom Webler in 1993 in the canton of Aargau (see Dienel and Renn in this volume). The problem was to find a place for a landfill in the eastern part of the canton Aargau. Some twelve potential sites were evaluated by the cantonal administration. Planning Cells modeled after Swiss commissions were established with members from concerned communities in order to evaluate the sites and articulate recommendations for selecting the best site(s). This project was successful for the following reasons:

- The commissions did not discuss the need for a landfill, but only its location and specific attributes. The problem was well defined, widely accepted, and could be resolved in isolation of interconnected issues.
- The formal competence to decide remained within the cantonal parliament, but the problem was manifested on the local level. From the cantonal viewpoint it does not matter where the landfill is located; the main concern was to have one sited within a given time period. Consequently, the cantonal authorities had no *prima facie* reason to deviate from a solution proposed by the concerned communities.
- Given the fact that the affected communities were represented in the Planning Cells, there was no strict difference between a general interest – represented by the Planning Cells – and the especially affected people.

For cases such as this, I believe the instrument of Planning Cell can be quite useful and legitimate.

## References


Dienel, Peter C., *Die Planungszelle* Second Edition (Westdeutscher Verlag: Opladen 1991).

Kant, Immanuel, *Metaphysik der Sitten*, 1. Teil, Anfangsgründe der Rechtslehre, edited by K. Vorländer (Hamburg 1959 [original 1797]).

Rhinow, René, "Grundprobleme der schweizerischen Demokratie," *Zeitschrift für Schweizerisches Recht*, N.F. 103, II. Halbband, (1984), 111-273.

Seiler, Hansjörg, *Das Recht der nuklearen Entsorgung in der Schweiz* Abhandlungen zum schweizerischen Recht, Heft 502 (Stämpfli & Cie AG: Bern 1986).

Seiler, Hansjörg, "Rechtliche und rechtsethische Aspekte der Risikobewertung," in: S. Chakraborty/G. Yadigaroglu (ed.), *Ganzheitliche Risikobetrachtungen* (Verlag TÜV Rheinland: Köln 1991), pp. 05-1 – 05-26.

Seiler, Hansjörg, "Halbdirekte Demokratie: Verfassungskonzept und Herausforderungen" (gemeinsam mit Pierre Tschannen), *Zeitschrift für Schweizerisches Recht* N.F. 110, I. Halbband, (1991a), 117-134.

Seiler, Hansjörg, "Die (Nicht-)Oeffentlichkeit der Verwaltung," *Zeitschrift für Schweizerisches Recht*, N.F. 111, I. Halbband, (1992), 415-440.

Uebersax, Peter, *Betroffenheit als Anknüpfung für Partizipation* Basler Studien zur Rechtswissenschaft (Helbing & Lichtenhahn: Basel and Frankfurt/M 1991).

*Chapter 8*

# *Citizens Juries: One Solution for Difficult Environmental Questions*

*Ned Crosby*

## Introduction

This paper presents a particular method of citizen participation, the Citizens Jury[1] process, as a novel way of obtaining citizen input on environmental questions. The Citizens Jury process uses random selection to impanel a group of citizens to study specific public policy issues or to review candidates in an election. The paper starts with an overview of the process, moves on to a discussion of using Citizens Juries on nuclear waste facilities and concludes with a review of how the process meets some general criteria of fairness and competency derived from the work of Jürgen Habermas.

## The Citizens Jury Process

### *A summary of the process*

A Citizens Jury is a group of randomly selected people, gathered in such a way as to represent a microcosm of their community, who are paid to attend a series of meetings to learn about and discuss a set of candidates in an election or a specific public policy issue and make public their conclusions. The process is similar to Peter Dienel's Planning Cells, which were invented two years before the Citizens Jury process (cf: Chapter 6). Fourteen Citizens Jury projects have now been conducted, only one of which was on an environmental issue. There are some key aspects of the process which help differentiate it from focus groups and other similar processes.

- A Citizens Jury should be designed and conducted so as to serve the general public and not a special interest. The process works better on value questions than technical issues.

---

1 The Jefferson Center has been granted a service mark by the U.S. government for the term "Citizens Jury." This was sought in order to prevent improper use of the method.

- The jurors are selected at random through a quota system which makes them a microcosm of the community from which they are drawn. Typically, the quotas include demographics such as age, gender, education and race, or else rely on balancing the attitudes of the jurors on the question at hand so that they resemble their community. Sometimes both demographic and attitudinal quotas are used.
- The jurors are paid to attend hearings where they become informed about the topic at hand. Hearings typically run for four or five full days. If a two-tiered system is used, where several regional juries meet and then select one to three of their members for further joint meetings, then those in the second meeting may spend up to 12 days in hearings. Longer meetings could be used, but the gain in understanding may be offset by a drop in acceptance rate among those who participate.
- The information presented to the jurors must come from several points of view and be presented in a way which is fair to the concerned parties. Usually this is done through witnesses. Sometimes the testimony of witnesses will be organized by case managers, analogous to litigators, so that two to four different points of view are presented to the jurors. Since there is no objective standard to use in achieving the proper balance between points of view, the views presented should be selected either by a group with a clear public record of holding fair discussions or done by the jurors themselves. A neutral moderator should facilitate all discussions, with the possible exception of the final deliberations.
- The findings and recommendations of the jurors should be made specifically in response to a "charge" given them by the sponsors of the project. This charge should be short, direct and clear and be made public at the beginning of the project. The jurors should be given adequate time to deliberate and should have the option of conducting their deliberations in private. The jurors must have the opportunity to review and approve all of their findings and recommendations.
- The jurors must be given the opportunity to evaluate the process and make public their views at the project's conclusion. This evaluation should include at least one standardized question on the fairness of the proceedings so that comparisons with other projects are possible. Also one or more of the jurors should be selected at the end of the hearings to attend an "oversight" meeting with jurors from other projects to review the rules and staff performance.

## *Past projects and their results*

The Jefferson Center, founded in 1974, is now the oldest of the six organizations which have been set up in the United States, independent of government and universities, to work to improve democracy without taking policy stands.[2] From 1974 to 1983 we experimented with two other new democratic processes besides Citizens Juries, only to decide they were unlikely to make major changes in the way democracy functions in America. Since 1983, therefore, we have concentrated on Citizens Juries.

Three projects were conducted in the mid-1980's. The first used five regional Citizens Juries in Minnesota to study the impacts of agriculture on water quality. Representatives from each panel were then assembled to come up with a set of recommendations for the state as a whole. This was conducted for a steering committee of 11 organizations, including four state agencies, two major farm groups, two environmental groups, etc. The second was conducted on organ transplants, for an advisory committee of people prominent in the health care system. The third was on school-based clinics, done for the Health and Human Services Committee of the Minnesota Senate. But the major problem was not overcome: the Citizens Jury process is at odds with the way governmental decisions are made in the United States. Except for a few idealistic legislators, the process was viewed as a bother at best and, at worst, as a real challenge to the way business is currently conducted. The process was not openly condemned, but few urged us to continue.

This led us to return to Citizens Juries on elections, something we tested in 1976. In 1989 we held a Citizens Jury on the St. Paul mayoral race and in 1990 on the Minnesota gubernatorial race, both done jointly with the League of Women Voters of Minnesota. These projects were praised editorially and received top marks from the jurors involved. The latter project received extensive media coverage around the state.

The 1990 project showed the process has good prospects for becoming an institution voters can trust. We sent the findings of the jurors to 450 randomly selected people. One third said that it helped them make up their minds how to vote. Among those in this group who told us how they voted, the vote was two to one for the candidate who had received the best marks from the jurors (the state as a whole split 49 percent to 51 percent on the candidates).

In 1992 we have moved beyond Minnesota. The League of Women Voters of Pennsylvania used a Citizens Jury to review the candidates in the 1992 Senatorial race in their state. The national media coverage given to this project encourages us to believe that indeed it can be turned into something which will be given strong political backing on the federal level.

In 1993 we conducted the first two national Citizens Jury projects. Both were held in Washington, D.C. with 24 jurors drawn at random from around the nation. In the first project the jurors developed their own version of the

---

2 The four, besides the Jefferson Center, which have lasted for more than five years are the Center for the Study of Democratic Institutions (which lasted from the early 60's to the mid-70's), the Roosevelt Center (1981-1988), the Public Agenda Foundation (1976 to present) and the National Issues Forum (late 70's to present). The latter two are run under the auspices of the Kettering Foundation.

federal budget; in the second, they evaluated the Clinton health care plan. Especially the latter showed the growth in respect for the process from the powers that be: testifying at the hearings were three U.S. Senators and Ira Magaziner, the chief health care planner for the President and Mrs. Clinton. The process has been praised in the *Washington Post*, with David Broder naming it one of the two most interesting voter reform projects in the nation (December 27, 1992) and William Raspberry calling it a "paragon of representative democracy" (October 19, 1993).

## *Designing citizens juries*

The main goal of the process is to enable a microcosm of the community to exercise both reason and concern for others to answer the question: "How should we live together?"[3] Such a broad goal is not easy to turn into reality; many of the specific aims of the process conflict. Also, the details of the process depend on common sense as well as past experience. For this reason, there is not a simple "cookbook" which lays out how a Citizens Jury should be designed and run.

The difficulty of how to set up a Citizens Jury process properly lies with the age-old problem of how to bring fairness to the exercise of power. In 1993 the Jefferson Center set up an "Oversight Committee" to enable former jurors to review the rules and staff procedures of the method. The decisions about design are made by the staff of the Jefferson Center and then evaluated by the jurors who participate in the project, as well as by the Oversight Committee. A few examples will show the kinds of dilemmas we face.

A critical matter in designing any project is insuring that the jurors will be able to make good judgments on the issues before them. Note that competence and good judgment are somewhat different terms. In 1990 in Minnesota one of the big issues in the race for governor was the budget shortfall facing the state. Governor Perpich insisted that it was no more than $200 or $300 million, while his opponent Arne Carlson insisted it was going to be $1 billion. When the jurors in our Citizens Jury project on the gubernatorial race voted overwhelmingly in favor of Carlson on the issue of taxes and spending, the *Star Tribune* criticized them strongly for not understanding the good work which Perpich had done. There is no doubt that in the normal meaning of competence, the finance planners working for Governor Perpich and the editorial writers for the *Star Tribune* were much more competent than the jurors in their understanding of state finances. But three weeks after the election, it was announced by the Governor's office that

3 This approach should be compared to that of Bernard Williams who suggests that the primary question of moral philosophy should be Socrates' question, "How should one live?". See *Ethics and the Limits of Philosophy*, Harvard University Press, 1985.

the shortfall for the biennium was estimated to be $1.6 billion, considerably larger than even Carlson had estimated. Our claim for the Citizens Jury process is that a group of randomly selected citizens, when exposed to good information presented by witnesses from differing points of view, is able to make good judgments on public policy matters even though in terms of training and experience there are many people more competent than they. One anecdote does not prove the point, but we would submit that a careful study of the jurors' recommendations over the years would show that they have by-and-large made good judgments on complicated issues.

But the ability of any group to make good judgments is strongly affected by the time they have and the size of discussion group. It is hard to know how long a Citizens Jury should be run if judgment were the primary goal and costs were of no concern. Certainly if we were to run the process for several months, the judgments of the jurors would likely be better, but the representativeness of those attending would drop off considerably, unless the process were much better known or we were to pay jurors a great deal more than the $100 a day now paid.

But if an improvement in judgment could diminish representativeness, the reverse is also true. Surely the representativeness would be improved if we were to have 600 jurors, as opposed to the 96 for the school-based clinics project or 18 for some local projects. But with this many people it would be almost impossible for a good dialogue to take place. Of course each jury could have been broken down into smaller units for discussion purposes, but then there is the problem of how to bring their views back into an effective large group discussion.

This raises a key point: the Citizens Jury process is not based on the analogy of a public opinion poll, but on the analogy of a jury. Juries of 12 are widely accepted to decide matters in the areas of criminal and civil law. The assumption is that 12 people who are well informed on the facts of the matter are much better than 600 or even 6,000 people who are not well informed.

Another important decision which the staff of a Citizens Jury project needs to make is with regard to the "charge" that is presented to the jurors at the beginning to guide their work (and shape the testimony of the witnesses) throughout the hearings. This charge usually contains a clear statement of the question which is to be examined as well as a few brief follow-up questions to be answered by the jurors during their deliberations. The aim of the staff in setting the charge is to frame a question which is satisfactory to the sponsors of the project, fair to the parties affected by the issue, and which will provide a framework within which jurors can make good judgments. These goals are not easy to meet simultaneously.

The most obvious problem is that a charge which is satisfactory to those who currently have the power to decide may not be seen as fair by those affected by the decision. If the Minnesota Department of Transportation wants to build a new freeway between Minneapolis and Duluth, they might be tempted to frame the question as: "Should the new freeway between the two cities be built on corridor A, B or C?" But others may feel that the most important question is not where to build the freeway, but whether a new freeway should be built at all. In such a case a Citizens Jury is inappropriate.

Occasionally the dilemma is not so clear. In the national Citizens Jury conducted in October 1993 on health care, we decided to concentrate on the health care plan presented by President Clinton. The charge, therefore, asked jurors whether they thought there was a need for health care reform in America and, if so, whether the Clinton plan was the way to get it. The first two days of the project concentrated on background information on health care, but most of the rest of the time was taken up with advocacy presentations pro and con the Clinton plan. We realized, however, that the jurors might want to learn about plans other than the Clinton plan or the Republican plans in opposition, so we asked Senator Durenburger to present a bi-partisan plan and Senator Wellstone a proposal for a single payer approach. The jurors liked the latter so well that they invited Senator Wellstone back twice to make further presentations. When they came to answering the charge, however, they limited themselves to the original charge, voting 19 to 5 against the Clinton plan. Several of the jurors wanted to vote on the single payer approach, but the jurors decided not to by a vote of 14 to 9, even though an informal vote after the project was completed showed that 17 of them liked the single payer approach. In deciding to limit themselves to the original charge, several jurors said that they liked the single payer approach, but had not heard enough about it to be willing to make a formal vote.

The obvious question for us in retrospect is whether we should have had a broader charge which would have asked the jurors to review several plans, as opposed to simply the Clinton plan and the Republican opposition. Clearly the "conventional wisdom" in the national media was that the single payer approach was "dead." If staff had insisted on its inclusion, it might well have not been possible to get any cooperation from the White House in the presentation of the Clinton plan, something which would have made the whole project less credible. Also, there is the question of what plans the staff should have selected in an expanded agenda. The single payer approach was only one of many possible additions and it is not at all clear which of these should have been added to enhance the fairness of the project.

A different problem with the charge arose in the 1993 Citizens Jury on the federal budget. There we had a seven point charge which asked the jurors to come up with statements about the largest budgets and smallest budgets they wanted and to say in each case who they thought should be making sacrifices for the public good. The reason for including these questions was that we thought that the actual construction of a federal budget was a task too difficult for citizens in the five days available to them and they would find the broader questions easier to answer. As it turned out, the jurors found it easier to construct their own budget than to answer the broader questions. Part of the reason they were able to do this was that they could borrow from the information given them by the liberal and conservative advocates in order to come up with their own set of figures. The art of creating an agenda and charge is to set up information in such a way that the hearings are fair to the parties affected and also so that the jurors can reach closure on a complex issue in the time available. As the project on the federal budget showed, sometimes the staff makes poor judgments regarding the tasks jurors are capable of assuming. The charge must be flexible, but not unfair to the advocates and witnesses whose commitment to appear was made in light of the original charge.

## *Maintaining the integrity of the process*

There are two essentials in maintaining the integrity of the process: minimization of staff biases and maintenance of effective citizen control. Only a time-proven commitment by Citizens Jury staff to minimize its biases will persuade the public to believe this is a method for letting a microcosm of the public suggest what is good for all. Since it is rare for a staff *not* to impose its own values on the organization it runs, it will take exceptional work on the part of the Citizens Jury staff to maintain these standards and convince the public of their commitment.

The Jefferson Center has taken steps to promote this, but more remains to be done. In the first place, the staff discusses potential bias and tries on its own to eliminate it. For example, staff monitor the moderator of a project to insure that his/her facial expressions and body language does not indicate a preference for one point of view over another.

The most important step is that every project conducted gives the jurors an opportunity to rate the project and comment on it. These comments are always made public at the same time as the final recommendations are released. We are proud of the results we receive (see *Table 1).* Once a staff gets good evaluations, it becomes highly motivated to maintain public credibility. That is why we require these opinions always be made public. If the process continues to grow as it has in the past two years, we are considering establishing a system of peer review to enhance staff performance.

But juror evaluations and peer review will not be enough to maintain the integrity of the process. There must be effective citizen control: the process must be under the ultimate control of the informed wishes of average citizens. The Oversight Committee set up by the Jefferson Center in 1993 is a significant step in this direction. It currently consists of 16 people selected from the two most recent statewide projects and the two national projects

*Table 1:* Juror Evaluations of Citizens Jury issue projects. ("Project" refers to satisfaction with the project as a whole; "Bias" refers to satisfaction with how staff has minimized biases")

| | 1980-81 | | 1984 | | 1987-88 | | January 1993 | |
|---|---|---|---|---|---|---|---|---|
| | Project | Bias | Project | Bias | Project | Bias | Project | Bias |
| Very Satisfied | 67% | 33% | 60% | 40% | 54% | 88% | 78% | 96% |
| Satisfied | 33% | 67% | 40% | 60% | 46% | 12% | 22% | 4% |
| Neutral | 0 | 0 | 0 | 0 | 0 | 0 | 0 | 0 |
| Dissatisfied | 0 | 0 | 0 | 0 | 0 | 0 | 0 | 0 |
| Very Dissatisfied | 0 | 0 | 0 | 0 | 0 | 0 | 0 | 0 |

conducted in 1993. At their first meeting, they decided their mission should be as follows, "To monitor the Citizens Jury process to judge its integrity as an institution citizens can trust. Both the Citizens Jury process itself and specific uses of it must be evaluated. Outside resources will be brought in as necessary to enable the Oversight Committee to prove to the public that they are making fair and objective judgments."

The first Citizens Jury project they were able to evaluate was the project on the Clinton health care plan. They sent two representatives (selected at random) to observe the hearings and then had an extensive discussion of the reports by those two people. Their ratings of the project were interesting, running from a rating of "poor" in terms of juror preparation (materials should be sent out in advance and a better job done describing voting procedures) to a rating of "excellent" on the performance of the Jefferson Center staff. Interestingly, they rated the performance of the long term Jefferson Center moderator as "good," but that of a well-known Dean of a School of Communications "moderate", commenting that when well-known people are brought in to moderate, a better job of training must be done to be sure that the moderator does not dominate the proceedings.

## A Citizens Jury on Nuclear Waste Facilities

From a rational perspective, decisions about disposing of nuclear wastes resulting from energy production are fairly straightforward, as long as sound empirical studies have been done. One estimates future energy needs, examines the relative costs of energy production from different energy sources, takes a look at the environmental and health risks of the different methods of production, and then chooses the method which leads to the greatest utility in light of these factors.

Of course this is not the way things work, given the roles played by money and fear. The investments needed in energy production (and the potential profits available) mean that very powerful corporations and governmental bureaucracies become deeply invested in producing energy through nuclear power. On the other hand, an emotional issue such as nuclear waste disposal is an organizer's delight, since it is relatively easy to get a large group of people to attend major public meetings and get good media coverage. It is a rare group of local officials which is able to withstand pressure of this variety. Hence, the power of NIMBY (not in my back yard) has exerted itself in a very strong way on the question of locating nuclear waste facilities. Caught between these two sources of political power, the public officials who are charged with coming up with finding a rational solution to the problem are frustrated.

This is the kind of problem to which Citizens Juries are well adapted. Everyone is at potential risk from the production of nuclear wastes and virtually the whole population would benefit from a solution. But the two obvious solutions (cut down on production or dispose of the wastes in a carefully controlled way) mean that either industry or some particular community bears the burden of a program which benefits the large majority. This is a classic situation where a social contract is needed. Representatives of the population should come together, listen carefully to each other and the evidence about the nature of the problem, determine whether indeed there is a need for nuclear power and then decide where to locate the waste facilities (with compensation to those nearby, if necessary) so that no one ends up carrying an unfair burden in order to benefit the majority.

Many people believe that citizen participation, if broad enough, could create such a social contract. This has often not been the case. In Minnesota in the early 1980's an attempt was made to locate a hazardous waste facility based on extensive public participation. Open meetings around the state were attended by several thousand people. Over 70 percent agreed there was a need for a hazardous waste facility. When the possible sites were narrowed to four, however, the resistance from the residents of those counties was so strong that in 1985 the process was brought to a halt and a search was started to find another way of locating a facility. To date, none has been found.

A Citizens Jury process can improve on this situation by insuring that the group of people gathered to discuss needs are also those who must decide how and where to locate the facility. When jurors play the roles of both tax-payers and residents of the county where the facility will be located, they are in a good position to decide how much compensation, if any, should be given to those who live where the facility will be placed. When potential local sites are identified, then another Citizens Jury can be called, using randomly selected residents of these communities. These people can be brought together with the jurors from the original project to discuss what ought to be done. In this way, representative members of the affected community are asked by other citizens to take the waste, as opposed to confronting governmental officials who appear to be foisting something on them.

The dilemma of risk management is that it is virtually impossible from a values point of view to have anyone but the affected people make the final choice, yet often these people can be whipped up emotionally and make what appear from the point of view of experts to be irrational choices. The position of the Jefferson Center is that it is easier to put citizens in a setting where they are likely to be rational than to put experts into a setting where they are viewed as legitimate substitutes for the public on risk management issues. This is especially true in the United States where skepticism over public officials runs very high.

Given these advantages to the Citizens Jury approach, why is it not used? The answer is that it is very difficult to convince elected officials that it is worth the effort. The process is expensive and there is no use in conducting it unless there is some reason to believe that public officials will pay attention to the recommendations.

The Jefferson Center had an interesting experience with this in 1989, when we tried to use the process to resolve a dispute over solid waste disposal. Winona county is located on the Mississippi in far southeastern Minnesota and was admired for being one of the first to develop a comprehensive plan for solid waste disposal. But if their plan was technically sound, it was politically foolish, proposing that an incinerator be located in a strongly ethnic community which traditionally resented the way others in the community treated them. The County argued that since 80 percent of the county's waste was created within the city limits, the method for dealing with it should be located there too. Not surprisingly, the City Council refused to allow it to be put there.

The Jefferson Center proposed a Citizens Jury made up half of city residents and half from the broader county, asking each governmental unit to pay $5,000. The County voted approval, the City did not. The mayor of the city claimed that our process was too expensive, yet a year and a half later, the two sides together had incurred legal fees of over $500,000.

After our failure to get a contract, a mediation group was called in to help. With pressure from the Governor, both sides did agree to try mediation, but it ended up going nowhere. But it was the mediation effort which showed us how culturally radical our process was for politicians. A good mediation process attempts to get all important players to sit at the table and pursue a compromise. Some sophisticated mediation groups even go so far as to try to create representation for those groups or interests which exist latently but are not organized. If mediation is to succeed, everyone must be at the table who has any interest in, or power over, the solution to the problem.

Our process, however, comes in and metaphorically asks all those with power sitting at the table to stand up and move away, since we are going to seat average citizens at the table to devise a solution. In terms of etiquette in dealing with elected officials, that would be like walking into the mayor's private dining room and telling everyone, mayor included, to get up so that we could let a group of people, none of whom they had ever heard of, sit down to eat. Given this cultural discrepancy of the Citizens Jury from normal political procedure, there has been a gut resistance on the part of most elected officials to the process.

This situation may be changing, however, as the public distrust of politicians in America continues to grow. As of December 1993, two Members of Congress (Reps. Peter Hoechstra, Republican, Michigan and Tim Penny, Democrat, Minnesota) have committed to using the Citizens Jury process to "center the dialogue" over a policy issue in their district. The aim in these projects will be to promote a dialogue between the constituents and their U.S. Representative in which the Citizens Jury process serves as the link between the two. Currently, it is often talk shows or the Representative's newsletter which serves as the main link. The Representative will be asked to attend the beginning and end of the Citizens Jury hearings to dialogue with the jurors about their initial desires and their conclusions, thereby enhancing the Representative's appreciation for the process and the likely coverage in the local media.

Opportunities will be given to the public to call in and register comments and also make suggestions about matters they think should be covered in more detail. The Representative has been asked to show a strong interest in the recommendations of the jurors, but emphasize that his/her ultimate commitment is to *all* of the citizens of the district. If they do not come around to sharing the same views as the Citizens Jury, then the Representative will be guided by their views unless s/he feels that there is some compelling reason to side with the jurors' recommendations.

Assuming that "centering a dialogue" through the Citizens Jury process becomes popular with Members of Congress, does this mean it should be the method used to site nuclear waste facilities? Not necessarily. One has to look at several factors in deciding whether to use a Citizens Jury or mediation or another participatory method. Even if the Citizens Jury process receives strong backing from Members of Congress, that is not to say they would not give equally strong backing to some other participatory method. Beyond that one must look closely at how the agenda and charge are set. If the only reason Members of Congress become interested in the Citizens Jury process is that they can set the agenda as they wish, then some other method might well be preferable. If funds were available, the best way to set the agenda for a

Citizens Jury is to hold an initial project simply to set the agenda for the main set of hearings. The initial project would give the jurors the opportunity to review a large number of witnesses and proposals and cull this down to those they thought the most worthy of consideration. But so far we are a long way from having the funds and the political support to do this.

## Citizens Juries and the Discursive Standard Criteria

### *Fairness criteria*

#### Making of the agenda and the process rules

Does the model provide everyone with an equal chance to put their concerns on the agenda and to approve or propose rules for discourse, to debate and critique proposals for the agenda and the rules, and to influence the final decision about the agenda and the discourse rules within the agreed upon decision making procedure?

In theory, the Citizens Jury process meets this criterion well, although not in the direct way the criterion is stated. As noted in the discussion of nuclear wastes, we would optimally convene a Citizens Jury for the sole purpose of setting the agenda for the main hearings of a newly formed Citizens Jury. Also, our Oversight Committee will review the setting of the agenda by Jefferson Center staff and insure that it has been done properly. This gives "everyone" an equal chance to be selected for a Citizens Jury project and thus to be voted onto the Oversight Committee, thereby giving them the opportunity to engage in the above activities. We would submit that no real world discussion could ever begin if this criterion had to be met in its presently stated form, since it would take years to allow "everyone" to engage in the above activities and reach a consensus about what should be done.

In practice, the Citizens Jury process does not do as well, given the practical need to set agendas that those who currently hold power will participate. Also, the Oversight Committee is in its first year of operation and has yet to demonstrate how well it works and to what degree the Jefferson Center follows its recommendations. But the increasingly good ratings given the staff by jurors on the degree to which they operated in a neutral way indicates that in practice the staff has performed well in this area.

#### Moderator and rule enforcement

Does the model provide everyone with an equal chance to suggest a moderator or debate proposals for the moderator, and comment on the facilitation style or debate proposals for how moderation should be carried out? It is standard procedure for staff to inform the jurors at the beginning of every Citizens Jury project that they are there to serve the jurors. Their style of moderating is described (we limit the length of time and frequency any given juror is allowed to speak) and jurors are urged to comment and propose changes in this style at any time. This means that the Citizens Jury process has always done well with the second half of the criterion. With regard to the selection of moderators, now that the Oversight Committee is in operation and rating moderators and their styles, we assume that we will improve on the opportunity for everyone to have a say in the kinds of moderators who are chosen. If more than this is needed, then when we get to the point of running an initial Citizens Jury to set the agenda, this device could also be used to select moderators if the jurors or Oversight Committee deemed it necessary.

### Discussion rules

Does the model provide everyone who is potentially affected by the decision proposal (positively or negatively) an equal chance to be present or represented at the discourse and to put forth and criticize validity claims about language, facts, norms and expressions? This criterion is met very well if one assumes that "everyone" means individuals in the community under consideration. It is not met perfectly, since those without telephones are left out of the jury selection process. This is not a problem in theory, since it would be possible to give "everyone" an equal chance of being selected if the funds were available to do a thorough random selection by dwelling, including methods for reaching the homeless in the proper proportion.

If, however, "everyone" is taken to mean all groups affected by the decision to be made, then the process does not do as well. Group representation is not provided for, other than those who are selected as witnesses or advocates. Even when a Citizens Jury is used to set the agenda for the main hearings, the number of groups that will have a chance to speak up will be limited to a few dozen. If one assumes that the most important policy discussions occur between group representatives and those who hold power, then the current legislative system does a good job of allowing "everyone" to participate. If one assumes that "everyone" refers to individuals in society, then the Citizens Jury process does a very good job of giving all an equal chance of participating in the discussion and allowing them to structure it in a way which is most fruitful to them.

The difference between representing interests (or stakeholder groups) and the public at large is such a basic one that it is difficult to know how to advance definitive arguments in favor of one approach over the other. Certainly one of the standard arguments against including the public at large in public policy decisions is that they are not competent to do so. We believe that the experiences of the Citizens Jury process to date shows this to be false, if average citizens are placed in a setting where they have time to learn about the issues and discuss them. The reason for not using volunteers in a Citizens Jury is just so that we maintain a microcosm of the public and avoid groups which are formed of those who feel empowered enough to donate their time. The question of how a group representative of the public should be constituted and conducted is discussed further under the third competence criterion.

## *Competence criteria*

### Rules for redeeming comprehensibility validity claims

Are the participants provided with equal access to sources for commonly-agreed-upon standards and definitions, an opportunity to confirm each others' terms, definitions and concepts, and the ability to take advantage of pre-established reference standards?

On the initial day of a typical Citizens Jury, the jurors are presented with a glossary of terms and given introductory information on the topic at hand. There has been a long debate among the staff of the Jefferson Center about the wisdom of sending written information on the issue to jurors before the hearings begin. The reason for not doing this is the fear that it would unjustly disadvantage those whose reading abilities are below average. On the other hand, the low rating given by the Oversight Committee to the preparation given jurors indicates clearly that they would like this to be done. The staff of the Jefferson Center does not think the preparation given jurors was as bad as the rating given by the Oversight Committee, but nevertheless are determined to improve on it.

### Rules for redeeming truth validity claims

Does everyone have equal access to available and relevant systematic and anecdotal and intuitive knowledge about the objective world? Are uncertainties regarding factual information considered? Is there a means of verifying whether factual claims are consistent with the prevailing expert opinion or consistent with the anecdotal knowledge of other persons not involved in the discourse? And are cognitive claims separated from normative claims?

This criterion is met rather well by the process. Jurors are exposed to a variety of points of view and the witnesses who appear do a fairly good job of critiquing each other's statements. However, there are difficulties which remain to be overcome. The witnesses and advocates do not do an especially good job of separating cognitive and normative claims. They find it tempting to highlight emotional claims, while ignoring basic facts. Several observers at the 1993 health care project felt this was being done by both of the principal advocates. The easiest way to overcome this problem is for staff to allocate more time to "neutral" experts to make things clearer. But it is very difficult to know who is "neutral." The more time allocated to such witnesses, the greater the danger that staff bias is introduced. The Jefferson Center has long wanted to use some sort of "technical review" method for clarifying disagreements between experts, but has not had the funding to do so. This would be done before the hearings begin and would be somewhat analogous to the "discovery" process used by the U.S. legal system, except that there would be a method for encouraging experts to be clear about their agreements and disagreements and whether these are over empirical or normative matters. So again, in theory the method does very well, but in practice there is room for improvement.

### Rules for redeeming normative validity claims

Does the model contain any explicit barriers that bias the distribution of interests that participate or prevent people from making subjective determinations? Does the model promote the discovery and development of mutual understandings among all the participants? And does the model make certain that the factual implications of normative choices are considered, consistent with themselves or the general will, and compatible with laws and present expectations?

The Citizens Jury process was created to deal with normative problems and it is here that its greatest strength lies. The Citizens Jury process was created on the assumption that there is no intellectual way to create a foundation from which an objective definition of normative concepts like "bias" or "general will" is possible.[4] If one wants an answer to whether or not the Citizens Jury process is biased or in tune with the general will, one should not turn to academia, but to the people themselves in a reflective mode.

Imagine, for example, that someone were to claim that the "general will" is best indicated through public opinion polls or through an "electronic town meeting." In such a case our position would be that people who have participated in either one of these should be brought together with people who have participated in a Citizens Jury to discuss which method works best. Great care would have to be taken in setting up such a meeting so that the agenda and rules were set so that the participants who have experienced the different methods agree that they are taking part in a fair discussion, such that none of the methods is given an unfair advantage over the other. If such a discussion led to a consensus that some method were superior to the Citizens Jury process, we at the Jefferson Center would give serious consideration to modifying the method to meet the suggestions made.

---

4 Crosby, Edwin L. *Concern For All*: (unpublished Ph.D. thesis, University of Minnesota, 1973).

In the absence of such an experiment, it is our belief that no current method does as well as the Citizens Jury process in providing a format for normative discussions. We know of no method which does as well in bringing a diverse group of citizens together under circumstances which "promote the discovery and development of mutual understandings." The hearings are conducted in such a way that the mutual respect of the jurors is enhanced and they are given an opportunity to listen to each other and learn about different points of view.

Of course other methods of discussion may provide for longer discussions and greater mutual understandings, but no method of which we are aware gives a virtually equal chance to "everyone" to participate. For example, mediation, when conducted in a sophisticated way, can give the participants more time to understand each other under circumstances where the participants are better informed on the matter under discussion than randomly selected citizens. The Jefferson Center is quite prepared to admit that there are good grounds for decision-makers to choose mediation over a Citizens Jury if the former can be done with a fairer agenda or with a greater chance of its recommendations being followed by those in power. But if one's goal is the normative one of seeking the common good, then we would challenge any other method to plead its case against the Citizens Jury process before a representative group of the public chosen in such a way that all citizens had an equal chance of participating in a deliberative forum.

## Rules for redeeming truthfulness validity claims

Does the model promote discussion about the authenticity of the speaker's expressive claims and an examination of the speaker's sincerity as well as the qualities of the situation? Are individuals given enough time to state and defend their expressive claims? Is a translation scheme used that is acceptable to everyone?

The hearing format of the Citizens Jury process means that the jurors are given a good opportunity to check out the truthfulness of claims by the principal participants. In the 1990 Citizens Jury on the Minnesota gubernatorial race, the jurors in the Western panel became interested in a concept mentioned by a witness, referred to as "zero-based budgeting." There were four witnesses on the panel where this was mentioned and all four agreed that the concept was interesting, but had never been used in Minnesota. The jurors wanted to learn why it had not been used and decided to ask this of all five candidates who appeared before them. When Governor Perpich, alone among the candidates, said that it was a good idea and that it was currently in use in Minnesota, the jurors concluded that he was less than truthful. It is rare that average citizens get an opportunity to learn about some complex method of budgeting from a variety of well-informed people and then pose a direct question to candidates for an important office.

On the other hand, there are a number of witnesses who appear for a short enough time so that the jurors do not find it easy to check out the truthfulness of some of their claims. Generally the jurors have the opportunity to check out only the most important claims which are made. Again, we face a situation where a longer time in the process would yield better results. Currently, however, we believe that the gains in doing this would be more than offset by the loss in representativeness resulting from the longer time.

## Conclusions

The Citizens Jury process does a good job of meeting the three fairness and four competency criteria discussed above. There is no doubt that there are some processes which are more representative (e.g., public opinion polling) or involve people whose backgrounds make them intellectually more competent to deal with the question at hand (e.g., mediation). We would submit, however, that there is no process which has the potential for combining fairness and competency as well as the Citizens Jury process.

The main problem with the method is the compromises needed to empower the process. Because it is relatively expensive to conduct a project, there is not much sense in doing so unless it is clear that the recommendations of the jurors will receive serious attention from policy makers. But often the only way to gain the commitment of those in power in advance of the project is to set an agenda which is not as open or fair as it ought to be. Only as the public dismay over politics in America is becoming very serious do we find any significant politicians willing to commit to the process at the beginning with the possibility of our setting a fairly open agenda. We hope that the projects with Reps. Hoechstra and Penny, where a Citizens Jury will be used to promote a dialogue between the officials and their constituents will be successful.

With regard to the example of siting nuclear waste facilities, we believe that the Citizens Jury process has considerable potential so long as there are powerful public officials willing to endorse it. So long as officials act as they did in Winona, Minnesota, the process is not at all helpful. But the longer that the problem of nuclear waste remains unresolved, the more likely it is that powerful officials may decide they want to resort to a method which has the following advantages:

(1) it does not require the consent of every concerned party, as is the case with mediation;

(2) it makes the NIMBY tactic much less appealing for local organizers, since the people calling for the siting are other citizens and not governmental officials; and

(3) it approaches the dilemma of risk management by helping a representative group of citizens be rational, as opposed to trying to make a group of government officials legitimate surrogates to citizens on value questions.

*Chapter 9*

# *The Citizens' Jury Model of Public Participation: A Critical Evaluation*

*Audrey Armour*

## Introduction

"Our political system has become so complex that the decency and common sense of average citizens have only a faint influence on our national politics."[1] This is the premise underlying the Jefferson Center's efforts over the past decade to promote the use of Citizens Juries as an alternative approach to public involvement in policy making processes. The intent is not to usurp the decision-making power of elected officials but to ensure that they have a thorough understanding of the general public's views when they exercise that power. More fundamentally, the aim is "to change the [policy] dialogue to form a partnership between average citizens and elected officials."[2] The Citizens Jury model, the Center contends, is an effective means of obtaining informed and representative citizen input to policy decisions and of fostering this partnership. As will be shown in the evaluation which follows, the model meets the first objective. But there are reasons to question the extent to which the informed input is truly representative and there are reasons to doubt that the model will lead to a partnership between citizens and policy makers.

## Key Features of the Citizens Jury Model

In his paper, Crosby presents a brief description of the Citizens Jury model. This information was augmented with that contained in written reports of the model's application in various U.S. states[3] in order to obtain a full abstract of

1 "We the People," *The People Themselves* 1:1 (1989).

2 ibid.

3 The reports included N. Crosby, J. Kelly and P. Schaeffer, "Citizen Panels: A New Approach to Citizen Participation," *Public Administration Review* 46:170 (1986), the *Final Report, Policy Jury on School-Based Clinics*, conducted for the Health and Human Services Committee of the Minnesota Senate, 1988, *The People Themselves*, Volumes One and Two, 1989 and 1990, and *Citizens Jury Update*, June 1991.

the model's key features. Basically, the Citizens Jury process is a fairly straightforward one:

- the project sponsor asks a neutral facilitator, such as the Jefferson Center, to set up a Citizens Jury;
- the neutral facilitator selects the Jury members from a "jury pool" obtained through a quota sampling procedure designed to ensure that the sample matches the demographic (e.g., age, gender, education, race) or attitudinal (i.e., opponents, supporters, undecided) characteristics of the population at large;
- the two-tier Jury process may be used with one or more Citizens Juries set up at the regional level and one or more set up at the local level in which case either members of the regional Citizens Jury(ies) decide who among their members should be delegated to represent them on the local Jury(ies) or vice versa;
- the jurors are paid a nominal amount (e.g. US $75 a day) to attend hearings which may run four to five days for local hearings and a week for regional hearings;
- in determining the hearing schedule, consideration is given to the need to balance the potential gain in understanding and the potential drop in the participation rate if more time were allowed;
- the project sponsor and/or the neutral facilitator gives the Citizens Jury its "charge" or statement of the problem to be addressed;
- the neutral facilitator and/or the Citizens Jury decide who to call as witnesses with the aim being to ensure that the different points of view on the problem are presented to the jurors; sometimes a case manager may be called upon to organize the testimony of witnesses;
- the Citizens Jury members decide what questions to ask the witnesses;
- a neutral moderator facilitates all discussions, with the possible exception of final deliberations by the Jury;
- Jury members are given instructions on the method of voting that they are to use (e.g., a two-step vote, the first vote being a weighted vote);
- the Jury members present their informed opinions and recommendations in writing to the project sponsor;
- at the end of their deliberations, Jury members are asked to complete a questionnaire as a means of obtaining their evaluations of the way in which the hearings were run.

Funding for each Citizens Jury has tended to come from a variety of sources. For example, the Citizens Jury which examined the impacts of agricultural run-off on water quality in Minnesota had eleven sponsors including four state agencies, two farm organizations, two environmental groups, two state wide associations, and a group affiliated with the University of Minnesota.

Similarly, the housing Jury which investigated the lack of low-to-moderate income housing in Rochester and Olmstead County was sponsored by a number of local organizations and the Jefferson Center. And the budget priorities Citizens Jury process in Hennepin County of Minnesota was jointly funded by the West Hennepin Human Services and a private foundation.

Needless to say, as with any public participation process, there are complexities which cannot be conveyed in a simple listing of key features. There is much involved in structuring the sampling procedure, setting the hearing schedule, developing questions for witnesses, facilitating discussion, deliberating facts and opinions, and writing and presenting advice to elected officials. Each of these tasks has its own set of procedural requirements and dynamics. And each has the potential to either reinforce or undermine the basic framework of the Citizens Jury model. Unfortunately, it is not possible here to explore the internal complexities of the model. And so, attention is focused on its key procedural features and the extent to which they combine to form a good instrument for public participation.

## Evaluation of the Model

### *Criteria for evaluation*

The normative model of "right" public participation espoused in Chapter 3 sets out two broad categories of criteria for evaluating the structure and conduct of public participation processes. The first category is fairness (measured against conditions of the ideal speech situation) and the second is competence (measured in terms of the performance of the participants in constructing the best possible understandings and agreements, taking into account the information and knowledge available to them). Does the Citizens Jury model embody fairness and communicative competence? Does it seek to attain a condition of fairness as described by the "ideal speech situation" advocated by Habermas? A third evaluative category is also addressed. It is efficacy and applicability. It was felt that these were important considerations, even though such criteria are not part of the normative model of "right" participation.

In evaluating the Citizens Jury model of public participation, there is a need to distinguish between the involvement of citizens as jurors in the application of the model to specific problems and the involvement of those citizens in the refinement of the model once the jury process has ended. In his discussion of the Citizens Jury model, Crosby addresses both aspects. However, since it is the application of the jury process itself that is of central concern, it is this aspect which was taken as the focus for evaluation. In any case, as Crosby points out, the Jefferson Center has never had the funds to implement its "oversight method" and to have members of Citizens Juries assist the Center in modifying its rules and procedures.

## *Fairness criteria*

### Making of the agenda and the process rules:

Does the model provide everyone with an equal chance to put their concerns on the agenda and to approve or propose rules for discourse, to debate and critique proposals for the agenda and the rules, and to influence the final decision about the agenda and the discourse rules within the agreed upon decision making procedure?

The model, as practiced to date, has only partially met this criterion. The "charge" or agenda for each specific application of the model and the rules for discourse are set by either the neutral facilitator (in this case, staff of the Jefferson Center) or by the project sponsor. They are not set by the Citizens Jury itself who is simply given the charge and the rules. If, either during the hearing or prior to their private deliberations, the jurors were allowed to negotiate a change in their "charge" with the project sponsor in order to allow them to address additional issues in their recommendations, then this limitation in the model would be overcome.

Occasionally the agenda is set with input from members of the groups or interests who might be potentially affected by the outcome of the decision-making process which would make up somewhat for the lack of input from the Citizens Jury. Unfortunately, whether this happens appears to depend on the inclination of the project sponsor to seek such input. For example, in 1990 the Rochester School District called upon the Jefferson Center to set up a Citizens Jury to help it evaluate directions for its Fine Arts program.[4] The School District put together a steering committee to assess whether the "charge" for the Jury was appropriate and how the Jury agenda and witness list should be developed. The steering committee consisted of four teachers and two community people as well as School District administrators. Thus, the representation of the steering committee extended beyond the project sponsor (although students were noticeably absent). It should be stated, however, that it is not clear whether the charge set by the steering committee was ratified by their respective stakeholders. Indeed, it was reported that "few teachers liked the charge, but most agreed that it accurately reflected the conditions facing the school."[5] At the other extreme is the approach taken in the case of the budget priorities Citizens Jury in Hennepin County, Minnesota. The purpose

---

4 M. Greenwald, "Painting a New Picture? Fine Arts Jury lays Groundwork for Change," *Citizens Jury Update*: 8-10 (June 1991).

5 ibid.

of the Jury was to provide county residents and officials with an independent perspective on the county's budget. The project was led by a steering committee consisting of the County Commissioners, County planning board members, a community planner, and an independent consultant – in other words, a project sponsor-dominated committee. Again, this limitation can be overcome by setting down a requirement that the steering committee be representative of the full range of stakeholders associated with the policy issue.

The rules for each specific application of the model are also set by the neutral facilitator in consultation with the project sponsor. The Citizens Jury itself does not have any input to rule-setting. Nor do those who might be potentially affected by the outcome of the decision-making process. This limitation can also be addressed by allowing Jury members an opportunity either at the start of the process or sometime during it to re-negotiate the rules with the project sponsor.

### Moderator and rule enforcement:

Does the model provide everyone with an equal chance to suggest a moderator or debate proposals for the moderator, and comment on the facilitation style or debate proposals for how moderation should be carried out?

The model does not meet this criterion. The moderator is selected by the neutral facilitator and/or the project sponsor with no input from the Citizens Jury or those who might be potentially affected by the outcome of the decision-making process. Although jurors are asked at the end of their deliberations to evaluate the performance of the moderator, this evaluation comes after-the-fact.

Given the role played by the neutral facilitator in selecting jurors and helping to define their "charge", it will be difficult to overcome this limitation. Although it is possible to conceive of the Jury being given the opportunity to select its own moderator, it is unlikely that this change in the model's procedures would be supported by project sponsors. By the time the Citizens Jury is in place, the project sponsor is likely to have made a substantial commitment (both financially and psychologically) to the neutral facilitator it hired to help it and would be inclined to rely upon the facilitator's expertise to select an appropriate moderator rather than turn the task over to non-professionals.

### Discussion rules:

Does the model provide everyone who is potentially affected by the decision proposal (positively or negatively) an equal chance to be present or represented at the discourse and to put forth and criticize validity claims about language, facts, norms and expressions.

The model partially satisfies this criterion. The sampling procedure used to select jurors theoretically provides all people in the greater affected population with an equal chance to participate or to be represented. However, much depends upon the criteria used to structure the quota sample and the extent to which the selected jurors can be said to be truly representative of the interests at stake. It is not clear from the available descriptions of the model who has input to the development of these criteria and on what basis the criteria are set. Which demographic or attitudinal characteristics are important and who determines this? And even if the random sampling ensures that everyone has an equal chance to be represented in the discourse, it does not allow potential affected interests to select their own representative. Rather, a representative is chosen for them. Furthermore, given that jurors are selected solely by virtue of meeting certain demographic or attitudinal characteristics and that they are given no opportunity to interact with their "constituents," it is open to question whether the views they put forward actually reflect the views of the affected interests they are supposed to represent. Do all senior citizens think alike? Do all people who oppose nuclear power do so for the same reasons or share the same views on non-nuclear energy sources? The Jefferson Center would contend that Citizens Juries are intended to represent the views of "average citizens" but this only brings us back to the question of what is "average" and who decides this.

The opportunity given to the jurors to question witnesses should enable them to question the validity claims of the witnesses which come before the Jury. However, the public hearing format may constrain the dialogue considerably. Rather than a multi-way communication, the interaction among participants is most likely to be characterized by one-way exchanges (as witnesses try to educate or persuade the jurors) and at best two-way (as jurors ask questions of witnesses in an effort to solicit or elucidate facts and opinions). And there is no guarantee that the deliberations of the jurors, especially those conducted in private, will be any better. The potential for coercion in juror interactions amongst themselves is high.

Finally, in the interest of time, the number of witnesses who appear before the Jury has to be limited and consequently "all relevant parties" may not be present. For example, the Minnesota electoral Jury heard testimony from 25 witnesses, the Rochester housing Jury (which rated candidates) heard from 43 witnesses, and the Jury on school-based clinics (for the prevention of AIDS and teen pregnancy) involved 22 witnesses. Another key deficiency with respect to who appears before the Jury is that policy makers, the people who will make the final decision, are not required to appear before the Jurors. This is not to say that they never take part – in the case of the agricultural run-off Citizens Jury, officials in state agencies responsible for water quality made presentations to and answered questions from the statewide Citizens Jury. However, this appears to be the exception rather than the rule. When policy makers do not take part in the Citizens Jury process, there is no opportunity for dialogue between them, the jurors and the witnesses. Thus, the validity claims of each group cannot be explored in the question-answer exchange with the result that both the Citizens Jury and the policy-makers proceed to make their decisions in the absence of potentially important information. Moreover, lack of involvement of policy-makers in the Citizens Jury process may contribute to their discounting the output of the process.

These limitations of the model are inherent ones. Some are a direct consequence of the question-answer format of hearing processes and of the group dynamics of juries. Others are a direct consequence of the need to balance hearing length and participation rates.

Of the three fairness criteria, the discussion criterion is the most important. Input to rule-making and the choice of moderators amount to little if the discussion is constrained, as it is with the Citizens Jury, to at best two-way communication.

In many respects the Citizens Jury model is simply another, albeit advanced, form of public opinion poll or focus group. The advantage of the Citizens Jury model over these other methods of surveying public opinion is that the model provides the citizens involved with an opportunity to learn and reflect on the policy matters at issue before being asked to express a point of view. However, it does not facilitate the kind of multi-way communication needed to achieve an effective resolution of policy issues. The written input provided by Citizens Juries is, when all is said and done, just one more input policy makers have to take into account. Whether and how policy makers go about factoring the views of "average citizens" into the decision equation is not part of the Citizens Jury model.

## *Competence criteria*

### Rules for redeeming comprehensibility validity claims:

Are the participants provided with equal access to sources for commonly-agreed-upon standards and definitions, an opportunity to confirm each others' terms, definitions and concepts, and the ability to take advantage of preestablished reference standards?

It is not clear the extent to which the model satisfies this criterion of competence. There is no discussion in Crosby's paper or in the other reports of the Citizens Jury model as to whether time is taken in advance or at the start of the hearing process to review standards and definitions, resolve comprehensibility problems, and reach agreement regarding reference standards such as a dictionary or textbook. It is possible that, in the question-answer exchange between jurors and witnesses, standards and definitions are probed. However, since the intent of the question-answer format is to draw out information from the witnesses rather than to reach a consensus with them, whatever probing of standards that might occur is likely to be limited.

### Rules for redeeming truth validity claims:

Does everyone have equal access to available and relevant systematic and anecdotal and intuitive knowledge about the objective world? Are uncertainties regarding factual information considered? Is there a means of verifying whether factual claims are consistent with the prevailing expert opinion or consistent with the anecdotal knowledge of other persons not involved in the discourse? And are cognitive claims separated from normative claims?

The model has the potential to satisfy this criterion. Depending on the range of witnesses selected to appear before the Citizens Jury, the jurors should have equal access to both systematic and anecdotal knowledge. Uncertainties regarding the factual information are likely to become evident in the course of witness presentations since it is highly likely that, in presenting their case, witnesses will point out such uncertainties. Furthermore, the model's reliance on witnesses to provide information and points of view also enables the jurors to verify that factual claims are consistent with prevailing expert opinion. And the survey process used to select potential jurors (people in the jury pool are interviewed) should provide the Jury members with data to verify that factual claims are consistent with the anecdotal knowledge of other persons not involved in the hearing. Reliance on the results of the quota sample is not the most effective technique for this purpose but, given that public opinions are gathered in the sampling process, it offers a cost and time saving means of accessing the anecdotal knowledge of persons not involved in the hearing.

It is not clear from the available descriptions of the model whether any effort is made to separate cognitive and normative claims. This may happen in the course of juror questioning or as a result of interventions by the neutral facilitator, but there does not appear to be any explicit intent to ensure that it happens.

### Rules for redeeming normative validity claims:

Does the model contain any explicit barriers that biases the distribution of interests that participate or prevent people from making subjective determinations? Does the model promote the discovery and development of mutual understandings among all the participants? And does the model make certain that the factual implications of normative choices are considered, consistent with themselves or the general will, and compatible with laws and present expectations?

The model partially satisfies this criterion. Theoretically, the selection of jurors through a random sampling process and the adherence to quotas to ensure that the sample reflects demographic and attitudinal characteristics of the jurisdiction should prevent bias in the structuring of the Citizens Jury. This is not to say that gaps in representation will not occur – if the jury process is to function effectively the number of jurors has to be kept small, which means there is the potential that not all interests will be represented. In addition, as noted above in the discussion of the third fairness criterion, the extent to which Citizens Juries can be counted on to represent affected interests is open to question.

The use of a neutral facilitator to select witnesses and the procedural rule that the witnesses must represent a broad range of points of view on the policy issue also prevent a biasing of the distribution of interests that participate in the hearing process.

Where the model falls short is in regard to promoting a mutual understanding among all the participants. No doubt the jurors, in their deliberations amongst themselves, could develop such an understanding, given that the jurors are often together for four to six full days and assuming the dialogue they have is characterized by consensus building and not by coercion on the part of more aggressive or dominant jury members. However, there is no similar opportunity to discover and develop mutual understanding between the jurors and the witnesses or between these two groups and the policy makers. As noted earlier, the question-answer form of dialogue between jurors and witnesses has little potential to foster mutual understanding. And the absence of policy makers from the dialogue that occurs either during the questioning process or during the jury's deliberations significantly reduces the potential for a meaningful exchange of ideas between the "average citizens" who serve on the jury and the people with decision-making responsibility.

With regard to the third question above, it should be pointed out that to date the Jefferson Center has not selected cases that focus solely on cognitive dimensions. Indeed, some of the cases seem to be strongly normative, such as the case involving school-board clinics and the one concerning the Minnesota 1990 election process. In any case, there are no apparent constraints which would prevent the jurors from considering the factual implications of normative choices and whether these choices are internally consistent and consistent with laws, other than the willingness and ability of the jurors themselves to undertake the task. The first constraint could be removed by making this task part of their "charge" and the second by providing the jurors with training and appropriate information.

More problematic is the matter of whether the normative choices deliberated and recommended by Citizens Jury are consistent with the general will and compatible with present expectations. On the one hand, it has been shown that the opinions of Citizens Juries and the general public are not consistent. For example, in the case of the Citizens Jury for school-based clinics for the prevention of AIDS and teenage pregnancy, the initial survey of 800 people from around the state showed 77 percent who thought school-based clinics were "generally a good idea." Yet only 54 percent of the statewide jurors voted in favor of the idea.[6] On the other hand, it has been argued that Citizens Juries provide a better reflection of general public opinion than polls. "Public opinion polls only show what the average citizens think without reflection [while] policy juries allow them to learn, reflect and then express themselves."[7] However, without a way to test this assertion, it must be accepted on faith.

This is perhaps the most significant limitation of the Citizens Jury model. It does not provide for ratification of the normative choices of the Jury. This could be done by convening public workshops in which the Jury members could explain to other "average citizens" how they came to the choices they made, answer questions, exchange ideas and obtain feedback. This would allow the Jury to test the consistency of its choices with the general will while at the same time allowing other average citizens an opportunity to learn more about the policy issue and how their peers addressed it.

### Rules for redeeming truthfulness validity claims:

Does the model promote discussion about the authenticity of the speaker's expressive claims and an examination of the speaker's sincerity as well as the qualities of the situation? Are individuals given enough time to state and defend their expressive claims? Is a translation scheme used that is acceptable to everyone?

The model has the potential to satisfy this criterion. It is reasonable to assume that, in the course of their deliberations, the jurors discuss the authenticity of the witnesses' expressive claims and their sincerity. It is also reasonable to assume that the witnesses' presentation could be scheduled to afford them the time they need to put forward their expressive claims. Finally, it is also reasonable to assume that the witnesses could be given the opportunity to verify the translation of their expressive claims.

Overall, the Citizens Jury model appears to address the matter of competence with respect to two of the four forms of discourse – explicative and theoretical. As noted above, however, in terms of therapeutic and practical discourse, the model has some serious shortcomings.

---

6 *Final Report, Policy Jury on School-Based Clinics*, 2.

7 ibid., 1.

## *Efficacy and applicability*

Although these issues are not part of the normative model of '"right" public participation, the evaluation of the Citizens Jury model would not be complete without some reference to its efficacy and its potential applicability to different problems. The issue of the efficacy of the process raises several questions: Is the process a reliable source of citizen input, that is, representative of the general public? Are the outcomes of Citizens Juries (i.e., the recommendations they make) better than what would have resulted from more traditional processes of public participation? And have Citizens Juries proven effective in influencing the decision-making of the sponsors? It was noted in discussing the rules for redeeming normative validity claims, the model, as currently defined, does not include a ratification procedure which would confirm that the views of the Citizens Jury are supported by their supposed constituents. It was also noted that this limitation can be easily addressed. As to the merits of the Citizens Juries recommendations and degree of influence the Jury process has had on decision-making processes, there is a noticeable lack of research which would enable any assessment to be made. This is unfortunate because, without such assessments, the model is likely to remain "unique" rather than a standard approach to public involvement. Public policy-makers have to be confident that the Citizens Jury represents an improvement, not just in terms of its procedural features but also in terms of the quality of the product. For example, a key concern for policy makers will be whether the Jury process yields recommendations which are more than a "wish list" and reflect considerations of implementability. Until this is demonstrated, persuading policy-makers to adopt the Jury process will be difficult.

The model fares better in terms of its applicability to different problems. As can be seen from the cases cited in Crosby's paper and in this review, the model has been applied to a variety of substantive issues, including economic (setting of budget priorities), environmental (agricultural run-off), social (school-based clinics), and political (rating of political candidates). Still, it cannot be assumed that the model is applicable to any issue. It is unlikely, for example, to be an effective approach to resolving the issue of siting "unwanted facilities" since the problem is not simply one of arbitrating the competing interests affected by a specific siting decision (is community A a better site than community B?) but reflects fundamental flaws in siting procedures and methodologies.[8] In addition, the model has only been applied in the North American context and to mainstream cultural groups. Whether it is applicable to other contexts and cultures remains to be seen. Its cross-cultural

8 Institute for Environmental Negotiation. *Not-In-My-Backyard! Community Reaction to Locally Unwanted Land Use* (Charlottesville, VA: IEN, 1986).

applicability is questionable – the selection of jurors through quota sampling, the question-answer form of dialogue, examination of witnesses, and voting form of decision-making – may not be transferable to aboriginal cultures where decision-making is entrusted to elders and consensus-based processes. It would be premature, however, to reach any firm conclusions on the matter without considerably more "testing" of the model.

## Conclusion

On balance, the Citizens Jury model represents an improvement over more traditional forms of public involvement in policy making. If care is taken in the structuring of the sampling procedure and in the selection of witnesses, it has the potential to be an effective means of obtaining *informed* citizen input to policy decisions. This achievement is not to be underrated. Far too often, the input of citizens is ignored by policy makers simply on the basis that such input is ill-informed.

However, the Citizens Jury model needs to be strengthened if it is to become a more significant instrument of public participation. First, there is a need to guarantee the representativeness of the Citizens Jury. The model now lacks a reliable and open procedure for obtaining consensus on the demographic or attitudinal characteristics to be taken into account in setting up the Jury (at the start of the process) and for ratifying the recommendations of the Jury with those they are supposed to represent (at the end of the process).

More significantly, the model as currently structured has no potential to achieve its intended goal which is to "change the policy dialogue to form a partnership between average citizens and elected officials." There is no interaction between the Jury and elected officials save for the submission of a written report. Providing input to the policy making process is not the same as taking part in it. The Citizens Jury model, like so many models of public participation, still leaves citizens outside the policy arena, talking among themselves and hoping that their input will somehow be taken into account. The Citizens Jury model does not enable citizens to dialogue directly and in a meaningful way with those who govern them. Furthermore, the input of the Jury to the policy process is one among many and, although this input may be an improved form of input (in that it is more informed), it would be naive to think that this alone will make a substantial difference in the forces that are at play in policy making. Requiring policy makers to be active participants in the Citizens Jury process, to ask and be asked questions, to put forward and receive points of view, in short to be accountable and responsive, would go a long way to making the Citizens Jury model a much stronger and potentially powerful instrument.

Finally, there is a need to provide rigorous assessments of the model's efficacy and applicability. While its procedural characteristics are laudable, potential sponsors will be equally (if not more) concerned about the potential for citizen juries to improve, in a substantive way, the outcomes of decision-making processes and the extent to which citizen juries can be applied to different issues and contexts. Answers to these questions will reveal the model's full power.

*Chapter 10*

# *The Varresbecker Bach Participatory Process: The Model of Citizen Initiatives*

*Frank Claus*

## Introduction

This chapter describes a public participation model that has been developed and tested in the case of soil contamination in the residential area of Varresbecker Bach, which is located in Wuppertal, Germany. The case is ongoing in the sense that decisions have yet to be made concerning the actual sanitation of the area. But an effort has been made to involve the public in a meaningful way in this decision process. This chapter describes this effort, and the following chapter analyzes the model according to the criteria set out in Chapter 3.

Wuppertal lies in the heart of the most industrialized area of Western Europe, the Ruhr valley, where the industrial revolution on the continent is said to have begun. As early as 1900, there were over 100 chemical factories operating in Wuppertal, which later gained the title of the Cradle of the Chemical Industry. Wuppertal is a middle-sized city covering an area of some 150 square kilometers. For its size, it has a large number of contaminated waste sites: 430 registered sites and an expected additional number of 1,500 to 2,000 (see Gerdts, et al., 1992).

The contamination of residential or other areas presents one of the most difficult challenges to collective choice processes. The issues are usually emotionally charged, involving the health and safety of the residents, their children, and generations to follow. The cost of cleaning up contaminated sites is often very high. Moreover, the scope of the contamination problem throughout the industrialized world is truly daunting. Very difficult choices must be made concerning which contaminated sites should be selected for remediation, and how clean is clean enough? Because the health risks from contaminated sites usually affect an identified and localized group of citizens, the distributional questions also present a complex set of issues.

No country has done more than only begun to address the problem of contaminated waste sites. Even in the US, after more than a decade of the Superfund program, this country is still in the beginning stages of any sort of meaningful cleanup program. This is equally true for countries in Western

Europe, and one can only speculate about the dimensions of this issue in Central and Eastern Europe.

## The Case of Varresbecker Bach

This case study describes one of Wuppertal's contaminated sites in the residential area of Varresbecker Bach. It can be viewed as a test case for a much larger number of such cases that will undoubtedly follow.

In September 1990, construction workers in the area came across a number of buried barrels. Subsequent testing revealed that the construction site was seriously contaminated with arsenic, lead, copper, mercury, cyanide, and PCB's. Further testing by the Wuppertal Environment Agency revealed that the contamination was not limited to the construction site, but that over a kilometer of a stream bed (the now rerouted Varresbecker stream) had been used as a dumping ground for one or more now nonexistent and unknown chemical firms. The creek bed extends through a residential area of single family houses, perhaps affecting up to 1,000 residents.

The Wuppertal Environment Agency has official responsibility for the contaminated site. Yet, according to the state law of Nordrhein-Westfalia, if the original polluting firms cannot be identified, then the owners of the property have full responsibility and liability for the costs of any measures deemed necessary to reduce the environmental risks of the contamination. The Environmental Agency's task is to assess the seriousness of the contamination and determine the extent of the necessary remediation measures.

The Agency makes a recommendation to the Wuppertal City Council, specifically to the Wuppertal Environment Council, which has ultimate decision authority. Seldom, however, is a recommendation by the Agency not accepted. The Agency has no legal obligation to include the public or property owners in this process. They wisely chose to do so, however, and this chapter describes the participatory model that was developed and tested at Varresbecker Bach.

## Key Features of the Varresbecker Bach Case

### *The issue*

Contamination at Varresbecker Bach appears to be of concern primarily to the residents and property owners of the community and the City Government, that is, the issues appear to be local in scope. This is reflected by the official political agenda which contains two main questions:

- Is there a need to clean up Varresbecker Bach? (In other words, what are the public risks involved?)
- If so, how should the cleanup proceed? (Implicitly, how clean is clean enough?)

Not originally on the official agenda, but very much on the minds of all involved, was the issue of *who* will bear the costs. This raises some fundamental questions concerning the fairness or justice of holding the victims financially responsible. As mentioned above, if the polluter cannot be identified, then the property owners have full legal liability for the remediation costs as deemed necessary by the Environment Agency. This legal provision would appear to constrain any further negotiations on the issue of financial responsibility, but for a number of reasons this law was viewed as adaptable. Only a short distance to the south is the State of Hesse, where property owners are not legally responsible for such costs. Moreover, the public outcry surrounding this issue could lead to political pressure for government support for the citizens of Varresbecker Bach. Both a change in the law and/or an exemption for Varresbecker Bach were viewed as politically feasible by all involved.

## *The stakeholders and their concerns*

There are two main groups of stakeholders in this case. The first are the residents and the property owners of Varresbecker Bach, most of whom are one and the same (around 90 percent of the residents own their homes). Their main concern appeared at first to be the health risks from the contamination, but later other concerns arose. Many residents have fruit trees and small vegetable plots in their back gardens. The risks result not only from contaminated food, but also from children playing in the contaminated areas. There has been little discussion of ground water contamination.

To date, the residents have voiced sincere concerns about these risks, but the arguments have not had the intensity witnessed in many other similar cases. The explanation may lie in several factors. The citizens of Wuppertal have a long involvement with the chemical industry, which has provided many jobs in the area. The Wuppertal branch of the well-known BAYER firm, where some of the residents work, is less than 100 meters away. Chemical risks have thus been a part of the lives of many residents, and most families have lived in their homes for many years without any noticeable affects. At this time, there are few families with small children.

Perhaps the most important reason for the seemingly languid response of the residents to the potential health risks is that health concerns are in conflict with their concerns as property owners over the high costs of sanitizing the area as well as the losses in property values if perceived risks persist. This conflicting situation – where the residents are interested in reducing their risks but also in keeping costs to a minimum – is an important characteristic of this particular case. It renders the case more analogous, for example, to radon in private homes than to cases of hazardous waste cleanup in most other countries where property owners are not responsible for the financial burden.

This feature of the Varresbecker Bach case can also lead to conflicts within homes and between neighbors. A notable feature of the case is the lack of involvement on the part of women, who often are more concerned with the health risks than their male counterparts. A concern that arose at a later stage in the process, particularly on the part of the elderly residents, is the disruption that the testing and sanitation will cause to their gardens and their everyday lives.

The City of Wuppertal, more precisely the Wuppertal Environment Agency, constitutes the other main stakeholder. The Environment Agency has legal responsibility for determining the extent of the cleanup. Although it appears that the Agency can objectively weigh the risks and the costs in carrying out its mandate, a closer look reveals that it also has a substantial interest in the outcome. It is deeply concerned with maintaining its legitimacy as an independent regulatory agency. Knowing the potential for public outrage and conflict that often accompanies these issues, and which often damages the credibility of public officials, the Agency's interest is in a mutually agreeable and amicable solution with the other stakeholders.

The Environment Agency is not the only stakeholder in the City of Wuppertal. The entire political administration has a stake in the outcome of this case. If the Wuppertal Government makes an exception to the law and assumes the costs of cleaning up Varresbecker Bach, there will surely be demands for further exceptions. Alternatively, public demands for a change in the legal responsibility could also shift enormous potential costs to the government. Given the very large number of contaminated sites within the city limits, this case will clearly set an important precedent for the avalanche of cases likely to follow.

The Wuppertal Environment Commission is one of several commissions set up by the Wuppertal City Council. This Commission has the final word (or it recommends to the City Council) on all decisions for sanitizing Varresbecker Bach. In practice, it generally accepts the recommendations of the Environment Agency.

## *The steps*

This case is still in the early phases. According to the Environment Agency, the process and actual cleanup of the site should continue at least until mid-1995. During this time, the following four tasks have been set out by the Agency:

(1) Assessment of the Risks. The task is to determine the type, extent, and seriousness of the contamination.

(2) Evaluation of the Sanitation Alternatives. Given the many possibilities for dealing with the risks – from sealing off the site to removing the contaminants – the task is to examine and decide upon the alternatives.

(3) Planning the Technical Sanitation. The task is to choose the contractor and to plan the technical procedures.

(4) Implementing the Plan.

At the time of this writing, the process is in the first phase of assessing the type, extent and seriousness of the risks to the public. The Wuppertal Environment Agency has compiled evidence indicating the possibility that the entire stream bed is seriously contaminated. With permission of the residents, hundreds of soil samples were taken, which revealed that at least 100 private properties are contaminated with heavy metals, cyanide and PCB's. Another large area on which apartments owned by the City are built is also affected. Overall, about 1,000 persons are directly affected. After a long process, the residents have granted the Agency permission to excavate their yards in order to determine more precisely the extent of the contamination. Currently, these excavations and testing are near completion.

## The Process

### *The environment agency initiates a participatory process*

From the outset, the Wuppertal Environment Agency recognized the value of full public disclosure and participation in the decision process. The Agency feared that any decision it might take regarding the sanitation of the site might confront strong public opposition, particularly since the victims of the contamination are financially responsible for its remediation. The Agency, therefore, decided to proceed along the steps mentioned above with full cooperation of all the stakeholders. This would mean devising an *extra legem* (beyond legality) process with novel forms of public participation. As fundamental to this process, the Agency decided to proceed by

(1) disclosing all available information to the affected parties;

(2) openly revealing all information gaps; and

(3) engaging an outside mediator to communicate with the public, advise the Agency, and generally organize the participatory process.

This initiative by the Environment Agency is rather unusual for Germany. Unlike the United States, which has a slightly longer tradition of public disclosure, Germany has a tradition of administrative secrecy in public policy making, and the rights of the public for formal participation are only starting to emerge. The decision to engage an outside mediator is even more unusual, and the success of this particular model may have rather far-reaching influence in the City of Wuppertal and the State of Nordrhein-Westfalia.

## *An independent mediator is brought in*

The mediator in this case is the Institute for Communication and Environmental Planning (IKU), a small and young organization located in Dortmund (of which the author is the President). In the words of the Environment Agency, the tasks of the IKU are to inform the affected citizens about the contamination, organize and moderate citizen and working group meetings, serve as an independent advisor for the affected citizens, promote strategies to avoid conflicts, and generally assure the transparency of the process to all concerned. The IKU sees its role as one of communicating the risks to the citizens and promoting a consensus based upon open communication and mediation of any disputes that arise. The IKU's stated objectives are: (1) to provide open and easily comprehensible information to the citizens; (2) to minimize any negative psychological and social problems by involving the citizens in a meaningful way; and (3) to keep the process focused on the "objective" problem at hand. Although the tools and procedures are selected in an *ad hoc* manner to fit the changing nature of the process, the following fundamental principles should, according to the IKU, remain unchanged:

- open and unrestrained information and access to all official files;
- early information before decisions are taken on the official level; and
- cooperative procedures for arriving at solutions.

## *The citizens are informed and involved*

The first task before the Agency and the IKU was to inform the affected citizens of the contamination. The IKU prepared an address list of approximately 30 households. These households were selected because they bordered on the Varresbecker Bach, and therefore it was thought that they were the only households affected. The IKU sent letters informing these households of the potential problem and inviting them to the first public information meeting. At this meeting, the public's additional insights on the geography and history of the area resulted in an expansion of the address list to approximately 100 affected households, almost all single family houses.

These public meeting were the first of many to follow. The meetings are organized by the IKU, which also plays a role in moderating conflicts and assuring that all participants have a chance to speak and voice their opinions. The IKU also advised the Environment Agency concerning its role at these meetings. Specifically, official spokesmen were advised to be frank and open, and to admit to any possible mistakes that they might have made. Not only does the IKU distribute information and organize public meetings, but it also serves as advisor and consultant to the residents. For this purpose, it has open office hours and publicizes its readiness to discuss the problems with any of the stakeholders.

Already during the first large public meeting, where information was distributed concerning the possible health risks, the residents and property owners formed the Varresbecker Bach Contaminated Site Interest Group (hereafter referred to as the Interest Group). The Interest Group has seven active members. It does not, however, officially represent all the property owners and residents since its constellation and mandate have not been the subject of a general vote. Yet, its role in the early stages of the process was not challenged, and it appeared to be the *de facto* authority in this case.

The spontaneous formation of the Interest Group appears to have resulted from concern about the health risks, but also about the sanitation costs and property values. The stated priority of the Interest Group is to find satisfactory solutions to the following questions:

- Are the health risks really acute, and what short-term measures can be taken?
- Who is responsible for protecting the residents from the risks, that is, who will pay the costs of the sanitation if it is shown to be necessary?
- How can a deterioration in property values of the affected residents be avoided, and what possibilities exist for negotiating financial agreements with banks?

It is noteworthy that this is the first explicit mention of one of the core issues of this case, that is, who will pay the necessary sanitation costs? Since this question does not appear on the agenda of the Environment Agency, there seem to be two different problem definitions and agendas that will have to be reconciled.

An important tradeoff is also alluded to here in the sense that the costs of reducing the risks (the sanitation costs) should be balanced with that of the benefits (reducing health risks). A striking difference between the language used in issues of this type in Germany compared with the U.S. is that the cost/benefit tradeoffs are not explicitly mentioned in German cases. There has been no suggestion in this case that the "necessary" extent of the cleanup should depend critically on the costs and health risks reduced.

In the beginning, the Interest Group concentrated on sensitizing the relevant politicians and the general public to the plight of the residents in this area. This was done by writing letters to the political representatives in the Wuppertal Parliament and to the Wuppertal Environment Commission, and drawing repeated attention to the issue in the local press. The activities of the Interest Group are constrained, however, by the private finances of those actively involved since the group receives no financial support from the government or from the residents and property owners it represents. This makes it difficult to carry out extensive mailings or even extensive telephone conversations. The services of a lawyer are apparently out of the question. To improve communication with those it claims to be representing, the Interest Group has established a centrally-located information board, where announcement about the regular meetings of the Interest Group and the public meetings are posted.

The Interest Group began its activities by reviewing the results of the soil tests which had been carried out by the Environment Agency. The soil in many gardens appeared to be seriously contaminated with lead, arsenic, mercury, and, in some cases, PCB's. Working together with the Environment Agency, information and recommendations were sent to the residents regarding the use of their gardens and the consumption of the vegetables and fruits. It was agreed that the public playground would be closed.

## *The first consensual decision is taken*

The first step in the lengthy process to follow was to assess the extent and seriousness of the contamination. Up until this point, only the surface soils had been sampled and tested, a procedure which had not required approval by the residents. It is not yet known how deep the contamination extends. Far more extensive testing of the area, involving extensive digging and drilling on each of the surrounding properties, will be necessary. This excavation of the private property can, in turn, only be done with the permission of the residents. Some were disinclined to give this permission because of fears that public knowledge of the contamination on their private properties would affect their property values. They have some legal protection, however, from the German Data Protection Law, which states that data on the contamination of private property cannot be made public without the consent of the property owner. Another concern that arose at a later stage in the testing process was that the testing would seriously damage backyard gardens. Those opposed to any cleanup activities felt the testing was unnecessary.

The Environment Agency agreed that the property owners and residents should participate in the choice of the expert institution that would carry out the envisaged testing program, and that the Agency would pay the full costs of the testing. The IKU devised a procedure for this participation. At the suggestion of the Agency, the IKU invited three certified expert agencies (Gutachterbüros) to a public meeting, where each was asked to present its concept of a testing and risk assessment program. The citizens present at the meeting decided upon the following three criteria on which to judge the three competing concepts:

- the proposed involvement of the affected citizens in all steps of the testing program and assessment of the risks;
- the degree to which the information would be made understandable and transparent; and
- the manner in which the risk estimates for individual properties would be reported.

Only the residents and property owners present at the meeting were permitted to participate in the actual evaluation of the expert groups. At this time, the invitation list to the public meeting had expanded significantly to include owners and residents of all potentially affected properties. This included about 1,000 persons living in (or owning) around 100 private homes as well as a large public housing complex owned by the government. Each resident and/or owner present was asked to weigh the three concepts according to the above three criteria. The results were very evenly spread over two of the expert groups. Both groups were then recommended to the Environment Agency which, exercising its decision authority, decided to employ both to carry out a joint plan. This consensual selection process appears to have created a great deal of trust on the part of the Varresbecker Bach residents. Subsequent to the decision, 98 percent of the owners of the around 100 affected properties gave their permission for testing on their property. At this writing, the testing is well underway.

## *A consensus-building round table forum is established*

Shortly following the choice of the expert teams to carry out the testing and risk assessment, the stakeholders formed a Working Group to assure that adequate information from the expert teams would flow among the stakeholders and to anticipate potential conflicts. Participating in this Working Group are the Environment Agency, the Interest Group (and later a second interest group of citizens against cleanup), the District Government, the Wuppertal Environment Commission, and the two expert teams.

This Working Group was initially organized on a voluntary and informal basis. However, shortly after its constellation, a blunder by the Environment Agency – in which the Agency failed to invite members of the Interest Group to a meeting where some early test results were discussed – motivated the Working Group to establish a more formal working arrangement. The now so-called Varresbecker Forum drew up a charter which was then approved by all its members, including the Wuppertal Environment Commission.

The Varresbecker Forum Charter calls for regular "roundtable" discussions of representatives from all the stakeholder groups. The expressed goals of the Charter are as follows:

- to promote consensus among the stakeholders on the testing of the properties, the interpretation of the test results, the actions necessary, and the control of the sanitation process;
- to clarify open questions and to anticipate and avoid conflicts;
- to work towards the rapid and comprehensive reduction of the health risks from the contamination;
- to avoid unproductive conflicts by focusing only on the facts;
- to make recommendations to the politically-responsible bodies; and
- to work towards a socially and environmentally acceptable solution to the contamination.

According to the charter, the voting members of the Varresbecker Forum include:

- two members of the Interest Group;
- two representatives from the Wuppertal City Government;
- two representatives from the expert teams (one from each);
- one representative from the Environment Commission;
- one representative from the District Government.

Non-voting members include representatives from the Düsseldorf Office for Water and Waste Management and the Government of Düsseldorf. Inclusion of additional voting members requires a 2/3 majority vote. The Forum meets every month and when deemed additionally necessary.

The charter declares that the IKU will continue to serve as mediator/facilitator for the Forum and will notify the participants of the meetings. The Chairperson of the Forum can rotate among the participants and the IKU. The duties of the Chairperson include calling on speakers and declaring voting issues. Meetings are generally open to the public, but can be closed with a majority vote. The protocol is set by the facilitator, who should collect items for the agenda at the end of each meeting (for the next meeting). Recommendations by the Forum should, as far as possible, represent a consensus of the members, as confirmed by a majority vote. At the request of 1/3 of the members, the Forum participants can be required to keep certain information confidential. A member can be expelled from the Forum with a 2/3 majority vote, but he or she must be replaced by the respective group he or she is representing.

The Varresbecker Forum as a consensus-forming group is rather unusual in Germany, and at first the local politicians viewed it skeptically. This appears to have changed, to such an extent that recent meetings have actually been attended by several local politicians.

## *A second citizens' initiative is formed*

The latest and most surprising development with respect to the Varresbecker Bach case is the entrance of a second citizens' initiative. The group taking this initiative calls itself the Citizens' Initiative for the Environment: Against Cleanup. This group opposes completion of the testing of the private properties and all types of possible cleanup activities. What is most remarkable is that this group does not oppose cleanup activities on financial grounds. Rather, they oppose having their back gardens excavated and sanitized because of the damage this will cause to their gardens and the disruption this will cause to their daily lives. This group is mainly elderly persons, who have spent years nurturing their gardens and understandably see little benefit in massive excavations.

The IKU has managed to integrate this citizen group into the participatory process. By visiting individual members of the group, and spending time discussing their concerns (what is called in Germany "sofa talks"), the IKU now has their cooperation in the Participatory Model. This initiative enjoys the same status of the Interest Group, and members regularly attend the scheduled meetings.

## *The financial liability issue is resolved*

Another milestone in this case is the recent resolution of the financial-liability issue. A state fund for cleaning up contaminated sites has recently been created in Nordrhein-Westfalia. The source of this fund is an annual licensing fee for all generators of hazardous waste in the state. In 1994, this fund contained approximately 50 million marks.

It became apparent to all stakeholders in the Varresbecker Bach case that an application to this fund would be in everybody's interests. The application was submitted, and the state authorities agreed to pay 80 percent of the Varresbecker Bach cleanup costs, which are now estimated at approximately 35 million marks.

The state authorities are willing to allocate a large part of the cleanup fund to Varresbecker Bach for two reasons: First, Varresbecker Bach is one of few seriously contaminated sites that is a residential area. Therefore, the state need not be overly concerned that a flood of similar requests will follow. Secondly, Varresbecker Bach is one of the few sites that is far along in the process of assessing the contamination and risks. Therefore, Varresbecker Bach fits nicely with the state's requirements to allocate its 1994 budget. Naturally, this state financial aid is a windfall for both the residents of Varresbecker Bach and also for the Wuppertal City Government.

This extensive financial aid would appear to be the happy ending of a process that has been non-contentious from the start. However, the Citizens' Initiative against Cleanup remains opposed to any cleanup activities even with such favorable financial arrangements. The Agency has proposed that those property owners who do not wish to have their properties cleaned up be excluded, and this will be arranged through contracts with each property owner. If the property is not cleaned up, however, the owner will have to accept certain legal restrictions on its use.

For those agreeing to a cleanup of their property (and agreeing to pay 20 percent of the costs) there are definite benefits. First, the residents can enjoy their gardens without fear of contamination. Moreover, their property values are greatly increased because the property is canceled from the state list of contaminated sites. Furthermore, they will be given a sum of money to restore their gardens.

## Summary

The main feature of the Varresbecker Bach Participatory Model is an outside, independent mediator – the IKU – which was asked to develop a participatory model for making consensual decisions regarding the eventual cleanup of the Varresbecker Bach contamination. With the support of the IKU, a number of participatory forums have evolved, including public meetings, individual consultations, and a roundtable forum for making recommendations to the City of Wuppertal. A first important decision has been made, that of selecting the expert teams for the comprehensive testing program. Subsequently, nearly all of the property owners have agreed to allow excavations on their properties, which indicates a significant trust in the process on the part of the residents and property owners.

The Varresbecker Bach case is characterized by two main groups of stakeholders, the residents and property owners of Varresbecker Bach and the City of Wuppertal. The residents are concerned about the health risks from the extensive contamination of their community, and, as property owners, about the financial implications arising from their legal responsible for the sanitation. Another concern arising late in the process is that of elderly residents who view the cleanup as disadvantageous in that it will ruin their gardens and disrupt their lives. The City of Wuppertal, and especially the Environment Agency, has legal authority to require the sanitation. Because of the Agency's concern about a possible public outcry implicating the City Administration for past pollution practices, it wisely decided to initiate a participatory decision process with full disclosure of all relevant information.

The process has proceeded with extensive public meetings, information dissemination, and the establishment of a forum for sharing information and reviewing the process. The IKU has managed to direct the process towards a consensual decision concerning the selection of the experts, and has integrated the newly formed Citizens Initiative for the Environment: Against Cleanup. Moreover, the financial liability issue has been resolved to the satisfaction of all the stakeholders via the availability of a state fund. Overall, the Varresbecker Bach Participatory Model shows great promise in resolving the contamination issue in a cooperative, consensual fashion that involves the public in a meaningful way.

## References

Gerdts, D., Claus, F,. and E. U. Lohrmann, *Risikokommunikation und Partizipation bei der Sanierung Bebauter Altlasten,* Draft Paper, Institute for Communication and Environmental Planning (IKU), Dortmund (1992).

*Chapter 11*

# *The Varresbecker Bach Participatory Process: An Evaluation*

*Joanne Linnerooth-Bayer*

## Introduction

Faced with the possibility of serious contamination at the residential area of Varresbecker Bach, the government of the German City of Wuppertal initiated a unique participatory process for making the difficult decisions involved in cleaning up the site. There are two main features of the Varresbecker Bach Participatory Model: the early involvement of the public in a consensual decision process and the contribution of an outside, independent mediator in facilitating an agreement among the stakeholders. While the Varresbecker process is not yet fully resolved, the model has had its first successful trial in the form of a consensual decision on the selection of the experts responsible for assessing the extent and seriousness of the contamination. If the model continues to prove successful, its features will undoubtedly be a key factor of many similar cases that will follow.

This chapter evaluates the Varresbecker Bach Model as an approach to public participation and conflict resolution in environmental disputes. The fairness and competence of the model will be assessed according to the criteria set out in Chapter 3. As a part of this evaluation, it is important to explore the extent to which this model can be applied in other environmental policy contexts, both within and outside of Germany. We must ask if the apparent success (to date) of the model is due to any unique characteristics of this particular case, or, more generally, of environmental policy making in Germany.

## Environmental Policy Making in Germany

Comparative policy studies reveal rather striking differences between American and European styles of environmental policy making. In the words of Sheila Jasanoff (1990:63), the U.S. process impresses observers as "costly, confrontational, litigious, formal, and unusually open to participation." European processes by comparison are more "uniformly cooperative and consensual; informal, cost conscious, and for the most part closed to the public." In Europe, this style is evolving towards more open information and participation, as the Varresbecker Bach and other German experiences indicate. Throughout Europe, there is a discernible trend towards the liberalization of participatory processes, and some countries, including Germany, have institutionalized participation by major interest groups and the lay public. Still, there remain rather fundamental differences between the European and American political contexts of participation, and an understanding of these differences explains the significance of the exceptional procedures at Varresbecker Bach.

The political culture of the U.S. can best be described as individualistic and competitive as compared to the more hierarchical and consensual political traditions of Germany. Both countries instituted strong reforms throughout the 1980's. However, unlike American legislation of this period, European laws did not go as far in formally expanding citizen participation in environmental policy making or in allowing public access to expert documents. The influence German citizens have on environmental and risk policy making remains much weaker, and for the most part the decisions are the preserve of the administrative authorities.

This difference in public influence is reinforced by the contrasting role of the courts. Unlike Europe, administrative decisions in the U.S. are not finalized until they have undergone judicial review. This provides opportunities for interest groups to delay and reverse environmental decisions made by government agencies, and greatly enhances the significance attributed to public participation. This form of judicial review does not exist in Germany, which means that decisions made by a regulatory agency, such as the Environmental Agency in Wuppertal, cannot be challenged.

The eventuality of judicial review also shapes the way in which expertise enters political arenas. Predictably, the Americans have gone much further in adopting formal analytical procedures and technical justification to satisfy judicial demands. Environmental conflicts are necessarily framed as technical conflicts, and political battles are often settled by appealing to quantified estimates of risks and benefits. To be effective in this arena, both regulatory agencies and public interest groups have developed sophisticated scientific capabilities.

The scientific pluralism characterizing U.S. environmental risk management has developed to a far lesser extent in Europe, where regulatory agencies have not had the same pressures to defend their policies with scientific evidence and interest groups have had limited opportunities to challenge administrative decisions. Not only are German administrative decisions considered final, but most legislated participatory forums (e.g. public hearings) do not permit cross examination of expert witnesses. As a consequence, risk and other environmental assessments continue to be performed in closed expert committees, shielded from public review and criticism. Moreover, public authorities can appeal to a closed set of experts in justifying policy decisions, and public access to technical documents continues to be limited.

From a U.S. perspective, it is puzzling that the European public tolerates this government/expert community monopoly on the generation of public knowledge. This reflects the political culture of European countries, where experts have traditionally enjoyed more authority than their American counterparts. As Jasanoff (1990:75) puts it, the prescription to "consult the experts and do as they say" has relatively more meaning on the European scene than in the more scientifically plural American context. Of course, as mentioned earlier, this is changing as European structures yield to more individualistic demands for citizen access and influence.

In evaluating the Varresbecker Bach Participatory Model, it is important to recognize these differences in political traditions. The significance of the participatory innovations at Varresbecker Bach can only be appreciated in the context of traditional, hierarchical policy procedures in Germany, where there are few opportunities for challenging administrative decisions. A coalition of administrative authorities and their expert networks have operated with limited public access and influence. The Varresbecker Model was developed at a time of transition to more open processes, but with the traditionally closed structures still yielding considerable influence.

## The Problem Context

To date, the Varresbecker Bach Participatory Model appears successful to the extent that a first important decision concerning testing of the properties of many private residents has been resolved in a consensual manner with meaningful public participation and general public acceptance of the process. Moreover, the issue of whether to make an exception to the law requiring the property owners to bear the financial burden has been resolved to the mutual satisfaction of all the parties involved.

This success must, however, be viewed in the context of the circumstances of this particular case. To date, Varresbecker Bach is *not* a case of environmental conflict or dispute resolution, but rather a case of cooperative, joint problem solving. The only contentious issue, namely that of shifting financial responsibility to the City of Wuppertal or the surrounding industries, has now been resolved by the State of Nordrhein-Westfalia assuming most of the financial burden. At the outset of the process, however, this option was not obvious, and the City of Wuppertal was understandably concerned that the public might demand financial support at the City level. Whereas this demand did present itself on the agenda of the Interest Group, the Environment

Agency did not mention this issue in its original definition of the Varresbecker Bach problem.

Policy making differentiates itself from decision making in that the constraints – even those set in law – are often viewed as flexible (Majone, 1989). Nonetheless, during the initial stages of the process, the Interest Group appeared to tacitly accept the exclusion of the financial question. This was apparent from the Interest Group's endorsement of the Varresbecker Bach Forum Charter with its expressed goal "to promote consensus among the stakeholders on the testing of the properties, the interpretation of the test results, the actions necessary, and the control of the sanitation process." No mention was made of examining the issue of financial responsibility.

If we accept the formal agenda, then initially all the stakeholders shared the same problem definition and interests, namely, finding a solution (to the question of to what extent the site should be cleaned up) that was acceptable to the residents and property owners who would (at that time) bear both the benefits and costs of this solution. Once again, it may be helpful to compare the initial situation at Varresbecker Bach to the analogous case of radon contamination of private homes. If an environmental authority had the mandate to require homeowners to reduce radon exposure (which is not the case), it would appear to be squarely in the authority's interests to allow the homeowners to decide the level of radon with which they wanted to live given the costs of its reduction.

Whereas officially the problem confronted at Varresbecker Bach is one of joint problem solving in the absence of conflicting interests, this is certainly not a complete characterization of the problem. Hidden agendas are as much a part of political bargaining as the issues on the official agenda, and the crucial hidden agenda at Varresbecker Bach was resolution of the financial question. The Environment Agency appeared to have the advantage in this regard insofar as it could block this issue from the official agenda, which was certainly an important strategic issue for the Wuppertal Government. But as the unfolding of the Varresbecker case revealed, political agenda setting is a complex, dynamic process where many factors, not the least, public outrage, can force issues onto seemingly closed agendas.

Even accepting the official agenda, there is one internal conflict situation that deserves mention. (A second conflict situation arose later in the process.) Just as in the radon case, the homeowners of Varresbecker Bach might wish to downplay or even cover up the contamination. Clearly, their property values are affected by public attention to this problem. On the other hand, in the interests of their own health, it is important to generate information on the extent of the contamination. This can lead to internal conflicts among the residents. Those who are more concerned about the health effects of the contamination have opposing interests to those who are more concerned about the expenses of the cleanup and the reduction in property values that might occur if the area becomes stigmatized as a toxic dump. These issues also raise the possibility of conflicts with the Environment Agency if the data is not dealt with in a confidential manner.

Recognizing the sensitive nature of the case, it is still noteworthy that at the outset of the process all the stakeholders appeared to share common interests, with the single exception of the financial liability question. It is not surprising, therefore, that the Environment Agency sought a participatory, consensual model to resolve the contamination problem. The lack of conflicting interests does not mean, however, that the mediator was not needed. To the contrary, even given few conflicts of interest and good intentions, processes of this sort can easily stray from a mutually beneficial path – as was illustrated by the blunder of the Environment Agency in failing to include the Interest Group in its deliberation on the initial sample results. The mediator certainly plays a difficult and important role in this case, but this role is fundamentally different when the process is characterized by joint and mutually beneficial problem solving rather than conflict resolution.

## The Varresbecker Bach Participatory Process

The Environment Agency of the Wuppertal Government was the initiator of the Varresbecker Bach Participatory Model. Given the Agency's mandate to decide on the extent of the required sanitation of Varresbecker Bach, and given that the property owners would shoulder the costs of cleanup, the Agency wisely chose to initiate an *extra legem* process where the residents would be directly involved in this decision. The central feature of the model is the outside mediator, whose task is to promote a smooth and consensual process in deciding the fate of the Varresbecker Bach contamination.

As a procedure for cooperative problem solving, the Varresbecker Bach model has many interesting characteristics. Public participation was initiated by the government, in this case, the Environment Agency, and ultimate authority for the outcome lies fully in government hands. The overall process agenda is set by the government, at least as indicated in the steps of the process set out in the last chapter. Within this overall agenda and government authority, the public has been authorized to make recommendations in a cooperative way. At the initiative of the mediator, many different communication and participation forms have evolved throughout this process. These include information distribution and personal consultations by and with the IKU, public meetings, meetings of the Interest Group, and meetings of the Round Table Forum (see Table 1).

The most interesting procedural initiative taken by the Environment Agency was that of allowing the Varresbecker Bach residents to choose the experts that would carry out the testing and risk assessment. As discussed above, the designation of experts in processes of this sort has typically been

*Table 1.* Forms of disourse and prticipation in the Varresbecker Bach Case

| Form | Example | Main/*other* participants |
|---|---|---|
| Direct Communication | •Education of Environment Agency Staff<br>•Consultation hours for citizens<br>•Discussion and agreement with authorities and experts<br>•Support for the founding of the Interest Group | Local environmental agency<br><br>Citizens<br><br>*Politicians* |
| Public Meetings | •Information presented to stakeholders<br>•Selection of Experts to carry out testing program | Citizens<br><br>Local environmental agency<br>Experts<br>*Politicians* |
| Published Material | •Information pamphlets<br>•Documentation of the public meetings<br>•Letters to public authorities<br>•Press releases | Politicians<br><br>Citizens<br>*Media* |
| Negotiations | •Charter<br>•Round Table Forum | Citizens<br>Experts<br>Politicians<br>Local environmental agency |

the domain of the government administration. Relinquishing this authority is both unusual and important. The experts have a decisive role in the outcome of the case since the findings of the "official" experts go virtually unchallenged and are the main criterion on which decisions of this sort are based. Since the mode of operation is to choose the experts and "do as they say", clearly the choice of the experts is critically important, all the more so in evaluating policy paths in the face of controversial scientific evidence.

The significance given to the expert choice suggests a general recognition that there will be conflicting scientific evidence and advice depending on the expert chosen. An analogous situation might be the choice a seriously ill person makes for a doctor. If only one doctor will be consulted, and if the patient intends to follow the doctor's advise, then the choice of the doctor may be viewed as critically important, all the more so if the patient believes that different doctors will have different diagnoses and prescriptions. The authority awarded the doctor's opinion, and the restrictions on consulting multiple doctors, would suggest extreme prudence in this choice. Similarly, the Wuppertal residents and the other stakeholders viewed the choice of the experts in making and interpreting the tests of the contamination as a critically important step in the overall process.

## Procedural Fairness

Fairness with respect to the procedures and discourse refers to the opportunities that each and every person have to express his or her legitimate individual interests and to contribute to the development of the collective will. Criteria for fairness can be evaluated both on the level of the overall process as well as on the level of the specific forums in which discourse occur. Referring to the principles discussed in Chapter 3, an accepted fundament of a "fair process" is the notion of equality, or that each person has an equal footing in the participatory procedures. This means equal opportunity in determining the agenda, the rules for discourse, and in raising issues and contributing to discussions. At the most fundamental level, a fair process means that each stakeholder has the right to take part in the discourse, to initiate discussion, to argue by challenging and defending claims, and to influence the decision or collective consensus.

At the macro-process level, the Varresbecker Bach Participatory Model does not provide equality among all the stakeholders or assure the rights of a fair process. The overall agenda is set by the Environment Agency in the form of a series of questions to be resolved through the process, and the question of financial responsibility for the sanitation was not initially on this list. The rules for social discourse have also been, to some extent, predetermined by the Environment Agency, which chose the mediator before the other stakeholders were informed of the problem. Consistent with German political tradition, the Agency's intent was to involve the public but to retain control over the agenda and ultimate decision authority. Therefore, participation was initiated after the agenda and the mediator were in place. While the citizens of Varresbecker

Bach can (and did) influence the collective decision, the government authorities reserve the right to ignore the recommendations of the public.

The mediator (the IKU) was given the task of carrying out the Agency's agenda in a way that would promote consensual problem solving among all the stakeholders with as little conflict and social disruption as possible. The first step was to inform the public of the contamination and to communicate the available information on the health risks. This step was followed by a joint decision on further testing arrangements.

At the micro-process level, equal opportunity or fairness was a stated objective of the IKU in setting up a participatory process. One of the early forums for public involvement was the public meeting, of which there have now been many. In contrast to the overall process, participants at the public meetings have been encouraged to set the agenda and rules, and even to change the moderator if desired. The IKU has promoted a fair process by (1) inviting all the relevant stakeholders to the public meetings, (2) discussing and deciding on the agenda of the meeting with all stakeholders present, (3) allowing the participants to discuss and agree on the procedural rules, (4) designating speakers at meetings in such a way that disparities in articulation skills are compensated for, and (5) granting each participant an equal vote or say in making the decisions. Moreover, early on in the process the IKU briefed the participants from the Environment Agency in order to improve their skills in risk communication. They were advised to admit government mistakes and to be responsive to citizen concerns. The press is invited to the public meetings as a way of assuring communication with the general public.

Fair process also requires equal access to the public participatory forums. Here, too, it seems that the IKU has made every possible effort to engage all interested persons. Some 450 people are now invited to the public meetings, and from 50 to 100 usually attend. Participation is by invitation. The address lists are prepared by the IKU, which has made an effort to include all residents and property owners of the contaminated area. Local environmental groups were also invited to participate, but with no response. This is rather significant, since this case could set a local and even state-wide precedent. Bound to local interests, the broader social questions concerning the ultimate responsibility and the role of industry in sharing the burdens are not addressed. The non-participation of local and even state environmental organizations serves to reinforce the status quo.

Regarding access to the political discussions, the issue arises whether the Interest Group was considered by all or most of the residents to adequately represent them. The Interest Group arose in an *ad hoc* manner, and its role in the process appeared to be unchallenged until the late arrival of the second initiative of residents opposing cleanup. The latent interests of elderly persons apparently were not taken account of in the early stages of the process, which seems to indicate that the Interest Group may have been too readily accepted as the legitimate representative of the residents and property owners. More efforts might have been made by the IKU to identify all the interests from the very start. For instance, an election of members to represent the residents and property owners would have allowed more open consideration of all possible interests.

Another important procedural initiative involving public participation is the Varresbecker Forum. Of course, this consensus group does not give each and every person an opportunity to express his or her individual views. The Forum operates as a representative body, and it consists of representatives from all the stakeholder groups (the residents are represented by the Interest Group and, more recently, by the Citizens' Initiative Against Cleanup). While the Forum has no direct decision power, it would be difficult (but not impossible) for the City Government to make decisions contrary to its recommendations.

In the Forum's charter, the members expressed the precise rules and conditions for their discourse. The main goal expressed in the charter is "to promote consensus among the stakeholders on the testing of the properties, the interpretation of the test results, the actions necessary, and the monitoring of the sanitation process." It is noteworthy that there is no mention here of allocating the financial burden. Recalling that the issue of financial responsibility is a stated priority of the Interest Group, apparently this group dropped this issue by agreeing to adopt the government agenda in this consensus forum. It may also be significant that originally the Interest Group (with two voting members) was outnumbered by the City Government and Environment Agency (with altogether four voting members). This ratio has changed with the arrival of the Citizen' Initiative Against Cleanup.

## Procedural Competence

Procedural competence refers to the extent to which the participants of a process are provided with the information and tools needed to make the best possible decision, or the best possible understandings and agreements given what is reasonably knowable to the participants. In the overall process, the "best possible decision" would appear to be finding an agreeable tradeoff between the risks of the contamination and the costs of reducing these risks, an agreed upon distribution of the financial burden, as well as acceptable procedures for the sanitation process. Yet, political processes are incremental, and decisions are made all along, many of which are irreversible. If earlier decisions close off possibilities, the parties become tied into certain paths which may preclude an overall "optimal" decision (for illustrations, see Kunreuther and Linnerooth, 1984).

One such incremental decision is the choice of the experts who have the task of assessing the seriousness of the Varresbecker Bach contamination.

Only with the benefit of hindsight can one judge if the choice of the experts was a good one, and then only if the alternatives can be known. But the conditions for an *a priori* good decision, and thus of a competent process, are that the participants to the decision are provided with sufficient information and tools for the choices at hand. The competence of the Varresbecker Bach process is discussed below with respect to four types of discourse: explicative, theoretical, practical and therapeutic.

## *Explicative discourse*

A competent process for resolving disputes, or even for arriving at joint decisions when little conflict is apparent, should be based upon agreed definitions of the terms of the debate and procedures for defining these terms. In the Varresbecker Bach case, there has been little discussion about the technical terms defining the issue, for example, the risks of the contamination, the types of technologies available for the sanitation, and so forth. The reason for this relate closely to the traditional role of expertise in German policy procedures. Since the public does not question the authority of the experts – and the experts have been judged to be competent – then there is little need to inform themselves of the technical intricacies of the case.

The IKU has taken the role of translating the expert reports into terms understandable by the residents. In this regard, extensive information has been distributed to the public in the form of information sheets. As an example, one early information sheet (INFO 1b) that appeared before the (extensive) testing was carried out is titled, "How Dangerous in the Dump?" It contains the results of the soil sample analysis shown in Table 2. Since there are no standards for soil contamination in Germany, the sampling results are compared with standards in the Netherlands.

In this early information sheet, the IKU also provides a cautious interpretation of the results. While explaining that the contamination does not pose an acute health risk to the residents, several precautionary measures are advised, including the avoidance of inhaling or eating the dust from any excavated areas. Children should not play in the dirt, but they may play on grassy areas. It is not clear if the fruit and vegetables are contaminated, but caution suggests that until more testing is done, they should not be eaten. Other information sheets contain information on the legal liability, data sources, and the availability of professional consultation.

Overall, the IKU plays a constructive, independent role in providing timely and open access to the work of the experts, and in interpreting the expert evidence in terms easily understandable by the residents. The residents have not questioned the content or interpretation of this information provided by the IKU. This is not surprising in light of the traditional authority given to expertise and in view of the fact that there were no interests to misrepresent or

*Table 2.* Toxic Substance Concentrations in Varresbecker Bach Soil Samples

| Toxic Substances | Concentration in Soil Samples in mg/kg | | | | | Netherland Standards | | |
|---|---|---|---|---|---|---|---|---|
| | Surface | | Underground | | Soil Samples | | | |
| | Minimum Value | Maximum Value | Minimum Value | Maximum Value | Varresbecker Strabe | A | B | C |
| **Heavy Metals** | | | | | | | | |
| Arsenic | 3.8 | 11.0 | 16.0 | 80.0 | 22.3 | 20.0 | 30.0 | 50.0 |
| Lead | 33.0 | 92.0 | 61.0 | 9700.0 | 1862.0 | 50.0 | 150.0 | 600.0 |
| Cadmium | 0.2 | 0.5 | 0.1 | 3.7 | 0.68 | 1.0 | 5.0 | 20.0 |
| Chrome | 33.0 | 39.0 | 26.0 | 310.0 | 162.0 | 100.0 | 250.0 | 800.0 |
| Copper | 22.0 | 45.0 | 25.0 | 580.0 | 41.4 | 50.0 | 100.0 | 500.0 |
| Nickel | 30.0 | 46.0 | 23.0 | 105.0 | 32.0 | 50.0 | 100.0 | 500.0 |
| Mercury | 0.0 | 0.1 | 0.0 | 2.6 | 0.62 | 0.5 | 2.0 | 10.0 |
| Zinc | 86.0 | 180.0 | 130.0 | 1700.0 | 439.0 | 200.0 | 500.0 | 3000.0 |
| Cyanide | | | 0.6 | 1.2 | | 5.0 | 50.0 | 500.0 |
| **PCB:** | | | | | | | | |
| Total | 0.6 | 7.3 | <0.0 | 1297.0 | 9.0 | | | |
| Benzo(a)pyren | 0.0 | 0.7 | <0.0 | 87.0 | 0.41 | 0.1 | 1.0 | 10.0 |
| Total acc: to US EPA | 0.6 | 8.2 | <0.0 | 1593.0 | 12.0 | 1.0 | 20.0 | 200.0 |

exaggerate the sample results in any particular direction. The IKU appears to have built a relationship of trust with all the stakeholders, including the experts. This has been helped by the fact that the IKU members also have technical backgrounds. The IKU recognizes the importance of trust in its dealings with the residents. In its own words:

> The (Participatory) Model, which was proposed by the City of Wuppertal, is being developed by the IKU. It is based on an open information policy and the inclusion of the stakeholders in the decisions. The success of this process depends crucially on your trust in the IKU. Distributing information in violation of the Data Protection Regulations will not be acceptable. Nor are we the spokespersons of the City of Wuppertal. (p.2)

This trust in the IKU, the authority attributed to expertise, and the general lack of conflicting interests, underlies the type of explicative discourse that has developed. The affected parties have no perceived need to inform themselves on the technical details and possible conflicts in the expert community. This is, after all, the responsibility of the experts. Indeed, the residents took little advantage of the information sources that were provided by the IKU. And, correspondingly, the IKU did not find it necessary to provide in-depth explanations of terms, definitions, assumptions, and concepts.

## *Theoretical discourse*

The most interesting aspect of the Varresbecker Bach process at the time of this writing is the importance placed on the consensual choice of the experts. The IKU's strategy of letting the three contending expert groups present their concepts to the voting members at the public meeting appears to have been successful. The decision procedure was also an interesting form of discourse and choice. The participants at the meeting were asked to articulate a list of criteria that were important to the choice at hand. Then each participant was awarded equal weights in matching the criteria to the three groups present. This "voting" procedure allowed the participants to weight their preferences.

The two consulting expert companies were thus chosen in a consensual manner, although the government stakeholders reserved the right of veto. Had there been a great deal of conflict between the property owners and the government, then it is not clear how this model would have fared. There is little tradition in Germany for the different interests to hire their own expertise, and this model has no provision for financial support to the various stakeholders for this purpose. In fact, there is no financial support to the property owners and residents of any kind, and this has constrained the activities of the Interest Group.

The expectation at this time is that the results of the testing program will not be challenged. Since institutional arrangements and forums for challenging government expertise are not a part of German political culture, no provisions have been made in the Varresbecker case for a peer review process or for independent verification of the scientific results. There has been no discussion of bringing in other expert teams or of selecting a panel of experts. It is not clear how uncertainty in the results will be presented. There appears to be a tacit assumption that if the experts, themselves, are selected as trustworthy, then the results will also be trustworthy.

## *Practical discourse*

The IKU has made numerous efforts to eliminate barriers to participation. The stakeholders are informed early on of meetings and encouraged to form their own working groups. The press is regularly briefed on the status of the procedures. Still, there may be important interests that are under represented in this model. It is troubling that most of the women in the community are not present at the public meetings (although several are active in the Interest Group). Because the process will ultimately have to address difficult tradeoffs between protecting the health of the community and paying the cleanup costs, the non-participation of women – who often are more concerned about the health issues – may endanger the competency of this outcome. Additionally, conflicting views on this tradeoff within the households might create difficult domestic tensions. No attempts have been made to form women discussion or support groups or to encourage more participation from women. One way this could be accomplished is through random selection of voting participants.

Other potential stakeholders, who have shown little interest in participation, are the publics of Wuppertal and Nordrhein-Westfalia. They also have an interest in the outcome of the Varresbecker Bach case insofar as it sets a precedent for government involvement and responsibility for the many similar cases that will undoubtedly follow. In fact, the Varresbecker Bach case could serve as a platform for a broadly based public discussion on the role of industry and government in the current contamination crisis in Nordrhein-Westfalia. The IKU has informed environmental groups of the meetings, but invitations are targeted mainly to the affected community.

The public meetings are carefully structured by the IKU. The setting is a medium-sized room with a large table for some 20 participants. The seats at this table are reserved for the members of the two citizens' groups and other residents and property owners who attend. A second row of seats around the table are reserved for the government participants, the experts present, and guests. The moderator from the IKU sits at the head of the table. The meeting begins with the IKU stating the discussion items on the agenda for the meeting, and asking those present if there are additions or changes. The IKU keeps the discussion from wandering from the agenda by means of a chart visible to all with a pointer marking the particular agenda item being addressed.

On only one occasion were small groups formed. This was at an early meeting to set out the aims of the cleanup. Since the goals were in some cases conflicting, for example, to reduce contamination risks and to keep costs to a minimum, the tradeoffs involved difficult value judgments. Building on the goals articulated by the small groups, the IKU devised a small questionnaire to elicit the values of the citizens. This questionnaire asked those present at the next meeting (the engaged public) to rank the various goals. This revealed differences in values among those present, some valuing the health benefits, others more concerned with property values, and others wishing to keep costs to a minimum. The goal of minimizing social disruption emerged at a later time in the process.

### *Therapeutic discourse*

According to the IKU, its main objective is to prevent negative psychological and social consequences from the sanitation process to the residents of Varresbecker Bach. Yet, the IKU does not consider its role as therapeutic in a narrow sense. Feelings and personal reflection are not, in their words, part of the participatory model.

Still, there is a great deal of therapeutic value in the now existing discussion forums. The main forum for discussion, the public meeting, is too large to accommodate a lot of personal reflection, but it does provide the citizens a chance to voice their fears, concerns, and opinions. Since the basic issues require value tradeoffs, the meetings must accommodate feelings and concerns that cannot be expressed entirely by facts.

Nonetheless, it is important to keep in mind that traditionally issues of this kind are framed as technical issues. At Varresbecker Bach, the tacit agreement is to follow the advice of the experts about the extent and seriousness of the contamination. Therefore, there is little room in this first phase of the process to explicitly discuss differences in values separate from the "factual" issues of the case. This will arise more explicitly at a later stage when the technologies for cleanup, and implicitly the issue of "how clean is clean enough", are discussed.

The public meetings are not the only forum for therapeutic discourse. The IKU also provides personal counseling by telephone or appointment. The way in which the IKU brought in the dissenting concerns and values of the Citizens' Initiative Against Cleanup is also notable in this respect. This was accomplished through personal communication in the living rooms of the members. Certainly, this was a delicate issue since the value of "leaving me in peace" was not voiced at earlier stages of the process. To the credit of the IKU, these concerns were not ignored or discounted, but taken seriously and eventually integrated into the participatory model. A solution has now been negotiated where property owners can choose not to have their properties further tested nor eventually cleaned up.

The IKU has gained a position of trust among all the stakeholders, which the Environment Agency does not enjoy to the same extent. Questions by the affected citizens are almost always directed to the IKU, and the citizens appear to be comfortable in disclosing their views and information to the IKU.

## Further Applications of the Model

With the apparent success of the Varresbecker Bach Participatory Model, it is important to reflect on its generality. Is the model applicable to other more contentious issues? Is it applicable outside of the political culture of Germany?

With regard to the latter question, cultural transplants of procedural innovations are almost inevitably doomed to failure. Certainly, in the adversarial climate in the U.S., the public would not be satisfied with an approach that gave such deference to expert opinion, that restricted opportunities for challenging this expert opinion, and that gave ultimate authority fully to the government authorities. Still, there are interesting cross-cultural lessons, and the U.S. is moving in many ways towards more consensual and less adversarial procedures. The idea of framing issues to exploit joint interests and involving the public in a structured, mediated process is certainly an exciting possibility. In this regard, the Varresbecker Bach case can provide useful insights.

A more difficult issue is whether the Varresbecker Bach model can be useful for coping with more contentious issues. For insights on this question it may be useful to examine the application of a comparable model in Austria, a country which shares very similar political traditions with Germany. The Austrian model was developed and tested in the context of finding suitable and publicly acceptable sites for badly needed hazardous waste facilities. Specifically, the State of Lower Austria developed and applied this model in a process to locate two land disposal facilities. Whereas the final decision on the site selection is legally the domain of the federal government, it will usually accept the governor's choice. According to recent constitutional amendments, the local communities have no legal voice in this choice.

Within these constraints, the Governor of Lower Austria initiated a two-stage process. First, a team of experts selected by the government screened potential sites and recommended the two required sites, Blumau and Enzersdorf. This procedure was conducted entirely by experts with no public input or access to the information. After the sites were selected, a second permitting stage was initiated. In a clear departure from legally required procedures, environmental impact assessments (EIAs) to determine the environmental and social suitability of the preselected sites were carried out in close cooperation with the selected communities. The Governor committed himself to this process by assuring that the hazardous waste facilities would be built only if the expert assessments on site geology, hydrology, ecology, public health, etc., showed the site to be suitable. As with the pre-selection stage, this decision would ultimately be based on expert evidence, but in contrast to the pre-selection stage, the communities would be informed and involved.

For this purpose, a participatory model similar to that at Varresbecker Bach was developed. The concept of an EIA team consisting of the developer (a joint government and private company) and members of an elected Citizens

Advisory Council from each community[1], was put forth. The essence of the participatory model is that all decisions taken by the EIA team must be reached by consensus. To facilitate consensual decisions, each team selected an EIA coordinator. This coordinator is very much analogous to the role of the IKU in the Varresbecker Bach case. The coordinator prepares agendas, moderates team meetings and acts as liaison between the team and the experts undertaking the environmental impact assessments. The coordinator also documents all steps of the EIA, reports progress to the responsible authorities and publishes a quarterly newsletter. The coordinator is, therefore, responsible for running the EIA and ensuring that all measurements and tests are completed.

In a striking analogy with the Varresbecker Bach case, a key decision in Lower Austria was the selection of the experts to carry out the technical assessments for the EIA. Expert opinion commands comparable authority in Austria as in Germany, and this opinion is ultimately the sole criterion for qualifying or disqualifying the candidate sites. Only one expert or expert group was asked to carry out each assessment, and this expertise cannot be challenged in a legal court. For this reason, the choice of the experts is viewed by all involved to be a key aspect of the outcome. Interestingly, in both Blumau and Enzersdorf, it was possible to find experts who were trusted by both the community and the project developer, although this trust was stronger in Enzersdorf than in Blumau.

There are then important similarities between the Varresbecker Bach and Lower Austrian participatory models. Each strayed from strict legal requirements and gave the citizens a forum for participation. Both were initiated by the city and state governments, respectively, which retained ultimate decision authority. Both engaged an outside mediator to conduct the consensual process, and this mediator was obliged to make all relevant information public. An important and crucial decision in both cases was the selection of the experts to carry out the investigations. There are also important differences. Whereas in the Lower Austrian case the citizens voted on their representatives, in the Varresbecker Bach case an *ad hoc* interest group was initially formed and later challenged by a second citizens' initiative.

---

1 In Blumau, a 16 member council was elected (10 with voting authority and 6 advisors) by the "Peoples' Platform." Membership in the Peoples' Platform was open to anyone who was interested in the matter (around 1,000 persons) and 77 percent of the members voted. Eight of the ten voting members of the council were from neighboring communities to Blumau. In Enzersdorf, the Citizens Advisory Council was chosen by citizens of Enzersdorf and nearby Margarethen with around 40 percent voter participation. All 10 elected members of the council were members of the "Citizens' Initiative" (a citizen action group formed in response to the planned landfill), 7 from Enzersdorf and one each from the neighboring communities of Arbesthal, Gallbrunn and Stixneusiedl.

At Varresbecker Bach, there was more direct participation by the citizens, also in the selection of the experts.

Perhaps, the most significant difference from which we can glean some insights on the generality of the participatory model concerns the nature of the issues. The Varresbecker Bach issue was essentially consensual with no real competing interests. In contrast, neither Blumau nor Enzersdorf share the same interests with the government/industry proposers for a hazardous waste landfill in their respective communities. Moreover, there are significant differences between the two communities. The citizens of Blumau are strongly opposed to the facility, whereas many citizens in Enzersdorf are willing to consider the facility if the community is sufficiently compensated for the burden.

How, then, does this participatory model fare for the resolution of a contentious issue with different stakeholder interests? Predictably, the outlook for the Enzersdorf process is considerably more optimistic than for Blumau. Whereas negotiations are continuing in Enzersdorf, the consensual approach appears to have lost all legitimacy with the citizens of Blumau. Indeed, when the expert assessments were presented at a public meeting, the meeting was interrupted by a bomb threat. The Blumauians were unwilling to accept the expert results, even in view of the fact that they had selected the experts. The Blumau Council commented that the facts of the case could be interpreted in widely different ways.

As the Blumau case suggests, the participatory model designed for Varresbecker Bach may be inappropriate for resolving more contentious issues characterized by opposing interests. Speculating on the failure of the model at Blumau, it is natural to ask if the citizens are willing to accept a model which gives no real decision authority to those who bear or will bear the burdens. If expert opinion is a key element of the process and tantamount to deciding the outcome, then this model depends on the acceptance of expert authority. This requires a deference to expert opinion that is nonexistent in many western countries, and as the Blumau case indicates, is waning in Austria.

## Summary

Although we have not examined enough cases to generalize authoritatively about other applications of the Varresbecker Bach Participatory Model, the German and Austrian experiences suggest that the model is most appropriate for issues with strong mutual interests among the stakeholders. There are many such issues, including such topical problems as radon contamination of private homes, zoning industrial areas, and community development plans.

The Varresbecker Bach Participatory Model was a government initiative. A mediator, the IKU, was asked to establish a procedural model that would share information with the public and involve the public in a way that would resolve the contamination issue in a non-confrontational, consensual manner. To date, the model appears to be appropriate for this particular case, and the public and other stakeholders appear to be satisfied with both the procedures and the outcome.

In many respects, the model can be considered to be procedurally fair, which refers to the extent to which each participant feels comfortable about advancing comments and sharing and debating different views. At the public meetings, the participants are encouraged to set the agenda and rules, and even to change the moderator if desired. In addition, the IKU designates speakers at these meeting in such a way that disparities in articulation skills are compensated for, and the IKU has coached the government participants to support rather than hamper the public discourse. The Varresbecker Forum, which is a representative body of all the stakeholders, has set its own rules for discourse and consensual decision making.

One troubling aspect of the fairness of the procedures, however, is the extent to which all the affected parties participate. There is seemingly little interest on the part of the more broadly-based environmental groups, which seems surprising since this case could set important legal precedents. Moreover, the women of the community are under represented at the public meetings. An important question is whether the Interest Group, which was formed in an *ad hoc* way at the outset of the process, was truly representative of all the interests. The subsequent formation of a second interest group at a later stage indicates the contrary. This raises the question of how a more representative body could be formed, perhaps through an election process.

While the agenda and moderator of the public meeting is set by the participants, the overall agenda for the process was initially defined by the Environment Agency. This means that the government defined the charge, selected the model, and identified the problem. This is especially significant because the Government's agenda was, at the outset, notably different from the resident's in that it circumvented the issue of financial responsibility.

Following the official agenda, there is no real conflict in this case since the interests of the community and government coincide, that is, both are interested in finding a tradeoff between the public risks and the costs of cleanup that is acceptable to the residents and property owners. This lack of conflicting interests has been reinforced by the recent decision on the part of the state government to assume a large part of the financial costs of cleanup.

It is also possible to evaluate the Varresbecker Bach Participatory Model from the perspective of procedural competence, or the extent to which the participants of the process are provided with the information and tools needed to make the best possible decision. Given the absence of conflicting interests, at least to date, the residents have placed full trust in the government experts and in the IKU to provide information on and interpretation of the available data on contamination. The deference to expert judgment was also an important factor in arriving at a consensual decision on selecting the two expert teams that would provide further risk assessments. There is no apparent need to fund separate groups of experts or to provide for a review process.

The selection of the expert teams at a public meeting, at which all the main stakeholders were invited, set the stage for a cooperative process for jointly solving the difficult problem of cleaning up Varresbecker Bach. The integration of the late arrival of a second interest group also proved the flexibility of the model. That a consensual solution appears on the horizon can be attributed to this innovative procedure and the efforts of the IKU in establishing a fair and competent participatory process. This process has many lessons for dealing with similar types of issues both within and outside Germany.

## References

Jasanoff, Sheila, "American Exceptionalism and the Political Acknowledgment of Risk," *Daedulus*, 11 (1991), 61-81.

Kunreuther, Howard, and Joanne Linnerooth. *Risk Analysis and Decision Processes*: (Springer Verlag: Berlin, 1983).

Majone, Giandomenico, *Evidence, Argument, and Persuasion in the Policy Process*: (Yale University Press: New Haven, 1989).

*Chapter 12*

# *Regulatory Negotiation as a Form of Public Participation*

*Daniel Fiorino*[1]

## Introduction

The growth of administrative power in the first two-thirds of this century posed a dilemma for democratic societies: How do they reconcile democratic process with the powers invested in non-elected administrative officials? A resort to administrative power to deal with the complex issues of modern society was almost inevitable. Legislatures simply lacked the capacity to develop the necessary technical expertise, establish the needed administrative routines, and concentrate the required attention on the narrow sets of issues that twentieth-century governments had to address – whether it was financial markets, food and drug safety, transportation, energy or environmental quality. Administrative agencies were an answer to one set of questions – of how to organize government to deal with the demands of modern society – but they raised an entirely new set of questions related to democratic control and accountability.

Administrative theorists have gone through stages in their attempts to reconcile administrative power with democratic accountability. Here we will distinguish two such stages – the early "rationalist" view and the "pluralist" view that came later. Agencies first were seen as neutral "transmission belts" that took the direction of the legislature, applied their technical knowledge and impartial view of issues, and acted according to the legislature's wishes (Garland 1985). Administrative power was not a problem under this view, because administrative discretion was limited by the legal delegations of authority from the legislature and by the agencies' assumed focus on technical issues. To ensure that agencies did not violate private rights, the U.S. Congress passed the Administrative Procedure Act (APA) in 1946. In addition to setting procedural limits on agency discretion, the APA authorized

---

1 Views and opinions expressed here are those of the author and do not necessarily reflect those of the U.S. Environmental Protection Agency.

the federal courts to review agency actions to ensure that they were consistent with the law and supported by the evidence (Bryner 1987; Verkuil 1978).

This view of agencies as neutral, fact-based implementors of the legislative will – what we can call a "rationalist" model – began to break down with experience. The administrative world was more complex, more laden with value choices and political conflict, than the rationalist model implied. A more realistic conception of the administrative process emerged. This "pluralist," interest group model recognized that administrative agencies perform a political function – that of reconciling the diverse interests in society. In the context of American interest group liberalism, this meant that agencies must balance the competing demands of different groups, much like a legislature, although in the form of the administrative process. Technical expertise, decision rules, and organizational routines – all are defining features of agencies and their decision processes. Agencies also take part in group politics, however.

This chapter examines the process of regulatory negotiation as a way to improve public participation in environmental decisions. Adopted by agencies in the 1980's to complement the conventional rulemaking process, regulatory negotiation brings together the representatives of different interests in an effort to agree on the content and often the language of regulations. Of the two models of the administrative process discussed here – the rationalist and pluralist – regulatory negotiation reflects the pluralist model; an agency is engaged more in reconciling diverse interests in society than in applying its neutral expertise to implementing the law.

The chapter begins with a discussion of rulemaking in the U.S. and the role of regulatory negotiation in the rulemaking process. It then examines negotiation as it has been used in one agency – the U.S. Environmental Protection Agency (EPA). EPA develops standards for regulating air, water, and land pollution in the U.S., making it one of the most active regulatory agencies in the world. Among the U.S. agencies that have used negotiation to issue rules, EPA has been the most active (Rushefsky 1991). The article concludes with a discussion of the strengths and weaknesses in the process. The negotiation process illustrates many of the tensions between the pluralist and rationalist models of administrative policy making. From a theoretical perspective, the agencies that use regulatory negotiation are building on the pluralist model of the administrative process. In addition to building on that model, however, these agencies are trying to improve upon that model, by designing the negotiation process in a way that will compensate for some of the weaknesses of pluralism.

Although this article focuses on regulatory negotiation at the EPA, the process has been used at other U.S. agencies, including the Department of Transportation and the Occupational Safety and Health Administration (Rushefsky 1991). In addition, the Clinton Administration has endorsed regulatory negotiation in two important ways. First, in an executive order on regulatory review, President Clinton urged U.S. agencies to use negotiation in their rulemaking (Clinton 1993). Second, in a memorandum issued the same day, the President directed each agency to select at least one rule that it would develop through a negotiation process. Both steps constitute important endorsements of regulatory negotiation.

## Regulatory Negotiation and the Rulemaking Process

In a regulatory negotiation, administrative agencies bring together the representatives of the interests that are affected by proposed rule before the agency makes decisions on the content of the rule. The goal of a negotiation is enable the representatives of these various interests (including the administrative agency) to reach agreement on the substance and, if possible, the language of a proposed rule. Its use in environmental policy reflects a trend toward resolving disputes through consultation, mediation, and negotiation rather than litigation or other forms of institutional confrontation (Bingham 1986). Negotiation is based on techniques of alternative dispute settlement, but it is applied to a different context. Regulatory negotiation is prospective in its orientation. Its purpose is less to resolve specific disputes than to define general rules. In this sense, it is more like a legislative than judicial process as a way to make policy.

Regulatory negotiation is a means for bringing in outside (i.e., non-governmental) parties to participate directly in agency decision making. An agency explicitly shares decision authority with representatives of industry, environmental organizations, state and local governments, trade associations, and other groups whose interests are affected by a regulation. The process offers a public forum for these outside interests to participate under conditions of rough parity with government decision makers.

The stated goal of a regulatory negotiation is to bring the representatives of "affected interests" together at the negotiating table. In environmental rulemaking, these interests typically will include representatives of the industrial firms who must comply with the regulation (both individual firms and trade associations that represent the industry); state and local pollution control, health, or regulatory agencies; environmental groups (national and regional or local, depending on the issue); other federal agencies; and sometimes people who can provide important kinds of industrial or scientific expertise that will support the negotiations. At a practical level, the "affected" interests are likely to include any organization or person that can be expected to challenge the rule later in the federal courts. The power later to oppose the rule in court is typically confers the status of "affected" interest. In addition, any other participants that the negotiating committee thinks can make a contribution to the process may be invited to participate.

To understand regulatory negotiation, one must first know something about the rulemaking process in the U.S. – the process in which administrative agencies develop regulations (or "rules") that apply generally to all the parties in a category. Rulemaking is a means for administrative agencies to define enforceable, broadly-applicable requirements for other parties in society. The scope and limits of an agency's substantive authority are defined in its authorizing laws, such as (in environmental regulation) the Clean Air Act or Clean Water Act. The rulemaking process in the U.S. takes place within the context of the Administrative Procedure Act (APA), passed in 1946. The APA requires that agencies provide the opportunity for comment on a proposed rule, that they consider and respond to those comments in issuing the final rule, and that they give fair warning before the rule takes effect. The APA also establishes the authority of the federal courts to review agency rules when a party whose interests are affected by the rule bring a lawsuit. In addition to the APA itself, judicial interpretations issued over the last fifty years have further defined agencies' rulemaking authorities and responsibilities (Bryner 1987).

The opportunities for public participation in conventional rulemaking in the U.S. are usually limited. Agencies may or may not hold public hearings; even when they are held, hearings usually do not allow interested parties much of a chance to engage in full discussions with the agency and other parties or to influence the outcome (Fiorino 1990). The APA requires that agencies publish a proposed rule in the *Federal Register* for public comments, then to consider and respond to those comments before issuing a final rule. It would be inaccurate to think of administrative agencies only as neutral, technical experts who simply implement legislative intent (the rationalist model). To grasp the role of regulatory negotiation in rulemaking, agencies should be seen as expressions of American pluralist politics (Lowi 1969). Pluralism presents politics as the competition among groups in society for influence in the open and decentralized circumstances of American politics. Because the decisions of administrative agencies have economic consequences, some of which are quite substantial for organized interest groups, rulemaking has become an important arena in the competition for influence over public policy. For a time, the chief participants in rulemaking were regulated industries. By the 1970's, the emergence of "public interest" environmental and consumer groups, the passage of laws designed to increase public participation, and a "reformation" in administrative law changed the rulemaking process in the U.S. (Stewart 1975; Reich 1985). As a result, public interest groups have played a larger role in rulemaking.

Regulatory negotiation reflects a conception of rulemaking as a pluralist process (Fiorino 1988 and 1991). It is based on the assumption that the role of regulatory agencies is to reconcile the interests of diverse groups in society and reach a result that is at least minimally acceptable to all of them.

Regulatory negotiation also responds to several perceived weaknesses in the conventional rulemaking process in the U.S. (Harter 1982).

One weakness of rulemaking is that it is adversarial. The parties strive to maximize positions, take extreme views for the sake of argument, and compete with the agency and other parties to marshal the data that will best support their position. A second weakness in conventional rulemaking is that it promotes litigation. Courts in the U.S. have adopted rules that give broad "standing" to various interest groups that want to challenge agency decisions in the federal courts. These generous rules of standing, a tradition of judicial activism, and the economic issues at stake in most regulations guarantee that parties unhappy with an agency decision will challenge it in court. Much of the conventional rulemaking process can be seen as a preparation for legal appeals rather than as a constructive search for good policy. A third weakness of conventional rulemaking in the U.S. is the extensive reliance on factual proof to justify as well as challenge decisions. Laws and courts apply legal standards of proof to probabilistic conclusions. It is difficult to prove or disprove the risk associated with a chemical's use, for example, yet much of the energy of rulemaking is focused on making just such a case. Facts are weapons in the struggle among the parties, not a basis for agreement on policy.

Regulatory negotiation was designed with these limitations in mind. It reflects a general dissatisfaction with the conventional rulemaking process and a trend toward greater use of alternative dispute settlement to resolve public policy disputes. Negotiation builds upon the pluralistic model of policy making. The goal is to identify representatives of the many interests affected by a policy decision and to enable them to achieve a consensus on what that policy should be. If negotiation in this context is successful, it will turn adversarial relationships into cooperative ones, reduce the chances of litigation, and substitute agreement for the demand for proof. How has this process worked in practice?

## Regulatory Negotiation in Practice

EPA's process for negotiating regulations has been described in detail in several publications (Fiorino 1988 and 1991; Perritt 1986; Rushefsky 1991; Susskind and McMahon 1985). For present purposes, three features are worth noting. First, negotiations complement but do not replace conventional rulemaking. They take place within the notice-and-comment format prescribed by the U.S. Administrative Procedure Act (APA). The objective in a negotiation is to reach a consensus that the agency can use in issuing a proposed regulation. Parties retain their rights to judicial review of the final rule. EPA charters the negotiating committees under the Federal Advisory

Committees Act (FACA); this requires public notice, open meetings, and publication of summaries of the negotiating sessions.

Second, EPA is a party-at-interest in the negotiations. It commits to publishing the committee's consensus as a proposed rule, as long as that consensus is consistent with the agency's statutory authority. Like any other party, the agency may block agreement or withdraw without prejudice at any time. The process also separates EPA's role as party-at-interest from a procedural role as convenor, facilitator, and project manager. EPA retains outside contractors to convene the negotiations and recommend a facilitator, subject to the committee's approval. Thus the *process* role is separated from the role of *substantive* negotiator. An EPA management official represents the agency in the negotiations and must obtain policy clearance from other offices in EPA before committing on issues. Although hired by EPA, the facilitator works for the negotiating committee as a whole; a loss of confidence can result either in the selection of a new facilitator or the end of the negotiations.

Third, once the committee is constituted, it has substantial control over its mode of operation, its composition, its use of resources, and the terms and timing of its dissolution. Each committee adopts its own protocols. Additional parties join only when the committee approves. Decisions are made by consensus, usually defined as "the concurrence of all interests represented" on the committee. The protocols allow participants to discontinue negotiations at any time if they are not productive. Participation does not prohibit a party from seeking judicial review of the final rule later. The committees are free to establish work groups, assign issues to the work groups for study and recommendation, and allocate funds from a resource pool provided by EPA to support the negotiation. The funds from this resource pool are hardly enough to offset the differences among groups in resources, but they at least provide some capability for the committees to investigate issues of common interest and to further the negotiations.

As of mid-1993, EPA had completed eleven negotiations, and three others were in progress. Of the completed negotiations, the committees reached a full consensus in seven cases and were able to agree on most, but not all of the issues on two others. Two ended before the work was complete. In the first (Farmworker Protection Standards for Pesticides), the negotiations ended when one set of interests – representatives of the farmworker unions – announced its intention to withdraw. In the second (Recycling of Lead Acid Batteries), EPA ended the negotiations after concluding that the rule would not meet the risk/benefit balancing test of the Toxic Substances Control Act. Although negotiations do not necessarily end if one of the parties withdraws, the withdrawal of a major participant often will lead the committee to end the process. The withdrawal of the agency, as in the lead batteries case, brings the negotiations to a halt, unless the other parties conclude that there is a chance to establish a consensus that would persuade the agency to formally reconvene the negotiations.

Three other negotiations still were in progress in 1993. The subjects of these negotiations were limits on disinfecting by-products in drinking water (under the Safe Drinking Water Act), emission standards for architectural and industrial coatings (under the 1990 Clean Air Act Amendments), and changes in rules governing requirements under the hazardous waste manifest system (under the Resource Conservation and Recovery Act). EPA's negotiators expect to be able to reach a consensus with parties on the committees on all or most of the issues in these three negotiations.

A brief case study of one of these negotiations illustrates how the process has worked in practice (Fiorino 1988; Funk 1987). Under the Clean Air Act, EPA was required to issue standards that would limit emissions from wood stoves into the ambient air. EPA had to set technology-based New Source Performance Standards (NSPS) that manufacturers would have to meet in future production. The negotiations began in March 1986 and concluded with a consensus on the core issues four months later. A work group (a subgroup of the full committee) drafted a preamble (the document explaining and justifying the rule) and proposed regulatory language, which the full committee approved and EPA published in February 1987. The wood stoves negotiation was a success, in that the negotiating committee agreed on the content and language of the proposed rule.

The negotiation process allowed the parties to make several useful tradeoffs in designing the rule. The most important was in balancing the interests of the government and environmental group representatives in an early compliance date against the interests of the industry representatives in avoiding unreasonable compliance burdens. State, EPA, and environmental representatives wanted the new emission limits to take effect quickly, so that any reductions achieved could help in meeting the national ambient air quality standards for small particulates. The wood stove manufacturing representatives wanted to be assured of sufficient time to test, certify, and market new models before the compliance date. In the give and take of the negotiation, the committee was able to agree on several measures that would allow the manufacturers to prepare for compliance. These included a one-year delay in compliance for small companies, a two-stage phase-in of the standards, and a waiver from testing for new models of wood stoves when they are identical to designs that had been previously tested. In exchange, the manufacturers agreed to an early compliance date for the first phase of the standards.

In interviews that were conducted after the negotiations were completed, several parties offered comments on the "competence" of the negotiations (Fiorino 1988). They cited the availability of practical information and the opportunity to consider it in open debate with other parties as advantages of the negotiation process. For example, representatives of the states of Oregon and Colorado--both of which had already established regulatory programs for wood stoves – provided valuable information to the committee. EPA was able to share information that it had collected on state practices, industry processes, and costs that otherwise would not have been available before a rulemaking proposal was published. An industry representative thought that the negotiations and the relationships that had been developed between the parties allowed for a thorough and effective use of information. The sources of much of data that was needed was at the negotiating table, and the committee could call on them for help in clarifying and resolving issues.

In most respects, the wood stoves case is representative of the negotiations in which the parties achieved a full consensus. It is, of course, less representative of the negotiations that did not lead to a consensus. The wood stoves case illustrates the many advantages of the process – in bringing practical information to the table, allowing for useful tradeoffs on the issues, and promoting better working relationships among the parties. These advantages were cited by many of the parties in other negotiations as well.

## An Evaluation of Regulatory Negotiation

So far, we have considered two conceptions of rulemaking in the U.S., the principal characteristics of conventional rulemaking, the weaknesses in the conventional rulemaking process that have led agencies to introduce regulatory negotiation, and the main features of the negotiation process itself. We now turn to an evaluation of regulatory negotiation. The focus of this evaluation is on the extent to which regulatory negotiation allows participants to "reach a non-coercive, rationally-motivated, cooperative agreement on collective action." When people decide it is in their interest to cooperate, which they do by definition when they participate in a negotiation, the question is whether the institutional forms they use to structure the deliberations are adequate to the task. How does negotiation stand up under our criteria?

The criteria for this collection follow the standard approach to rational policy making in the public policy field. One assesses the process as it allows people to define the problem, generate decision options, discuss and assess those options, and choose a preferred solution. Above all, the process should be *fair* and *competent.* Fairness raises questions about the capacity of the affected parties to influence the agenda and process rules, to be represented in the discourse, to put forth and criticize validity claims, and others. A fair process allows affected parties to assert and protect their individual interests and to shape the development of the collective will. Competence refers to the capacity of a process to provide procedural tools for making the best possible decision. A competent process permits access to relevant knowledge and interpretations and allows participants to use the best possible procedures for resolving disputes about them. Next is a discussion of negotiation as it fares under the criteria, in the form of eight generalizations about the negotiation process.

1. *Access to the process is highly selective and based on the representation of organized interests.* This point is related to the issue of fairness. Regulatory negotiation is highly selective about who is granted access to the decision process. Although participation is theoretically open to all interested and affected parties, in practice it has been limited to representatives of organized interests. The issues usually are national in scope, so limiting participation to organized interest groups is a practical way to identify and convene a workable number of parties that are representative of those affected by the decision. In this sense, the process builds upon the pluralist traditions of the American political system. As with pluralism, however, the negotiation process excludes interests and people who are unorganized, usually at the national level. It is the status of organized interest or, in a few cases, recognition as a source of technical expertise, that gives people and groups access to the negotiation process.

Under the protocols that EPA and other agencies have adopted, each committee determines who participates in the process. At least in theory, members of the lay public could participate. In practice, however, the costs and time commitments, the Washington focus of the negotiations, and the focus on the organizational representation of the parties make such participation unlikely. Regulatory negotiation is not an answer to the call for "strong democracy" that Benjamin Barber (1984) and others have issued.

2. *Once the parties have access, opportunities for influence are substantial.* The agency has announced, in advance, that any reasonable outcome will be adopted as a proposed rule. This alone constitutes a basic change in the agency's approach to rulemaking. Conventional rulemaking brings outside parties in for discussion one or two at a time, or limits them to filing written comments in response to public notices in the *Federal Register,* where all U.S. agencies are required to publish regulatory documents. But the agency retains final control over the decision. In a regulatory negotiation, an agency agrees to share decision authority with the negotiating committee. Those who get a seat at the table are empowered in many ways, but those who lack such a seat are excluded from access and influence.

3. *The process is a leveler yet reinforcer of inequities.* The definition of a consensus as "an outcome at least minimally acceptable to all parties" puts all the participants on the same footing when they make arguments, influence debates, and block or promote agreement. Once they have standing as members of the committee, the representative of the minor trade association or small environmental group has the same power to block agreement on a key

issue as the representative of General Motors, DuPont, or EPA. The threat of withdrawing from the discussions, thereby potentially dooming the negotiations, provides a constant source of influence. Despite this, the inequities in resources, time, and influence that the parties bring to the negotiating table have direct consequences for the ability of the participants to grasp the complexity of a problem, to formulate their position clearly and logically, to support their views and interpretations with data and argument, and to convince or win over the others to their preferred solution. Nevertheless, the process does have a kind of leveling effect, once the negotiations are underway. The small or poorly-funded interest group has far more influence than in the conventional rulemaking process, because of its equal power to influence group decisions as part of the negotiating committee.

4. *The parties have the opportunity to spend great amounts of time on an issue, but in turn must commit large amounts of time.* This point relates to the competence of the negotiation process. It is fair to say that in a regulatory negotiation, participants spend more time together working on a single issue than they ever have or probably ever will again. Negotiating sessions typically run for two or more days, and from four to six sessions are held. There are opportunities for informal discussions outside of the formal meeting as well as work group meetings on special issues. Still, participants complain that there is too little time to engage substantively in all the issues, to gather and assess all the data they would like to have, and to iron out unresolved issues that may remain near the end of the process. At the same time, the pace of the negotiations, especially as the parties reach the last ten percent of the process, amount to a large drain on everyone's time. Although negotiation may not take more time overall than a conventional rulemaking, the time demands are more concentrated.

5. *The parties in the negotiation have a great deal in common.* Because they represent organized interests and are professionals in their fields, the members of a negotiating committee are far more likely to share common backgrounds, expertise, and expectations than those in a public hearing or other such forum. Consider who is at the table in a negotiation: EPA and state/local officials, association executives, industry managers, technical experts from agencies or interest groups, and so on. All will not be technical experts in their fields, but they probably share membership in what Heclo (1979) calls "issue networks." They share a great deal in negotiating skills and an understanding of how to get things done in the policy community. People do not participate in negotiations as citizens who are taking part in a democratic process, but as professionals doing their jobs (Fiorino 1990). Although this raises questions about the overall fairness of the process, it does contribute to its competence. For the people at the table, there is rough parity in political and communications skills, if not always in technical knowledge.

6. *The process is well-designed to allow parties to challenge each other's factual claims.* Participants bring knowledge about the objective world to the committees in various ways – through reference to the professional literature, advice from recognized experts, special analyses or reports, and other ways. All such knowledge – if it is likely to affect the outcome of the process – is subjected to critical scrutiny from the other parties, all of whom have an incentive to challenge and debate the validity of factual information. Only if it is in the interest of all the parties not to challenge a piece of factual information – because they all benefit from an agreement based on its validity – will such claims go untested. In effect, parties can agree on what the relevant facts are as they relate to an issue if it is in their common interest to do so. The burden of proof for factual claims can be adjusted by the parties with their mutual consent. The advantage is that resources are saved and conflict is reduced. The disadvantage is that factual issues critical to the soundness of the decision may be ignored or misunderstood.

7. *The goal of the process is explicit from the start.* The negotiation process is designed explicitly to produce a statement of the collective will of the committee. The product of a public hearing may be little more than a transcript, that of an advisory commission a lengthy report. In a regulatory negotiation, the outcome, and thus the expression of the collective will of the group, is clear. The committee is formed to produce a level of agreement among the parties sufficient to lead to a notice of proposed rulemaking that the agency will issue for public comment. Some committees have agreed on the language for a rule, others on the conceptual basis. But a negotiation is successful only if the result is a consensus leading to a proposed rule. The objective is clear from the start. Indeed, the knowledge that the process will lead at some point to a product – assuming the negotiations do not break down – is an incentive to participate. As to the goal and intended policy significance of the negotiations, there is no doubt, even if there is about the ultimate substance of the rule.

8. *The process may cause the parties to lose sight of norms and values in favor of the practical issues in the negotiation.* The goal in a regulatory negotiation is to reach an agreement on the issues. It is a pragmatic goal. The problem is that, like the pluralistic politics of which it is an expression, the negotiation process may allow the parties at the table to place their needs and interests over those of society at large. Critics charge that the negotiation process is indeed pluralistic politics at its worst – a case of administrative agencies handing over governmental authority to representatives of private groups who make policy in their own interests (Funk 1987). Issues related to social norms and values then get lost in a process that emphasizes agreement as the test of good policy.

There is undeniable validity to this argument. In the wood stove negotiation, for example, the negotiating committee focused on practical issues – the time periods for phasing in the new, more stringent standards, the differences in compliance dates for large and small manufacturers, and so on. Were important social values neglected because of this emphasis on practical issues and because of the limited perspective of the parties at the table?

Several aspects of the negotiation process make it unlikely that important social values will be neglected (Fiorino 1991). First, the authorizing legislation (in this case the Clean Air Act) sets out the normative framework for regulation. Like any other New Source Performance Standard, the wood stove standard had to represent the best technology for the industry. Second, the EPA was a party to the negotiation; EPA's agreement was necessary for the committee to be able to reach a consensus. Third, the proposed rule that the committee developed was itself subject to the review and comment requirements that are set out in the Administrative Procedure Act. Fourth, once the rule was issued in final form, it was subject to review by the federal courts, just like any other final rule. These guarantees limit the possibility that the narrow interests of the parties will displace the broader public interest.

## General Observations

As a mechanism for the lay public to participate in making environmental decisions, regulatory negotiation is deficient. It allows for participation by representatives of organized interests in national-level policy debates, not members of the lay public selected on a random or other basis. It reinforces rather than mitigates the effects of pluralist policy making. For this reason, negotiation may involve bias in favor of the existing distribution of power and resources in society. Interests that are unorganized at the national level will lack access to the negotiations. Social values represented by these interests may be ignored. Diffuse, unorganized interests will go unrepresented and thus unrecognized in the policy debate. This limits the fairness of the process.

Yet the process has much to commend itself as a participatory model, once these weaknesses in representativeness are recognized. For parties that are admitted to the negotiating table, the process ranks high on many of the criteria for fairness and competence. It is especially important to compare negotiation to the conventional rulemaking process in the U.S. There, parties as a whole rarely have the opportunity to discuss issues together, relations among them are largely adversarial and mostly a prelude to litigation, information is used as a weapon more than a basis for sensible policy making, and exchanges of views take place through a written comment process that permits very little by way of constructive discussion, debate, and problem solving. For decisions where the representational issues can be dealt with adequately, negotiation offers a useful complement to conventional rulemaking.

Regulatory negotiation should be used only under certain conditions. First, it should not be used to make decisions about the acceptability of risks or distribution of costs and benefits across groups or among generations. These decisions raise value choices that negotiation is inadequate for addressing, on fairness and competence grounds. On the fairness criterion, negotiation is so selective and access so limited that the process cannot claim to represent all the social interests that should be considered. The parties at the table can claim to represent interests, but not values. It is least suited to "administrative decisions that are especially bound up with social values, or that are likely to have important effects on future choices" (Reich 1985:1640). On the criterion of competence, the lack of external standards on the quality of data used in making a decision means that data may be missing or unreliable. What is acceptable to the parties at the table may not be in the best interests of society or best reflect the prevailing social values.

Second, regulatory negotiations should be designed to meet the accepted tests of procedural legitimacy. In the context of U.S. rulemaking, this means designing the negotiation process to make it compatible with existing legal processes. In EPA's case, for example, the process is designed to complement conventional rulemaking. All of the usual rules apply – the opportunity for public comment; the obligation to respond to comments before issuing a final rule; the need to base the decision on substantial evidence; the right of the parties to challenge the decision later in court (Wald 1985). The same tests for judicial review apply as in conventional rulemaking. Given the limits in representation in the negotiation process, these tests of legitimacy are essential. Only if the process is firmly embedded in existing administrative and legal processes is it legitimate.

In many respects, regulatory negotiation may be seen as an attempt to make U.S. rulemaking procedures more like the processes used in the other Western democracies. Studies of environmental regulatory decisions in those countries suggest that their decision processes are based more on consensus, reflect more deference to technical experts in relevant fields, provide less opportunity for intervention by the courts, and are in general less adversarial than those in the U.S. (Jasanoff 1986; Vogel 1986). Because of these differences, the use of regulatory negotiation as a way to expand participation may be most relevant to the U.S. setting.

Regulatory negotiation relies on a pluralist conception of the administrative process. Agencies act more like legislatures than the technically expert, administrative rationalist organizations embodied on classic theories of administration. As a consequence, the process reflects the strengths and the weaknesses of pluralism. The close participation of affected interest groups, the search for consensus as a test of acceptable policy, the ability to stipulate as to facts, the higher likelihood of compliance and lower chance of litigation – all are strengths in the process. And yet there are weaknesses as well – the bias toward existing distributions of power in society, the exclusion of unorganized interests from influence, a risk of substituting the specific interests of the participating groups for those of society as a whole, among others. The goal with respect to regulatory negotiation should be to develop further criteria as to when it should and should not be used, and to apply those criteria to practical policy issues.

## References


Barber, Benjamin R., *Strong Democracy: Participatory Politics for a New Age* (University of California Press: Berkeley, CA, 1964).

Bingham, Gail, *Resolving Environmental Disputes: A Decade of Experience* (Conservation Foundation: Washington, D.C.,1986).

Bryner, Gary C., *Bureaucratic Discretion: Law and Policy in Federal Regulatory Agencies* (Pergamon Press, New York, 1987).

Clinton, William J., *Executive Order: Regulatory Planning and Review* (Washington, issued September 30, 1993).

Fiorino, Daniel J., "Regulatory Negotiation as a Policy Process," *Public Administration Review* 48:764-772 (1988).

Fiorino, Daniel J., "Citizen Participation and Environmental Risk: A Survey of Institutional Mechanisms," *Science, Technology, and Human Values* 15:226-243 (1990).

Fiorino, Daniel J., "Dimensions of Negotiated Rulemaking: Practical Constraints and Theoretical Implications," in: Stuart S. Nagel and Miriam K. Mills (eds.), *Systematic Analysis in Dispute Resolution* (Quorum Books: New York 1991), pp. 127-139.

Funk, William., "When Smoke Gets in Your Eyes: Regulatory Negotiation and the Public Interest – EPA's Woodstoves Standards," *Environmental Law* 18:55-98 (1987).

Garland, Merrick B., "Deregulation and Judicial Review," *Harvard Law Review* 98:505-591 (1985).

Harter, Philip J., "Negotiating Regulations: A Cure for the Malaise?" *Georgetown Law Review* 71:1-118 (1982).

Heclo, Hugh, "Issue Networks and the Executive Establishment," in: Anthony King (ed.), *The New American Political Establishment* (American Enterprise Institute: Washington, D.C., 1979).

Jasanoff, Sheila, *Risk Management and Political Culture: A Comparative Study of Science in the Policy Context* (Sage: New York 1986).

Lowi, Theodore J., *The End of Liberalism: Ideology, Public Policy, and the Crisis of Public Authority* (W. W. Norton: New York 1969).

Perritt, Henry J., Jr., "Negotiated Rulemaking in Practice," *Journal of Policy Analysis and Management* 5:482-495 (1986).

Reich, Robert B., "Public Administration and Public Deliberation: An Interpretative Essay," *Yale Law Journal* 94:1617-1641 (1985).

Rushefsky, Mark E., "Reducing Risk Conflict by Regulatory Negotiation: A Preliminary Evaluation," in: Stuart S. Nagel and Miriam Mills (eds.), *Systematic Analysis in Dispute Resolution* (Greenwood: New York 1991), pp. 109-126.

Stewart, Richard B., "The Reformation of American Administrative Law," *Harvard Law Review* 88:1667-1813 (1975).

Susskind, Lawrence and Gerard McMahon, "The Theory and Practice of Negotiated Rulemaking," *Yale Journal on Regulation* 3:133-165 (1985).

Verkuil, Paul, "The Emerging Concept of Administrative Procedure," *Columbia Law Review* 78:258-329 (1978).

Vogel, David, *National Styles of Regulation: Environmental Policy in Great Britain and the United States (*Cornell University Press: Ithaca, New York, 1986).

Wald, Patricia, "Negotiation of Environmental Disputes: A New Role for the Courts?" *Columbia Journal of Environmental Law* 10:1-33 (1985).

*Chapter 13*

# *Regulatory Negotiation as Citizen Participation: A Critique*

*Susan G. Hadden*

## Introduction

Techniques for facilitating citizen participation vary in utility and effectiveness in different nations and cultures. When the essential processes of democracy differ, naturally the means for participating will also differ. Regulatory negotiation, a means for drafting regulations required under law through consensus building among parties that are otherwise likely to disagree, is and will probably remain unique to the United States, because it is directly related to important characteristics of democracy in that nation: implementation of legislative policies characterized by adversarial relationships with government and among affected interests, the ability to challenge regulatory decisions of government agencies in court, and – both a consequence and a cause of the first two – the existence of powerful, well-organized national-level interest groups that are expected to act on behalf of their members without direct consultation.

In the adversarial environment of U.S. policymaking, regulatory negotiation is unusual in relying strongly on consensus-building. By bringing citizen groups into the regulatory process very early, moreover, regulatory negotiation empowers them. Finally, regulatory negotiation is a mechanism that can be duplicated and extended to other levels of government and other kinds of processes. For these reasons, it merits consideration, even though it may be inappropriate or inapplicable in many other countries.

This paper is organized into three parts: a brief discussion of why regulatory negotiation can even be considered a participatory technique in light of the very different experience in Europe; an evaluation of "reg neg," as many participants and proponents call it, using the criteria of competence and fairness; and a more general assessment of the overall utility and extendibility of the technique.

In this discussion, I rely upon the definitions and descriptions of regulatory negotiation provided by Fiorino (1992), Harter (1982), and others. Briefly, the purpose of regulatory negotiation (hereafter "RN") is to generate a Notice of Proposed Rulemaking, review and approval of which will then follow the requirements of the Administrative Procedures Act. To write a proposed rule, the agency convenes a negotiating committee; once constituted, the committee has substantial control over its composition, use of resources, protocols of operation, and conditions for continuance or dissolution. Most committees meet several times for two or three days over a period of one- to two-thirds of a year. The participating agency agrees to publish the rule so long as it falls within statutory guidelines. The U.S. Environmental Protection Agency has been the most active proponent of this technique. However, it is still in its infancy: EPA has tried to "steer 60 disputes into negotiation, but only twelve succeeded" (Wald 1991). (Fiorino, in this volume, counts eleven RNs.) EPA RN committees usually consist of a few representatives of environmental groups, affected industries, and the agency.

## Regulatory Negotiation in Europe?

Superficially, the technique of regulatory negotiation appears similar to long-established planning and consultation procedures in Europe. In reality, the two processes differ in important ways. Because the European procedures typically limit participation to three corporate interests – labor, capital, and the state; because interests often lack legal sanctions to induce competing groups to compromise; and because the activities are usually related to a social agenda, they differ from the narrower interests and issues included in most RNs, which develop very specific policies rather than broad social directions.

In his path-breaking book, *Modern Capitalism*, written in 1965, Andrew Shonfield identified economic planning procedures that differ in detail but not in essence in many western European countries – procedures in which certain designated interests work closely with government to influence or actually make a range of decisions that would be called "regulatory" in the United States. For example, the British Economic Development Council, comprising three members each from employers, organized labor, and government, along with two independents, was seen as a "forum for establishment of a new consensus among the main economic interest groups" (Shonfield 1965:151). Although different in its composition – civil servants and independent members were nearly half the total membership, and smaller industry – specific commissions were given considerable power to do the essential planning, nevertheless the French model also emphasized representation of standard corporate interests.

This pattern of close consultation among industry, employees, and government, although considerably altered by rapid economic development, diversification, and technological change, remained strong in the early 1980's. The Austrian "social partnership," for example, continues to comprise only capital, labor, and the state. As this name suggests, the focus of policy interest continues to be social welfare. Even regulation of risks to human health and the environment posed by chemicals is characterized in Germany and France by laws that specify the interests to be consulted or continued reliance on multipartite expert committees composed of peak organizations of workers and of employers in affected industries as well as government staff (Brickman, Jasanoff, and Ilgen 1985). Groups unhappy with resulting policies cannot file lawsuits challenging them, as they can in the U.S.

In contrast, RN in the U.S. developed in response to many years of adversarial policymaking, first in the legislature and then again as agencies attempted to implement legislative mandates. Interests are less corporate and more disaggregated, but even relatively small interests often have a well-organized association capable of sending a representative to a salient negotiation process; for example, the following industry groups participate in environmental regulatory policymaking: the American Wood Preservers Institute, the Compressed Gas Association, and the National Association of Solvent Recyclers.

Furthermore, participation early in the U.S. regulatory process is typically more effective than later entry. Working with agencies before regulations are proposed allows interests to help define the agenda, even suggesting the very words to which others less involved will only be able to react. Citizen groups seldom have the interest, the resources, or the expertise to participate at such an early stage; RN offers them the opportunity to do so. Thus in the U.S., RN does constitute a new form of participation; ensuring that environmental or other citizen groups are included in pre-publication discussions of proposed rules marks an important step towards opening up the process. On the other hand, RN can be criticized as institutionalizing a corporatist approach to policymaking, especially in its reliance on existing interest organizations, in contrast to other more genuinely open mechanisms for participation widely available in the U.S. political environment.

## Evaluating Regulatory Negotiation

Participation techniques can be evaluated according to a variety of criteria. As Webler indicates in Chapter 3, in this book we are focusing on two criteria: fairness and competence. A technique encourages fair discourse if participants have an equal chance to affect decisions setting the agenda and rules of participation and selecting the moderator and can join fully in the substantive discussions of the group. A technique is competent if it provides participants with procedures for making the best possible decision; that is, if it provides access to knowledge and interpretations and offers good procedures for resolving disputes about knowledge and interpretations.

Because RN is so severely constrained by the formal administrative process of which it is a variant, some of these factors (such as the need for the agency to set the limited agenda) appear to militate against fairness. Nevertheless, once we accept the constraints, RN does appear to be fair and competent, especially in contrast with the normal rulemaking procedures it replaces. To illustrate this point, I will divide RN into three stages – preliminaries; negotiation; and afterwards – and consider each separately.

## *Preliminaries: identifying issues and parties*

*Issue identification* is the heart of the policy process; whoever defines the problem can often determine the solution that is chosen. Normally, we would expect affected groups to identify issues of interest to them, organizing around the issue and/or using existing participatory mechanisms to put forward their agenda. The quasi-official status of RN changes this: the agency hoping to encourage RN must identify the issue.

Philip Harter, who first articulated the idea of RN, identified several criteria for identifying issues suitable for resolution using this technique:

- A limited number (15-25) of affected parties;
- Representable interests (I take this to mean there are already-organized groups, although Harter does not say this);
- A subject ripe for discussion whose outcome is in doubt;
- A realistic deadline;
- Agreement that no one will have to give up something essential; and
- An interested and committed agency (Harter 1986).

Other knowledgeable observers agree. Doniger (1986), for example, believes that a very limited set of issues is appropriate for RN, in part because industry does not always have an incentive to participate. In addition, there are other limitations on issues suitable for resolution through RN. Early in its trials of the format, EPA wanted to conduct a RN on radioactive waste disposal site selection: The agency was dissuaded from selecting this issue when it recognized that identifying interested parties (and their likely positions) would be extremely difficult, as siting decisions bring considerations into play that cause groups to take unpredictable and *ad hoc* positions. In addition, the strongly emotional nature of decisions involving siting and radioactive waste could well have precluded effective negotiation.

Neither Harter nor the other commentators mention one *sine qua non* of selecting an issue for RN; namely, that the agency must need to write a rule, which means there must be enabling legislation. Inherent in the process, therefore, is the assumption that the agency identifies the issue because a regulation must be formulated, and then it decides that it is interested in a RN on that issue. This gives the agency considerable power with respect to the other parties, compromising fairness at this point. As RN becomes more widely known, it is possible that affected parties will suggest its use to agencies, rectifying this difficulty in part.

*Identifying parties.* Once the issue is selected, the agency must identify parties to be invited to participate on the negotiating committee. It is this point upon which RN appears least fair; as Fiorino says, "Access to the process is highly selective."

In general, the agency identifies the parties rather than developing a means for parties to identify themselves. However, agencies must publish a notice in the *Federal Register* naming the groups already identified as parties. Presumably groups who came forward and asked to participate could be added if their interests appear legitimate. Members of the negotiating committee can also request inclusion of other groups.

The very nature of RN constrains identification of both issues and participants; that is, formal rule-making can never be anything but a prerogative of the statutorily designated agency. As such, the agency must play a lead role in both the beginning and, as we shall see, the ending phases of RN. It is also important to remember that rule-making is almost always a process characterized by limited citizen participation. Few citizens can or would care to follow the thousands of rules proposed each year, determine which are salient enough to merit further action, and then invest the resources in taking that action. Thus people and organizations are always delegating power, whéther implicitly or explicitly, to interest groups to follow rule-making proceedings and act for them.

It may be instructive in this context to contrast RN with alternative dispute resolution, or ADR. In both cases, affected parties agree to meet under the eye of a moderator; some forms of ADR call for the moderator's decision to be binding on the parties, but some are based on the kind of consensus-building that characterizes RN. Because such disputes tend to be quite specific, it is imperative to identify all the affected parties, or those left out will undermine the agreement. In a recent instance in Texas, several environmental groups joined together and negotiated effluent parameters with managers of a proposed copper plant that would have disposed a lot of toxic substances into Galveston Bay. Unfortunately, the groups failed to include the immediate neighbors of the proposed plant in the negotiations; when the negotiation became public, the neighbors became very upset and caused the agreement to dissolve. In contrast, in the successful RNs conducted so far by EPA, the negotiated rule has stood up well in the review process and has, in effect, been ratified by the public.

In sum, RN is a process intimately tied to formal regulatory procedures not subject to revision. As such, it is not intended to be a "complete" participatory process. The fact that the agency identifies issues and, to a large extent, participants makes the process less than fully fair in the initial stages.

However, the degree of empowerment it offers to those groups that do participate is so significant that these limitations may be acceptable.

## *Negotiating the proposed rule*

Negotiating the text of the proposed rule is the *raison d'etre* of RN. Because each negotiating team may set its own procedures, we can generally say that RN *could* be a fair and competent procedure, but we cannot say whether it *is* without examining the specific procedures adopted by each group. Nevertheless, the quotations from participants cited below and other descriptions of the process suggest that RN does meet the criteria at this stage – in part because it is a closed process where those who do participate can speak freely, explore the basis for differences with others, interact as much as needed, obtain and assess expert information, and otherwise allow members to define procedures and agendas.

The ability of each negotiating team to set its own rules offers each participant an opportunity to affect all important aspects of procedure and discourse within the parameters of the issue selected. Teams choose their leaders, rules, times of meeting, and set other procedural features of the negotiating process. In addition, they may establish working groups, subcommittees, or any other procedures for considering the substantive work of the group. Terms, definitions, and concepts, which lie at the very heart of rule-making, are the focus of considerable explicative discussion. To quote a participant in a negotiation concerning penalties for manufacturers of truck engines not meeting standards: "Our first task was to learn to talk *to* each other instead of *past* each other, to understand each other's vocabulary and concerns. We finally invented a language all our own. The resulting regulation is probably unintelligible to anyone outside the community of affected parties" (Doniger 1986:14).

Information is a critical component in the kinds of negotiations typical of RN, which typically concern a proposed regulation with a technical or scientific basis. In ADR, many discussions have failed or have simply used citizens to ratify the decision agreed upon by the industry and the agency, because the citizens lack either access to the same information or the expertise to understand it (for example, Reisel 1985). RN, in contrast, appears to encourage sharing of sound information – that is, theoretical discourse. Because most of the people in the room are experts, they can immediately decide whether the information provided makes sense, even if it takes some additional time to validate it fully. This equality of participants, which derives in part from the fact that participants are representatives of large interest groups, is a critical component of RN's success. It highlights the relative ease of consensus-building, even among opponents, if the decision making group shares many values and concerns, a point considered below.

Honesty is a critical component of any successful negotiation and a characteristic of a competent and fair procedure. RN offers several incentives for participants to be honest (engage in therapeutic discourse) in their dealings with each other. First, the proceedings are confidential but the outcome is meaningful for all participants. Second, the recognition that the proposed rule will be subject to public scrutiny discourages false assertions that others could later check. These themes are all implicit in the statement of one industry participant in a regulatory negotiation: "Honesty played an important role. Although initially we did not trust each other, we realized from the beginning that we had to develop and maintain absolute credibility. We succeeded in part because we realized that there was nothing to be gained by misleading or arguing in a way that might be misinterpreted. We knew that ultimately others would scrutinize our work" (Ferguson 1986:12)

Finally, the continuing relationship among the participants over several months of negotiations discourages dishonesty that might compromise later ability to work together. Similarly, it is likely that participants will also have to work together on other issues – either in future RNs, in working with Congress to develop new legislation, or in other less formal consultations about relevant policies. We know from the literature on risk communication and related work that trust can only be ensured if built up over a long-term relationship. Because dishonesty in the initial iteration of the RN would compromise the actor's ability to participate in any of these future activities, even a minimal level of commitment to the process gives actors an incentive to be honest.

To take advantage of the incentive toward honesty provided by the possibility of future interactions with the same parties, it would perhaps be useful to establish an ongoing negotiating team or standing advisory body for regulatory agencies. Although creation of standing committees would enhance the tendency for RNs to be exclusive, committees could always be modified for particular negotiations. Nevertheless, long-standing members would have some advantage over the ad hoc members in their familiarity with both the process and, more importantly, with each other.

Despite its openness and the incentives for honesty, RN retains some of the adversary character of non-negotiated rulemaking procedures. Human beings will always use the tools of discourse to achieve their goals. The same industry representative said, "Disarming candor was used as a tactical measure. It [was] hard to argue . . .that there is no problem. As the environmentalists began to document the problem . . .we said, 'Fine, we agree. Now let's help solve the problem.' This approach got us quickly on the right track" (Ferguson 1986: 12).

## *Afterwards*

Following the RN, the proposed rule drafted by the team is published in the *Federal Register*, where it is subject to the usual notice and comment provisions of the Administrative Procedures Act.

Our comments on this phase can be brief. It is commonly understood by students of citizen participation that publication in the *Federal Register* is not an egregiously participatory activity: the document is not widely read, and submitting comments is worthwhile only for those clearly affected by the regulations. Nevertheless, commenting does offer a means for refining the results of RN (not the process itself), by allowing people who were not party to the negotiations to comment.

However, federal procedures require agencies to solicit comments, but not necessarily to abide by them. Agencies are given considerable latitude to determine the extent to which they will revise proposed rules to reflect the comments they have received. We must hypothesize that the tendency to ignore comments received at this point will be stronger for rules resulting from RN because the agency has such a large stake in the proposed rule. Not only is it a party to the RN, and therefore has invested considerable resources (probably beyond the norm) in rules developed through RN, but the agency knows that the cost of changing the final rule will be to compromise its own ability to initiate or participate in future regulatory negotiations. Thus the notice and comment period for proposed rules that result from RNs could be, to a large extent, form rather than substance. If indeed this is the case, then the exclusiveness of the negotiations is enhanced in phase 3, which ought instead to provide a counterbalance – an opening up of a process that takes its strength from its initial limited size.

## *Summary evaluation*

Despite certain limitations, especially in the initial stages, regulatory negotiation can be accorded the status of a useful process of citizen participation.

Its virtues are especially clear in comparison to rulemaking as conducted without RN: First, it is widely believed that representatives of affected industries begin meeting with regulatory agencies about the language of proposed rules before most other people know that drafting has begun. Participation in the initial drafting has especially strong effects on policy outcomes, because commentators on the proposed rule can only react to it, while early participation may help frame the issue itself. Second, it is evident that in the drafting as well as the notice and comment phases, regulated parties have both the incentive and, usually, the resources to participate, while individual citizens with a more modest stake in the outcome have neither.

Thus participants in non-RN rulemaking procedures are mostly organized groups – representatives of the regulated industry and, in many cases, staff members from public interest groups.

RN does not change these features of the regulatory process; indeed it cannot change them, because they are inevitable within the constraints of the Administrative Procedures Act. The contribution of RN is to open up that portion of the process that is at once most influential and most secret: the initial drafting of the proposed rule. By bringing representatives of public interest groups into this early stage of regulation, RN empowers those groups. This is especially true since RN appears, during the closed negotiation portion, to offer participants an unusual degree of access in defining the issues on the regulatory agenda, obtaining and assessing the technical information on which regulation will be based, and in working with others to define terms and procedures of participation. While an ideal participation process would not give a government agency power both to identify the issue and the participants, nor would it limit participation to representatives of large, organized groups already known to the agency, these flaws in RN are largely offset by the level of empowerment it offers to those who do participate. The ability to design binding policies is an important benefit for which it is doubtless worth paying some price in reduced procedural fairness.

The limitations as well as the strengths of RN derive from its status as a variation on a mandatory process that provides for only limited citizen participation. Nevertheless, it is possible to imagine some refinements and extensions of the process that could increase its fairness, and some extensions that might strain the process to the limit. These topics are considered in the following section.

## RN in a Larger Context

Regulatory negotiation, like all other forms of citizen participation, does not take place in a vacuum. It is embedded in a cultural and political context that affects both its form and its likelihood of success. For example, the adversarial nature of regulatory policy implementation in the U.S. makes RN both a novel form (stressing negotiation and consensus-building) and a powerful form (bringing public interest groups into the process at a very early and influential point) of citizen participation. At the same time, the adversarial nature places limits on how citizens' groups can participate: they must not be perceived by their members or peer groups to be "selling out" by working with those who are usually opponents (and are likely to be opponents again). Similarly, RN depends strongly on the existence of organized interest groups that agencies can mobilize to participate in the negotiation. Extending RN to other levels of government may be difficult if public interest (or even regulated industry) groups are not as well organized at that level. This section elaborates these kinds of considerations as a prelude to a final brief evaluation of RN as a participatory technique.

## *Disincentives to participate in RN*

Perhaps the most striking feature of the context within which RN takes place is its adversarial nature. As noted, this may place some constraints on the ability of public interest groups to participate in RN. Two important disincentives are:

*Concerns of members.* Clearly, no group can enter a negotiation that will require it to compromise on issues of strong concern to members. The organization would be unable to bind the members to upholding the compromise; as a result the group would be weakened both by member disaffection and by engendering distrust in the other bargaining partners. In general, the public interest groups that participate in RN (such as the Environmental Defense Fund or the Audobon Society) are not true membership organizations; that is, people indicate general support of the organization's goals by contributing money and becoming "members," but the members expect the staff to keep up with policy, identify areas of concern, develop appropriate policy positions, and so on. Thus most of these groups would be free to participate in RNs on a wide range of issues. However, they must be careful to avoid RNs concerning topical "hot buttons" such as incinerators or radioactive waste. This constraint would be of more concern to state and especially local-level groups, where citizens can keep a closer eye both on the group and on the policy issues, should RN be extended to these other levels of government.

*Funding.* Participating in a RN is very expensive in terms of time if not money. Although social resources are likely to be conserved in the long run by avoiding litigation and developing regulations that are well-balanced between environmental protection and economic feasibility, in the short run participants bear the costs. This is particularly difficult for citizen groups, which often have limited staffs who must work on a wide range of issues. Small industry associations face the same difficulty. The ability to put forward the resources in terms of staff time is another limitation on the kinds of groups that can participate in RN: they must be large enough to spare a person for this purpose, which usually means having several full-time paid staff members. At the state and local levels, there are many fewer groups with this level of staff resources.

Many people think that RN would be more appealing and useful if methods could be found for providing funds to participating citizen groups. Industry, an obvious source of funds if only because of the money they might save in obviating very expensive regulations, is not acceptable to citizen groups because, however independent they remain, the perception of outsiders will be that they have been "bought." A pool of funds, perhaps industry contributions along with other sources, administered by a neutral group or the government agency might be acceptable where direct contributions from industry are not. As with so many forms of participation, creative attention to funding is imperative.

## *Extensions of RN*

As a relatively successful participatory technique, RN could be extended in several ways: it could be more widely used at the federal level, where the Environmental Protection Agency is the most aggressive advocate; it could be extended to rule-making proceedings at the state and local levels; and it could, with modifications, be extended to the writing of legislation to complement its present role in implementing legislation.

*Extension to state and local government.* RN is presently used almost exclusively at the federal level. However, state and even local governments also develop regulations. At these lower levels of government, the number of interests might be more circumscribed, allowing for use of the technique in a larger number of instances. Conversely, the interests at these lower levels, especially local, are less likely to be well organized, and unorganized groups cannot send binding representatives to the negotiating table. EPA might conduct workshops for state agencies to learn about the benefits of RN, perhaps starting with state environmental agencies that could learn directly from EPA's experiences.

*Extension to statutes/ordinances.* Developing statutes and, especially, local ordinances is often similar to developing regulations, although the scope of the problem is less well-defined. If state and local governments adopted some form of RN, it might well be extended to writing statutes. This practice already exists informally at the federal level as well: one documented case concerns the re-authorization of FIFRA (the Federal Insecticide, Fungicide, and Rodenticide Act). Legislators could agree to sponsor and vote for the compromises reached by the negotiating teams.

## *RN as a component in a hybrid model*

The weaknesses that I have identified of RN as a participation process include: limited number and kinds of participants; inclusion only of organized interests, primacy of the initiating agency in identifying issues and participants; disincentives for citizen groups to participate because of high cost and sensitivity to constituent disaffection; and the need to spend time in each new negotiation meeting and feeling out participants rather than developing a long-term relationship. Many of these problems can be overcome or offset if RN is considered as just one part of a hybrid model of participation.

I have already suggested standing multi-interest advisory committees as a complementary technique that establishes the kind of long-term relationship among the negotiating parties that can best form a basis for trust within the negotiations. An independent funding source might help relieve the burden of participation.

More important, other techniques that mobilize citizens and encourage them to organize will expand the number and diversity of organized groups upon which RN relies. Techniques designed to educate citizens and offer them background on issues with a strong technical component will also complement RN nicely. Citizens thus educated can both provide input into the RN through representatives and respond knowledgeably to the rule proposed by the negotiating team, enhancing the effectiveness of the post-negotiation comment period.

No single technique for citizen participation can meet all the different needs of citizens in a complex policymaking environment. RN admirably fits into a particular niche – one created by the present system in which agency rulemaking attempts become the focus of adversarial proceedings. Although it encourages fair and competent discourse within the negotiating room and it offers groups a chance to affect real policy outcomes, RN was intended to take into account only one of many different points at which citizens might wish to affect public policy. If other mechanisms for participation are available that offer citizens equally sound opportunities for effective discourse, RN can play a useful and unique role in empowering organized groups.

## Conclusion

Regulatory negotiation is an especially powerful form of participation because it focuses on setting general rules; in contrast, alternative dispute resolution attempts to resolve a particular dispute. RN empowers its participants by ensuring their equal participation at a point early enough in the policymaking process to have real effect. RN is also important because it is a good example of what Kelman (1992) calls "cooperationist" modes of decision making (in European literature the term neo-corporatist is used). Such modes are not only parsimonious of social resources in the long run, he argues, but they also result in better public policies.

The primary problems with RN are of two kinds: its limited application and the apparent narrowness of the groups represented. At present, EPA is one of the very few agencies that uses RN, and that only in a few cases. The cost of conducting a regulatory negotiation will always limit its use to important rules; nevertheless, it should be more widely deployed than at present. There would appear to be few if any problems with extending RN to lower levels of government and other substantive arenas. The interest in doing so must be fostered.

A more important concern is the narrowness and homogeneity of the groups represented. Because RN can only succeed if a limited number of people are on the negotiating team, and because team members must be recognized as representing a clear policy position, agencies have an incentive to identify existing interest groups as potential participants. Does this undermine the utility of RN as a participation mechanism? To answer this question, we must take two different kinds of concerns into account.

First, participants in RN must be well-versed in the technicalities of the issue area. In many cases, illustrated by the environmental RNs that have taken place so far, participants must be or become technical experts. It is unlikely that many members of the public at large will either recognize the issue as one of concern or have the expertise to participate effectively. In general, citizens must rely on trusted representatives to ensure that their concerns are voiced on highly technical matters; some citizens acquire expertise in one area of particular salience, but many do not. Thus the implicit delegation of representation to already-organized, professional interest group staff members possessed of appropriate technical expertise is not necessarily a limitation on citizen participation. Instead, it is a necessary and appropriate compromise in an age when detailed technical decisions have important ramifications for a citizenry that does not fully understand them.

Although representation, whether implicit or explicit, will continue to be a necessary component of regulatory policymaking and is therefore a tolerable component of RN, the lack of heterogeneity among RN participants is of more long-term concern. The diversity of views included in RN is not the issue: public interest groups can offer quite diverse approaches to the same issue. Rather, it is the homogeneity of the participants' experiences and activities. All the participants in RN are inside-the-beltway regulars – generally white males of middle age with considerable experience within government itself or as lobbyists. With such similar backgrounds, RN participants can readily communicate – "speak the same language" – no matter how different their views on a specific subject, and are thus able to arrive at consensus or compromise. More diverse participation, which is inhibited by minority groups' lack of organization and resources and their relative disinterest in environmental issues thus far, might make RN more difficult at the same time that its outcomes might be more legitimate or reflect different conceptions of the public interest.

Thus we return, inevitably, to the political culture within which RN is conducted. The legitimacy of U.S. public policy, for good or ill, rests in large measure on people's belief that all diverse interests were properly represented. The importance of RN is that it empowers formerly weak interests; at the same time, it does not and cannot empower people who are either unorganized or unaware of an issue. Given that other participatory techniques are in place to organize and mobilize affected parties, however, RN, especially if extended to other levels of government, offers an important means for providing citizens with true influence in a critical part of the policy process.

## References

Brickman, Ronald; Shiela Jasanoff; and Thomas Ilgen, *Controlling Chemicals: The Politics of Regulation in Europe and the United States.* (Cornell University Press: Ithaca 1985).

Doniger, David, "Negotiated Rulemaking at EPA: The Examples of Wood Stove Emissions and Truck Engine Emmissions," in: Standing Comittee on Enviromental Law (ed.), *The Private Assumption of Previously Public Responsibilities: The Expanding Role of Private Institutions in Public Enviornmental Decisionmaking,* (American Bar Association: Washington, D.C., 1986).

Ferguson, Scott, "Industry-Environmentalist Negotiation: the FIFRA Experience," in: Standing Comittee on Enviromental Law (ed.), *The Private Assumption of Previously Public Responsibilities: The Expanding Role of Private Institutions in Public Environmental Decisionmaking,* (American Bar Association: Washington, D.C., 1986), pp. 11-13.

Fiorino, Daniel, "Regulatory Negotiation as Public Participation," (this volume).

Harter, Philip, "Negotiating Regulations: A Cure for the Malaise?" *Georgetown Law Review* 71 (October 1982), 1-118.

Harter, Philip, "Overview," in: Standing Comittee on Enviromental Law (ed.), *The Private Assumption of Previously Public Responsibilities: The Expanding Role of Private Institutions in Public Enviornmental Decisionmaking,* (American Bar Association: Washington, D.C., 1986), pp. 7-9.

Kelman, Steven, "Adversary and Cooperationist Instutions for Conflict Resolution in Public Policymaking," J*ournal of Policy Analysis and Management,* 11 (Spring1992), 178-206.

Shonfield, Andrew, *Modern Capitalism:The Changing Balance of Public and Private Power* (Oxford University Press: London 1965).

Wald, Matthew, "U.S. Agencies Use Negotiations to Pre-empt Lawsuits Over Rules," *New York Times*, (September 21, 199), pp. 1, 12.

*Chapter 14*

# ***Mediation***

*Mike Baughman*

## Introduction

Environmental conflict is frequently borne of differences in individual and group values. Other important contributors to conflict include contrasting beliefs and differences in the distribution of costs and benefits. Societal choices made in an attempt to fulfill human needs and wants can eventually produce unanticipated outcomes, exposures, and undesirable consequences. The severity of these effects are often highly disputed by the scientific community and interpreted over widely varying measures of significance by affected parties.

Shrouded in conflicts of interpersonal values and lack of scientific consensus, environmental disputes are regularly brought before the judicial system for resolution. Not surprising, the courts are often unable to resolve differences in values nor sort through conflicting scientific evidence. When this happens, important environmental issues are frequently resolved along strictly legal process lines. While decisions of the court are likely to be procedurally fair, they are prone to being woefully insensitive to personal values and preferences and technical considerations. As a consequence, the effects of societal choices may not be effectively addressed. Mediation presents an alternative to the judicial process. When properly utilized, it maintains procedural fairness while offering substantial improvements in competence.

## What Is Mediation?

The rising popularity of mediation as an environmental dispute resolution technique has resulted in a variety of definitions of the process being put forth. The University of Washington's Institute for Environmental Mediation describes mediation in the following way:

*Mediation is a voluntary process in which those involved in a dispute jointly explore and reconcile their differences. The mediator has no authority to impose a settlement. His or her strength lies in the ability to assist the parties in settling their own differences. The mediated dispute is settled when the parties themselves reach what they consider to be a workable solution (Cormick 1987).*

Pruitt and Kressel (1985) have defined mediation simply as "third-party assistance to two or more disputing parties who are trying to reach agreement." Susskind and Ozawa (1985) describe mediation as a voluntary process distinguished from simple negotiation by the inclusion of a nonpartisan facilitator who serves at the pleasure of the disputants. A similar description is offered by Stephenson and Pops (1989): "Mediation involves intervention of an acceptable, impartial, and neutral third party who has no authoritative decision-making power to assist contending parties in voluntarily reaching their own mutually acceptable settlement." The direction that a particular mediation effort takes is largely dependent upon the skills of the mediator to guide participants through each phase of the process.

Common throughout these and other descriptions of mediation one finds in the literature are the presence of a dispute; the central role of an intermediary; the voluntary nature of party participation; and the consensual form to which agreements emanating from the process adhere. That parties choose to participate and can opt-out of the negotiations at any time is an important motivator to achieving consensus. Further, the focus of mediation upon collective rather than individual interests is essential to the formulation of consensually acceptable agreements. In this manner, mediation is pragmatic rather than idealistic. It is important to note that very few groundrules are applicable to use of mediation. The process encourages all parties to be highly flexible and adaptive in designing and implementing negotiating protocols appropriate to the dispute at hand.

Finally, it must be noted that mediation is one of the oldest and most common forms of dispute resolution. One need not look beyond the resolution of family or playground disputes to perceive the role that mediation has played in resolving conflict historically. Since time immortal there have been warring factions and peacemakers. Today's disputants and mediators are a continuation of this long lineage of conflict resolution.

Mediation is a process of integrative negotiation. The mediator will work with all parties to identify workable trades in order to reach a common-ground and consensus. Participants in the mediation then must have some bargaining leverage going into the negotiation process. Leverage can take many forms including financial resources; political strength; opportunities for legal challenge; and control of natural resources, among others.

Perhaps most important is the notion that mediation is largely negotiation behind closed doors. Mediators work with individual disputants to determine their ability and willingness to trade things each values differently. Mediators suggest alternative resolutions based upon the collective insights gained. Group discussions may, therefore, largely avoid discussions of differences, but focus upon common areas of interest. As a consequence, mediation facilitates public discussions of mutual gain, while issues of conflict may be reserved for discussions between the mediator and individual disputants. The ability to publicly cast the progress of a mediation process as "win-win" is an often important consideration for stakeholder representatives and can serve as justification to continue.

*Figure 1.* Progression of Alternative – Dispute Resolution Techniques

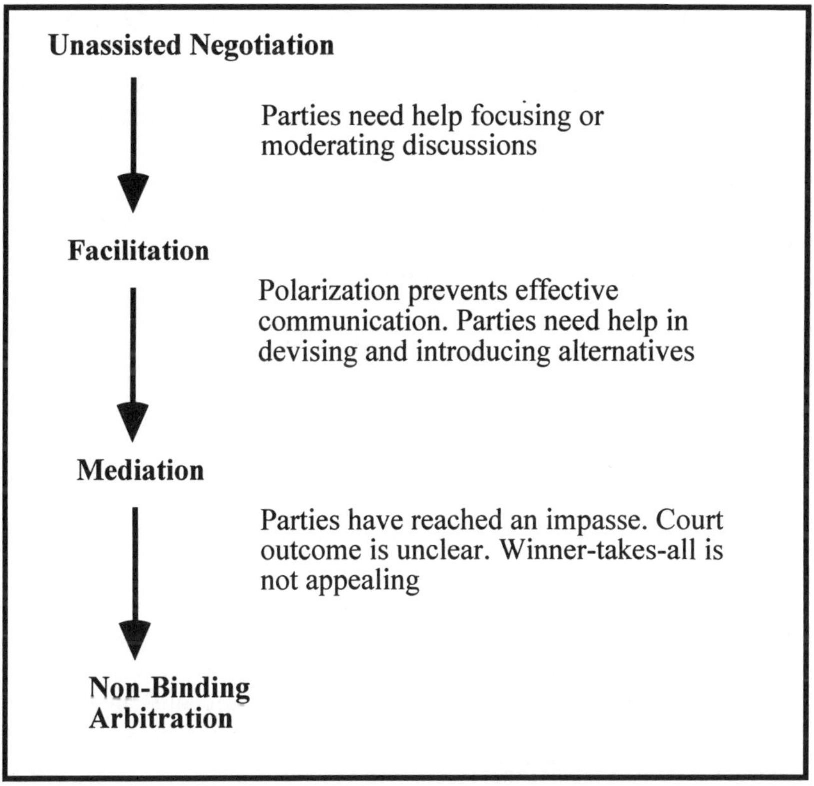


## When Is Mediation Used?

As is shown in Figure 1, formal mediation typically occurs only after the parties to a dispute have been engaged in an unassisted negotiation and perhaps have had the assistance of a facilitator. In many environmental dispute settings, the parties may initially begin a dialogue seeking to work out an amenable solution to the issue at hand. However, due to previous differences or other constraints such as political motivations, continued communication may be difficult. A facilitator may be brought in to assist the parties to maintain a dialogue. The role of the facilitator is largely limited to that of communicative referee, helping to focus or moderate discussions. When parties become polarized and unwilling to publicly pursue avenues of mutual gain, the derivation and introduction of resolution alternatives often benefits from a third-party mediator. Despite strongly held opposing views, resolution of the conflict through mediation may be desirable, as the disputants may be uncomfortable with the uncertain outcome associated with relegation of the issue to the court system. Once in the courts, the prospect for a "winner takes all" outcome becomes very real.

Mediation techniques are also well suited to complex disputes involving multiple parties and the potential for conflicting scientific information. In order to help ensure that negotiated agreements resulting from mediation enjoy broad acceptance over time, the mediator is encouraged to identify and seek to include in the negotiations all potential stakeholders to the dispute. In seeking to produce lasting and workable agreements, mediation tends to include rather than exclude interested parties.

## How Does Mediation Work?

Negotiations, whether unassisted or involving a mediator, can typically be thought of as moving through three distinct phases: pre-negotiation, integrative negotiation, and implementation. Each involve activities intended to promote the development of consensual agreements which are sustainable.

*Pre-negotiation:* At the initiation of any mediation effort there must be a dispute between two or more parties. The parties likely have attempted unassisted negotiations and may have engaged the help of a facilitator to keep such negotiations focused. For a variety of reasons, the disputants may have become polarized in their positions to the point that unassisted negotiations are no longer possible. Despite shared beliefs that a compromise might yet be possible, the parties have reached a point where they are no longer willing to carry on a dialogue which might lead to agreement.

The disputants may choose to engage the assistance of a mediator. In many cases, a third-party may suggest to the disputants that the services of a mediator may promote compromise and agreement. Alternatively, a mediator may offer unsolicited assistance to the disputants. A mediator attempting to sell the concept of mediation to disputing parties is not uncommon.

Agreement among the parties to use mediation embodies other decisions such as how a mediator will be selected; how the services of a mediator will be paid for; and how much each party will pay. Stephenson and Pops (1989) note that mutual stakeholder acceptance of an intermediary may help to encourage improved communication between disputants as well as broaden the spectrum of bargaining outcomes which each party is willing to consider seriously. Funding for mediation services may derive from governmental agencies, foundations, or non-governmental stakeholders.

Perhaps one of the most important strengths of mediation as an environmental dispute resolution technique is the extensive degree of foundational work which is done prior to commencement of negotiations. In a study of 164 cases of mediated environmental disputes, Bingham (1987) found that the conduct of a dispute assessment at the beginning of successfully mediated cases was a key factor in the success of the process. A dispute assessment seeks to define the disputants, the nature of the dispute, possible information needs of disputants, and the likelihood of the parties reaching accord. Dispute assessments conducted by the mediator may be essential to helping the parties decide whether to proceed with a voluntary dispute resolution process. Early efforts of the mediator to identify and include all affected and interested parties can also help ensure that resulting agreements will be sustainable. Assistance by the mediator to participants in understanding whether or not they can truly benefit from the negotiated process can help ensure that all eventual participants are participating in good faith. Good faith negotiations are prerequisite to consensual agreements.

One of the initial activities of the mediator should be to gain as full an understanding of the conflict as possible. Meetings with individual stakeholders to ascertain their view of the issues can also serve to help identify interests which may not be under-represented. It is important that missing groups be identified and strategies for their inclusion developed if subsequent agreements are to be sustainable. Representation then, becomes an important consideration to the mediation process. Certain interested and affected parties may be encouraged to join together to be represented. While helping to manage the numbers of participants, coalition building may be a strategic prerequisite for parties whose strength in negotiations may be political or legal leverage.

To justify participating in a negotiated agreement, stakeholders must conclude that alternatives would likely produce less satisfactory results. Administrative or judicial decisions may be less fair, less efficient, and less stable than solutions derived through a consensual process. The parties must also determine whether or not their interests, and those of other potential parties, can best be met through mutual agreement. Fisher and Ury (1981) have developed the term “BATNA” to explain the motivation behind a group having incentive to negotiate. BATNA stands for “best alternative to a negotiated agreement.” Susskind and Cruikshank (1987) note that no group should choose to participate in a negotiation if what it can obtain apart from the bargaining process (its BATNA) is better than what negotiation is likely to provide. Raiffa (1982) provides those interested with an excellent overview of procedures for estimating BATNA.

Prior to the initiation of negotiations, the mediator can assist parties in estimating their BATNA. Once negotiations begin, the ability of all parties to know that participants are participating in good faith can, in part, be based upon the knowledge that each group understands its BATNA.

Another early activity critical to the outcome of negotiations is the caucusing of stakeholder group members to select spokespeople or negotiating team leaders. Negotiations in good faith depend in part upon the ability of each party at the table to speak with authority on behalf of those they represent. That authority is best vested on individuals with the full consent and knowledge of stakeholder group constituents. It is unlikely that a participant in negotiations who does not have the support of his or her constituents will be able to obtain concurrence for agreements emanating from the mediation process.

Once matters of BATNA and representation have been resolved, the mediator should seek to develop and have the parties agree to protocols to govern the mediation process. Participants are free to structure the process in any manner consistent with their needs to achieve a consensual agreement.

The mediation agenda is also jointly developed by all participants. Topics to be discussed, the timing of deliberations, the extent to which outside experts will be called upon to provide information are all important elements of the agenda setting process. The mediator will seek to achieve consensus upon the agenda to be followed. It is important to note that mediation is intended to be highly responsive to the needs of the parties. At any time in the proceedings, the agenda can be revisited and revised to reflect the evolving progress of the group.

A final element of the pre-negotiation phase deserving consideration here is joint fact-finding. In joint fact-finding, parties agree to obtain a common body of information upon which discussions over implications can draw. Joint fact-finding seeks to avoid the problems of advocacy science, which often serve to frustrate rather than facilitate consensus. Rather than a "my expert versus your expert setting," joint fact-finding has all parties to the dispute agree to the selection and scope of inquiry of a common pool of expertise. To promote impartiality, the mediator may suggest a list of experts from which participants select. In addition, the mediator or some other third-party can manage the work of the experts and handle financial aspects of expert engagement.

As the pre-negotiation phase closes, a mediator should be in place; all appropriate parties to the negotiation should be involved and a spokesperson identified; each stakeholder group should have determined that they belong in and are committed to initiating the negotiation process; protocols and an agenda to govern the process should have been consensually developed; and joint fact-finding initiatives should have been undertaken as necessary. At this point, the parties are well prepared to begin the integrative negotiations phase.

*Integrative Negotiations:* In a mediated dispute, the mediator will typically work independently with stakeholders to develop "packaged trades" as proposed solutions. The basis of such trades will be found in solution options derived through group brainstorming processes. The mediator will seek to identify differences in the values parties place upon different characteristics of the issue in dispute. If both parties are willing to trade in areas they each assess differently, prospects for each party to emerge from the negotiations with a sense of having gained are very good.

The concept of trading to attain ones' objectives can be difficult to accept, especially for parties who typically find themselves in positions of opposition. Perceptions of selling out are easily formed and not easily overcome. To enable the integrative negotiation process to succeed, it is imperative that stakeholder representatives communicate their groups' BATNA with their constituency. Political constraints to negotiations can also impede initial face-to-face discussions between the parties. The mediator plays a key role in working between the parties to identify and communicate apparent opportunities for trading and agreement.

The behind the scenes work of the mediator can serve to bring the parties to the point where proposed solutions can be introduced for group discussion. The mediator's insights to each party's trading position serves to enable the formulation of proposed solutions, which, when presented for group discussion, can be fairly reflective of the consensual opinion of the stakeholders. Parties to the dispute can then debate the fine points of the proposed solutions and any alternatives thereto.

The process of negotiations can proceed for as long as the parties believe they are moving toward agreement. If at any point, a party to the process believes their interests can no longer be served through negotiations, the party can terminate participation. Obviously, if any group threatening to leave the negotiated process represents the power to block a subsequent agreement, all remaining parties may find it in their interest to discourage termination of negotiations.

Assuming that a group consensus can be achieved, written agreements are then produced for consideration and ratification by the participants. Drafting of agreements can be vested with the mediator or a subset of the disputants. Group review and debate of draft agreements is followed by preparation of a single preliminary draft agreement. Mechanisms to bind the parties to the agreements are then identified and established. Such mechanisms can range from simple interagency or inter-party memorandums of agreement to agreements adjudicated by the courts.

Once the participants to the mediation have achieved consensus upon a proposed resolution of the issue, each representative must then seek the concurrence of their constituency. To help ensure constituent backing, participants to the mediation process are encouraged to regularly report back to those they represent to ensure continued stakeholder support for the direction of the negotiations.

*Implementation:* Unlike a judicial decision, mediated agreements are not necessarily legal agreements which can be upheld in the courts. It is incumbent upon the parties to develop linkages between informal agreements and formal decision-making processes. For example mediated agreements can be submitted to the courts for adjudication. Alternatively, the agreements may take on the form of legally binding contracts between the parties. However, with governmental entities, such binding agreements which circumvent normal bureaucratic procedures can be problematic. The parties may agree to cooperatively seek statutory and/or regulatory authority for implementation of agreements.

Although conversion of mediated agreements into sustainable outcomes can be difficult, researchers have found that such agreements are characterized by very high rates of implementation. In a review of 100 site-specific environmental disputes, Bingham (1987) found that agreements produced through mediation were fully implemented in 80 percent of the cases. Another 13 percent of the cases were partially implemented. Bingham further found that disputes involving environmental policy produced agreements which were more difficult to implement. In these cases, agreements were fully implemented in only 41 percent of the cases studied (Bingham, 1987). According to Bingham, the most significant factor in determining the likelihood that an agreement resulting from mediation of an environmental dispute will be implemented appears to be whether those with authority to implement the decision participated directly in the negotiation process.

Once implemented, the ability for a mediated agreement to remain in effect over time may, in part, depend upon the establishment and functioning of an implementation monitoring group. A mediated agreement can stipulate that implementation be monitored to ensure that anticipated results occur. In the event that unanticipated or unacceptable outcomes result, the negotiations can be reopened.

## Is Mediation a Fair Process?

It can be said that a negotiated approach to environmental dispute resolution is a novel approach to public participation because stakeholders are afforded the opportunity to become decision-makers rather than simply decision-informers. The distinction is important. Decision-informers may largely be on the sidelines offering advice, having little control over the actual decision reached. The extent and capability to which stakeholders exercise their rights of participation in the mediation of an environmental dispute can vary considerably. Further, gaining entry to the dispute resolution process can be difficult if one does not possess bargaining power.

Elsewhere in this volume (Chapter 3), Webler suggests that evaluation of novel approaches to public participation fall along the lines of fairness and competence. Fairness is cast as including the extent to which participation in the process is open to all desiring entrance (attendance): the extent to which the process permits contributions to the discourse (initiation of speech acts); the extent to which the process encourages the free exchange and debate of ideas (participation in debate); and the extent to which individuals can influence group judgments (participation in group resolution). How does the mediation model measure up against these criteria?

*Attendance:* Mediation is not a process by which all interested persons are encouraged and invited to participate. Once a group is afforded a seat at the table, participants to the mediation proceedings are selected by constituents of affected groups or parties. While everyone is not afforded the opportunity to take part in the negotiations, representatives from every affected group are afforded opportunities to provide input to the process. Mediation encourages participation by proxy and necessarily seeks to limit the number of participants as a means to enhance the likelihood that consensus will be reached. To ensure that group interests are being fairly represented within the context of the mediation proceedings, participants to the mediation process are encouraged to regularly report back to and seek input from their constituents. Decisions regarding who to allow entrance to the negotiated proceedings is left wholly to the participants themselves.

*Initiation of Speech Acts:* Because a full airing of all concerns of each party is central to the establishment of mutually acceptable solutions, all forms of discussion are typically encouraged by the mediator – to the extent that they contribute to the agreed goals and objectives of the process. An important role of the mediator is to mitigate potential communicative inequities between participants. All participants are encouraged to freely suggest proposed agenda items, candidates for mediator, and process rules. Any issue or topic for which group consensual support can be obtained, is a candidate for inclusion in the mediation process.

*Participation in Debate:* Enduring agreements forged within a mediated framework have, as a prerequisite, the thorough debate of their various elements. However, the mediator will limit debate to the extent that it appears to serve the attainment of consensus. Debate is desirable as a means to consider pros and cons of various party positions. Debate is not desirable within the mediation framework if its purpose is to confuse rather than clarify.

Logistics of the discourse are determined jointly by all participants. For example, the schedule for the mediation process is typically flexible and driven by the needs of the group. Early in the process, a facilitator may be used to facilitate debate of pre-mediation issues such as selection of an intermediary.

*Participation in Group Resolution:* The survivability of negotiated agreements depends in large part upon a sense of ownership by participating parties. As with a market transaction, ownership will follow an expenditure of capital. In the case of mediation, such capital may be in the form of political, monetary, or other resources. The mediator is vested with the responsibility to ensure that all participants are afforded equal opportunity to participate in the resolution of issues. Any party feeling disenfranchised from the mediation process represents a possible dissenting opinion and a threat to consensus.

The mediator will typically work individually with parties to identify key issues and possible elements to a negotiated agreement. Group discussions most desirably focus upon the fine points of agreement or disagreement between parties. To enable this group focus, the intermediary will typically strive to assist participant groups to resolve internal issues.

## Does Mediation Result in Competent Decisions?

Minimizing known and unknown impacts should be a basic tenet of environmental conflict resolution. Many times, one solution to a problem can produce a new set of problems. The competence of decisions is dependent in part upon the level of knowledge applied to the formulation of dispute resolutions and to the extent that alternative interpretations of information are considered and understood. Mediation seeks to ensure competent decisions by facilitating the availability and access to information needed to understand issues and shape resolution alternatives.

*Access to Knowledge:* Joint fact-finding, often through the use of outside experts, is a central feature of mediation. Collectively, the participating parties agree to the scope and oversee the development of information needed to inform environmental dispute resolution efforts. In this way, information available to one party is simultaneously available to all other parties. Disagreements over data development methodologies are minimized, affording the parties the important opportunity to focus upon interpretation of results.

Regardless of whether the mediator is vested with the responsibility for selecting and managing outside experts or another third-party assumes this role, joint fact-finding promotes fairness through equal access to all information. Procedural equity with regard to access to knowledge is imperative if the information produced through the joint fact-finding process is to remain credible in the eyes of all participants.

*Differences of Interpretation:* Mediation offers a sufficiently flexible schedule. This provides participants with an opportunity to develop common points of reference. Participants are encouraged to take whatever steps are necessary to understand particular issues from the perspective of all interested parties. The ability for disputing parties to formulate mutually beneficial alternatives is recognizably dependent upon each party having some sense of what other parties value and why.

The mediator can be vested with the responsibility to confirm – through personal interactions – that all participants understand each others' terms, definitions, and concepts. Much of the mediator's early efforts will be to work individually with participants to understand individual or group values and to translate those values for other parties to the dispute.

Within the group, maintenance of interpretive consistency is ensured by drafting all agreements for review. Written agreements can be produced by a subset of the disputing parties, offering a good cross-section of participant values. Alternatively, the intermediary might be charged with the responsibility to prepare draft agreements. In both cases, the objective is to produce one version of the agreement which is as reflects individual participant values and concerns. The goal of drafting agreements is to produce language which is clear. (Not subject to diverse interpretation.) Outside of the negotiating group, participants are encouraged to take draft agreements back to their constituents. This helps to ensure interpretations are consistent among affected and interested parties. A formal written agreement, with enough elaboration to minimize interpretative inconsistencies, is key to the agreement finalization process.

## Can Mediation Work in Varying Geopolitical Settings?

The range of environmental issues to which alternative dispute resolution techniques have been applied is extensive. Bingham (1987) found nearly 50 differing contexts in which creative dispute resolution approaches had been applied. Cases identified in this study ranged from conflicts over land use to strategies for reducing quantities of hazardous wastes produced. Bingham found that mediation and other forms of dispute resolution were most frequently applied to situations involving site-specific conflicts rather than policy disputes. Application of mediation at the policy level may be difficult, because the parties to such deliberations may have few opportunities for "trading" in a negotiated setting. At the policy level of dispute, few tangible benefits or costs may have yet been accrued to any particular party.

If one considers environmental issues to be distributed along a continuum with policy decisions in the beginning and exposures and consequences resulting in the end, mediation might typically be found to be effective in dealing with the latter issues. It must be recognized that mediation of disputes about "downstream" environmental issues can and does have an influence upon policy. In this manner, mediation can be effective in shaping midcourse corrections to environmental policies which are shown to have unacceptable outcomes.

Mediation has a long history of international application. Use of mediation has been documented in China, Lebanon, and in Africa (Pruitt and Kressel 1985). Beyond traditional uses of mediation to resolve conflicts within nations, Dryzek and Hunter (1987) have documented several applications of mediation to the resolution of conflict among nations. A key factor in the success of mediation as a mechanism for resolving environmental conflicts which transcend regional boundaries is the decentralized nature of international governance. Dryzek and Hunter (1987) argue that the cultural and political diversity which characterizes international issue arenas actually facilitate effective environmental mediation.

## Summary

Use of mediation enables affected parties who may strongly disagree to work together to maintain control over impact inducing decisions. Effective mediation requires that participating parties negotiate in good faith, having a full awareness of their best alternative to a negotiated agreement. When carefully employed under the guidance of a skilled mediator, mediation affords the prospect of deriving win-win solutions to multi-party disputes. Mediation may be particularly useful in resolving conflicts over the distribution of costs and benefits of actions affecting the environment.

## References

Bingham, Gail, "Resolving Environmental Disputes: A Decade of Experience," in: Robert W. Lake (ed.), *Resolving Locational Conflict* (Center For Urban Policy Research, Rutgers – The State University of New Jersey, 1987), pp. 314-323.

Cormick, Gerald W., "The Myth, the Reality, and the Future of Environmental Mediation," in Robert W. Lake (ed.), *Resolving Locational Conflict* (Center For Urban Policy Research, Rutgers – The State University of New Jersey, 1987), pp. 302-313.

Dryzek, John S., and Susan Hunter, "Environmental Mediation for International Problems," *International Studies Quarterly* 31:87-102 (1987).

Fisher, R., and W. Ury, *Getting to YES: Negotiating Agreement Without Giving In* (Houghton Mifflin: Boston 1981).

Pruitt, Dean G., and Kenneth Kressel, "The Mediation of Social Conflict: An Introduction," *Journal of Social Issues* 41(2):1-10 (1985).

Stephenson, Max O., Jr., and Gerald M. Pops, "Conflict Resolution Methods and the Policy Process," *Public Administration Review* 3: 463-473 (September/October 1989).

Raiffa, Howard, *The Art and Science of Negotiation* (Cambridge:Belknap 1982).

Stulberg, "The Theory and Practice of Mediation: A Reply to Professor Susskind," *Vermont Law Review* (6): 85-88 (1981).

Susskind, Lawrence, and Connie Ozawa, "Mediating Public Disputes: Obstacles and Possibilities," *Journal of Social Issues* (41) 2: 145-159 (1985).

Susskind, Lawrence, and Jefferey Cruikshank, *Breaking The Impasse: Consensual Approaches to Resolving Public Disputes* (Basic Books: New York 1987).

*Chapter 15*

# *Environmental Mediation: Insights into the Microcosm and Outlooks for Political Implications*

*Werner Nothdurft*

## Introduction

In this paper I will argue that to view mediation as a means to resolve environmental conflicts is overly simplistic. It is simplistic for two reasons. First, it neglects the interactive features of mediation. Empirical studies have revealed that mediation consists of complex, contradictory, and even counterfactual elements of interaction. Insights into these features provide some sensitivity to the problems that are systematically associated with mediation. They also guard against overstrained expectations and provide a realistic basis for evaluating the usefulness and prospects of mediation. Second, it neglects the socio-cultural and political conditions necessary for mediation to function effectively. In the case of environmental conflicts the scientific understanding of what these conditions are are, as yet, unknown.

Therefore, I suggest we regard environmental mediation not as a useful instrument in environmental dispute resolution, but as a process which may change the political-communicative structure of our society and lead to new forms of communication about environmental problems.

## A Portentous Confusion: Mediation as a Procedure

The idea that we can resolve environmental conflicts by improving communication between disputing parties through intervention of a third party has received enthusiastic support. Often it comes in the shape of what is called a "procedural solution to dispute resolution." Such procedural methods have traditionally been used for conflict resolution with respect to labor relations, but they have gained renewed interest with the advent of alternative dispute resolution procedures (ADR). These borrow elements from routines of conflict-resolution in non-western cultural contexts and use the ideas and sentiments associated with these traditions. The protagonists label that appearance "mediation."

The verbal expression "mediation" functions as a catch-all to cover a great variety of cases of dispute resolution in environmental conflicts. There are many efforts to establish mediation as a regular part of the environmental decision making process. Land use disputes, management of land and water resources, siting of incinerator plants and energy facilities, disputes about air quality and toxics have been subjected to mediation. "[Mediation] has become a 'buzz word' used to describe a bewildering array of conflict-intervention processes and styles," writes the pioneer of environmental mediation (Cormick 1982: 15). Bingham (1986) counted about 160 cases of mediation of environmental conflicts. Furthermore the verbal expression "mediation training" is used to label educational activities undertaken by governmental agencies to enable practitioners to improve their skills in settling conflicts.

Faced with reports of success and observing the growing interest in institutionalization, one might wonder what is so special and magic about mediation that it can offer such overwhelming effects? Closer inquiry is disappointing and incomplete. Most evaluations are merely reports by the mediators or their home agencies. We cen expect little critical analysis from those with a vested interest in the process's success. A few studies include subjective evaluations of participants. Studies of mediation based on participatory observation are almost totally missing.

The absence of third-party evaluations is probably based on two reasons. One has to do with access. It is essential for mediation to take place behind closed doors. Mediation specialists agree that working in private is an important precondition for the participants. This excludes a non-participant observer. Even if access were to be granted notetaking and tape recording would more assurably be disallowed.

The second reason for the lack of empirical data resides in the way mediation experts write about their domain. Some authors present case-studies offering many idiosyncratic observations that do not allow generalizations (e.g. Nakamura et al. 1991; Susskind & Ozawa 1985). Others write about mediation as a procedure, but neglect the question of how it is performed and what the comparative merits are vis-a-vis other means of conflict resolution (e.g. Ozawa & Susskind 1985). This situation is rather typical for the state of knowledge in this field. The underlying reason for this unsatisfactory situation is based on the confusion between two analytical discourse levels on which mediation is described – the level of political or legal-political reasoning and that of interactional-structural rationalizations of interaction. At the legal-political level mediation is regarded as a technique, i.e. as a conception of dispute resolution which can be described in several phases and analyzed with respect to function and structure. A typical definition of mediation within this tradition is:

> The mediation process can be characterized as follows: it is (1) a non-compulsory procedure in which (2) an impartial, neutral party is invited or accepted by (3) parties to a dispute to help them (4) identify issues of mutual concern and (5) design solutions to these issues (6) which are acceptable to the parties. (Stulberg 1981).

At an interactional-structural level mediation can be described as a performance of face-to-face interaction. In other words it is a verbal-interactional event with characteristic interactive elements; relations between the elements; and the conditions, regularities, and constraints of the interactive performance.

Both levels – the legal-political discourse with its emphasis on procedure, and the interactional-structural discourse with its emphasis on complex interactional events – are crucial elements for a better understanding of mediation and its potential benefits and problems. This distinction is often neglected. Often authors talk about the interactive events in terms of the legal-political discourse. As a result, mediation is described as a process in which an autonomous machinery performs interactive results on the basis of specified procedural requirements. A quote can highlight this mechanistic view:

> *Mediation creates* a setting especially conductive to the scientist's information-sharing norms. ... *Mediation* also *has the virtue* of placing the decision makers in any dispute in a participating role. ... *Mediation can* easily *accommodate* joint fact-finding if the participants so desire. *Mediation can provide* a means of disengaging each party... (Ozawa & Susskind 1985: 32f; italics by author).

Such formulations invite confusion. The authors use the language of procedure, which belongs to the legal-political discourse, to describe interactive phenomena.

In this way, many authors change the main subject of their investigation. Instead of discussing features of procedures, they start to explore interactive phenomena, which they "magically" relate to the procedure of mediation. The subject of study is no longer 'procedure' in a legal sense, but interactive phenomena in a socio(psychological) or communicative sense. This change of perspective implies a semantic change. The discourse of reference is no longer the legal-political context (in which the concept of procedure has a reasonable position), but the discourse of interactive phenomena. It remains totally unclear if a conception of procedure has a legitimate position in this discourse and, if so, what the conditions for using the expression "procedure" in this discourse should be.

I do not doubt that disputing parties have cognitive concepts ("ideas") of legal procedures, but I do doubt that these concepts are identical with the components, phases and variables that underlie the discourse of legal-political reasoning. Empirical evidence suggests that these common-sense-concepts are used by people very differently depending upon the social interaction at stake.

## Misunderstandings: The Case of 'Fairness'

The misunderstanding is that the concept of procedure-as-such or procedure-as-pure-form (which makes sense when you discuss it in the legal discourse) has been taken as a concept for describing the interactive events in real cases of mediation or – even worse – mistaken as the interactional reality of mediation.

In his attempt to correct far too optimistic expectations of environmental mediation, Evan Cormick (1982) walks into this trap when he writes: "The term mediation [...] is nothing more or less than a device for facilitating the negotiation process." And, "Mediation does not lead to a resolution of the basic differences that separate the parties in conflict." And, when he writes about "application" and "misapplication of mediation."

At the same time, numerous phenomena which logically belong to the level of mediation as an interactional event are discussed at the level of legal-political reasoning. Fairness, distribution of control, process control, language, are all defined in terms of abstract procedural features instead of interactional phenomena and interactional dynamics. This can be illustrated with the concept of "fairness."

Fairness is said to be a feature of the procedure and is consequently defined in procedural terms. But this concept of fairness is not the way most people perceive it. Recent results of empirical studies on procedural justice show that judgments of fairness are connected with ideas such as "being treated respectfully," "concerns about the dignity and carefulness of procedure," and "how seriously litigants and the process are taken" (Lind et al. 1990:981).

All these phenomena associated with judgments of fairness have a lot to do with conceptions of personal and social identity, with social perception, and with personal dignity and respect. That means it is the subjective conception of oneself which becomes relevant when questions of fairness and justice are at stake. It is a matter of dignity, of intercourse, of interactional roles. Fairness is less dependent on a special procedure of conflict resolution condensed into a set of structural rules than it is on the way one is personally treated in a concrete conflict-resolution process. It is not "language or vote" per se, but the chance to contribute at relevant points in the negotiation process and to present oneself in a way consistent with one's own self-image. Having a vote or a right to speak at the wrong place in the course of interaction may induce a feeling of embarrassment rather than a feeling of fairness (Goffman 1967).

In essence, the interactional phenomenon "environmental mediation" cannot be subsumed under procedural components, phases, etc. They need to be analyzed separately. At the same time, however, interactive events are linked with the procedural frame defined as "mediation." The question, however, is just how this "interaction" can be specified.

## What Happens Inside? Characteristics of Mediation at the Interactional-Structural Level

From a political, legal, or educational point of view it might look as if the interactive events consist only of the peculiarities of the respective case or of the concrete individual-specific behavior of the participants. If so, they may be irrelevant for the scientific analysis, because descriptions of single cases do not allow general inferences. Neither the first nor the second assumption is true however. Neither are the interactive peculiarities irrelevant nor impossible to grasp. It is instead necessary and essential to describe and analyze mediation at the interactional-structural level. Only by so doing can we understand the communicative phenomena which systematically take place in mediation talks, and furnish insights into the interactive dynamics of mediation.

It is possible to describe mediation at the verbal-interactional level by means of conversational analysis. While Mike Baughman describes mediation at the procedural level, I would like to add a supplement at the verbal-interactional level. My point is that the working and the success of any dispute-resolution activity involving a third party essentially resides in verbal-interactional features of these activities. One needs to take these features into account in order to be able to evaluate the process of dispute resolution, to learn about problems systematically associated with this type of interaction, and to arrive at prudent expectations for what mediation can accomplish.

Investigating mediation at the verbal-interactional level furnishes the following insights. Procedural features certainly provide a macro-structural framework for interpreting interactive events, but interactive events transcend the mere realization of this macro-structural shape. Caused by the interplay of divergent interactive demands, the subjective interests of the participants, and the interactive dynamics of the negotiation process, many features emerge in the course of the interaction that are systematically associated with mediation-talk. These interactive events engrave the decisive shape on mediation. How these events manifest themselves depends on the activities of the participants – although not all events are necessarily intentional. They are the result of the ongoing interactional process and some may be accomplished even "behind the back" of the participants. These overt and hidden features of interactive events can be and are detected by conversational-analytical studies on mediation-talk in different institutional settings. That is, they are not derived from a specific model or conception of mediation, but are gained by empirical investigation of verbal (and non-verbal) transcripts of authentic mediation talk (cf. Nothdurft 1994a, Nothdurft 1994b).

Some of these features of mediation at the interactional-structural level are sketched in the following subsections. They seem especially suited to provide an understanding for the interactional reality of mediation-talk in a systematic way, to bring order to the wealth of empirical observations, and to represent the iridescent facets of mediation in an illustrative manner.

## *The verbal nature of mediation-talk*

Mediation is a communicative event of dispute-resolution. Whether a case of a neighborhood-dispute or an international conflict the dispute is not simply being acted out, it is processed in a meta-communicative way. The conflicting parties talk about how they want to talk about their dispute. One corollary of this is that the dispute is essentially present in symbolic way in mediation-talk. This holds true for environmental disputes, as well.

"At the heart of most environmental controversies lies a dispute over the likely future consequences of a proposed action... Making predictions about impacts that will occur well into the future necessarily involves a fair degree of scientific and technical analysis... Although we can predict a few natural systems quite accurately, our understanding of how most ecosystems operate is fairly limited. As a result, our predictions are necessarily approximations of reality" (Bacow & Wheeler 1984). Predictions, as well as statements of threat and danger, are verbal propositions without a definite reference to reality. They are constructed by means of language and acquire meaning only within a certain conceptual system.

The positions of disputing parties consist in the rhetorical efforts parties use to present their positions in and during the mediation-talk. In mediation-talk the dispute is present in the symbolic constructions of reality. This symbolic nature of the dispute is an essential condition for mediation to work. It is only through symbolic representation that the dispute becomes activated or present in mediation-talk. This quality provides the mediator with a full scope of potential interpretations and furnishes the development of a third perspective. Through this creative process, the participants develop suggestions for a mutually acceptable agreement. Such an agreement is also verbal in nature and is subject to the general regularities of symbolizations of reality, sharing their advantages and disadvantages.

In environmental conflicts even disputes over data or "facts" follow this pattern. Facts are only meaningful in the context of complicated technological procedures, interpretations, and decisions. That the dispute is only verbally activated does not mean everything is merely symbolic and has no impact on reality. The verbal rhetoric of an actor's presentations within the dispute aims at giving his or her own position full weight. This is achieved by making references to the desired goal of safety and by anchoring arguments in a discourse of mutually accepted or generally acceptable opinions and values. In addition, environmental disputes are characterized by claims to provide and challenge "hard facts." Nevertheless it is important to bear in mind that it is the rhetoric of verbal presentation by which a "fact" is established and its suggestivity accomplished.

As far as the question of fairness and competence is concerned, a corollary of the verbal nature of mediation is that the established competence of the participants (to a lesser degree, the impression of fairness) depends on the verbal competence of the parties involved in the discourse. The matter of competence in interactional situations is always a matter of how internal competence is articulated in an interactive setting. This depends on the verbal means and skills of the party who wants to prove or establish its claim of competence (cf. Nothdurft 1994).

## *The persuasive nature of mediation-talk*

The goal of mediation is to accomplish a consensus between the disputing parties, i.e., a solution of the dispute that all parties can agree upon. This requires a change of the attitude on the part of the disputants. The main task of the mediator is to facilitate this attitude change.

At the same time the disputing parties pursue their individual objectives. These they cannot force upon the group against the will of other participants or, most notably, the mediator. Their chance of influencing the whole group's decision increases when they try to cooperate with the mediator and to construct coalitions among the participants. Therefore the activities of all parties are subject to the condition that each party attempts to influence the other parties, particularly the mediator, to adopt its objectives or interests. These activities are grounded in persuasion. The activities of the participants do not follow the line of rational argumentation, but operate with common-sense-assumptions, opinions, and stereotypes. This is true for all parties. The need to use means of persuasion provides a constant threat to fairness. But the question of fairness cannot be answered in a global way. It depends on the respective activities and their relation to the roles and self-conceptions of the participants in the process (cf. Nothdurft, in press ).

## *The contradictory nature of mediation-talk*

The goal of mediation to resolve disputes by consensus requires many demands that the participants must fulfill, and many tasks they must perform. Some of these demands and tasks contradict each other. They refer to:

- contradictory demands of conflict-management: expansion versus confinement of the issue;
- contradictory demands of role-performance: neutrality versus expertise.

Agreement is always contingent on setting boundaries between the issue under consideration and the issues that are logically or intuitively associated with it. The stricter the boundaries, the easier it is to accomplish a task, but the more difficult it will be to find participants who will agree to work an such a limited mandate. At the same time, participants know that any resolution of the issue requires expertise and neutrality. Experts, however, are rarely neutral about the issue. And many who have no preferences with respect to the decision options lack the necessary expertise. These contradictory demands and tasks are partly derived from the procedural understanding of mediation, partly due to the institutional frame in which mediation is embedded. Participants who try to meet these demands and perform these tasks as best as they can create interactive conditions that are in opposition to other conflicting demands. Their contradictory nature cannot be resolved.

This contradictory nature of mediation-talk, however, does not necessarily result in a failure of mediation. Mediators are trained to deal with contradictions, to overcome them and to bring the mediation-talk to a successful end. A possible prerequisite for coping with contradictions is the non-transparency of the procedure, which makes it possible for the mediator to define the procedure step-by-step during the course of the interaction and to adjust the interactional frame in accordance with the respective state of discussion. Such a strategy and other similar procedures make it difficult to deal with the question of competence, because coping with one demand competently may lead to neglecting another. The management of the divergent demands in the course of mediation might well be understood as the best approximation of "competence."

## *The performative nature of mediation-talk*

The characteristic shape of the ongoing interactional events essentially resides in the rhetorical activities of the participants. In this way they mutually signal and ratify the definition of the ongoing event as part of the mediation process. Frequently a dissociation can be observed in mediation-talk between the characteristics of the ongoing interactional event and the definitional activities of the participants. The participants' common-sense knowledge of mediation and their attempts to formulate their expectations lead to certain interactive events and results which may transcend the conceptual frame of mediation and which may not be consistent with the original idea of what the mediation should accomplish. In such cases the nature of the interactive event as mediation is "performed" (cf. Schieffelin 1985). In other words, mediational activities are activated by a process of construction through performance. Features of mediation are unerringly performed in interactional sequences in order to engrave the definition of mediation into the interactional sequence.

This performance is rather effective. The success of many cases of mediation-talk relies in the fact that disputing parties act under the pretense of voluntary mediation – an inducement to participate which only works because the event itself is masked as mediation. It is exactly through the interplay of effective interactive events and performative actions that the intended goal of the mediation can be reached. In this sense the performative nature is essential (cf. similar observation on political discourse by Edelman 1964).

## *The fragile nature of mediation-talk*

For many reasons mediation is an extremely sensible enterprise. First, there is the dispute itself. Relying on interactive events, any mediation revives or even revitalizes the dispute and reminds the parties of their conflicts. once reminded of their opinon differences, participants may be enticed to escalate the conflict rather than resolve it. Second, reaching a final consensus is not always the most important objective of the disputing parties. They may regard other subjective goals as more important, e.g. to prevent their opponents from being better off, to receive more public support, or to push their own agenda on other issues. If these motivations dominate, consensus is unlikely.

Third, mediation is a complex process in which common-sense deliberations play a major role. In mediation-talk participants need to protect their social identity, emotional stability, anger, frustration, shame, face and other states of mind they value. These are aspects difficult to handle in normal conversations, let alone in situations of disputes and mediation. Considering these difficult circumstances, the most obvious question is: Why does mediation work at all? Certainly, it does not work all the time, but it works well in many cases. The reason may be that mediation (defined as an interactive event) takes place within a socio-cultural background which renders the mediator authority, his voice validity, his actions importance, and his values and orientations relevance.

## *The formal nature of mediation talk*

In many cases mediation takes place without changing the starting point of the dispute. The actions of the parties do not seem to be oriented toward the development of a consensus which can be integrated into their line of desired actions. In such cases, mediation is not functionally related to the end of mediation (i.e. consensus), but to the context of the dispute in which the mediation is embedded (cf. Nader & Todd 1978 for a transformational perspective on conflict). These situations require that mediation simply takes place, and do not demand the product of a consensus. Parties may participate to gain symbolic values (in the public eye) or strategic values (with respect to litigation). In such cases mediation is a mock fight.

## *The counterfactual nature of mediation-talk*

Mediation as a procedure is characterized by features such as the neutrality of the mediator, the voluntariness of participation, the consensual nature of the result, etc. These features provide mediation with its conceptual meaning, its relevance as a procedure of dispute resolution, and its rhetorical power in the public discourse. For parties who participate in mediation-talk these features have to be taken for granted. If they cannot assume that the mediator is neutral it makes little sense to participate. However, empirical evidence shows that: (1) the mediator more often than not is biased; (2) it is, at least, very difficult to decide whether he or she is neutral; and (3) that the predicate "neutral" is too global to be reasonably applied to the behavior of the mediator. Although empirical evidence supports the claim of bias, participants of mediation-talks need to stick to the fiction of neutrality if they want to legitimate their own decision to take part in the process. In this sense "neutrality" is a contrafactual assumption. The same holds for the other features of mediation mentioned above.

# Supporting Contexts: The Socio-Cultural Framework for Mediation-Talk

Mediation cannot be sufficiently explained by describing its intrinsic verbal-interactional nature. We need to improve our understanding of what a complicated and complex interactional endeavor mediation actually is. In spite of all the difficulties, the success of mediation relies on the complicated and complex nature of interactions that take place during mediation. It is exactly the flexibility, the openness, the vagueness of the interactional events that enables participants to deal successfully with conflicts. Success rests on a supporting and stable social context. The ethnographic literature on cultural studies of mediation shows that mediation only works when it is strongly related to a specific cultural context (cf. Nothdurft & Spranz-Fogasy 1986).

This is true for environmental mediation as it is true for any other application. Any application of mediation needs to consider phases of preparation, parallel ongoing processe, and implementation of results. It is essential that emphasis is given to preparing the process (cf. Nakamura et al. 1991), how it is "sold" (Kessel & Pruitt 1985), and that it takes place "in the shadow of the law" (Mnookin & Kornhauser 1979).

The relevance of the socio-cultural context for shaping the performance of environmental mediation can easily be shown using the features of mediation. A dispute is symbolically constructed and interactively accomplished based on social patterns and individual decisions of the participants. In all their utterances the participants rely on common-sense assumptions of how the world is (cf. Pollner 1987), on basic rules and idealizations of social situations (cf. Cicourel 1973), and on assumptions about "common language," "shared meaning" of verbal expressions, and identical comprehension of interactive events (cf. so-called meta-pragmatic studies, e.g. Kay 1987). It is exactly this cognitive-cultural web of commonsense assumptions and presuppositions of communication and understanding which guides the communicative behavior of participants in interactive situations.

Accomplishing a consensus or an agreement is not a sole product of individual persuasion. Making convincing arguments depends on culturally established systems of values, norms, "good reasons," preferences, and positively sanctioned inferences. Research on risk assessment shows that these systems vary considerably between different groups of participants, particularly lay persons and scientists (cf. Tversky & Kahnemann 1974).

Defining interaction as mediation rests partly on the definitional activities of the mediator, but also on the prestructured and shared concepts of mediation as a process and the issue in question among the participants. A good mediator has the ability to activate this reservoir of shared meanings, especially in the realm of political discourse and in the communication patterns of our society.

Issues discussed in mediation-talks may threaten the building of a consensus as they make people more aware of their differences in values and preferences. These differences, however, are never absolute. Participants share many beliefs and values within a socio-cultural context. Mediation-talk always takes place in the context of a stable background of defined relevances, interests, and constraints, which determine the superior goals of the participants. These refer to power structures, economic dependencies, the availability of alternatives, or "the shadow of the law."

## Back to the Legal-Political Level: Implications of Mediation for the Culture of Environmental Disputes

In the previous sections I placed most emphasis on two aspects of mediation-talk: first, on the difficult and complex interactional structure of mediation and second, on the solid socio-cultural context that has to be guaranteed for mediation to "work." Reports on and experiences of mediation show overwhelming evidence that these context-conditions vary among cases of mediation and that the resulting interaction is, in nearly all cases, a kind of an *ad hoc* coping with the respective conditions. In this sense mediation is what Bourdieu and others have called "praxis" (Bourdieu 1982). To the extent, however, that these contextual conditions can be systematically assessed and evaluated, mediation can evolve as a generalizable procedure or technique of dispute resolution. If this were to happen, praxis might become procedure. It was the pioneer of environmental mediation, Gerald Cormick (1982) who stressed this point in a remarkably reflective paper. He argued that it is not yet appropriate to call mediation a procedure and that, "a number of specific steps must be taken if mediation is to evolve from an interesting idea to a more broadly available option" (p. 38)[1]. Cormick elucidated the necessity of these "steps" or "required actions" by drawing analogies to the domain of labor-management disputes. He regarded the following goals laudable:

- To develop an encouraging attitude towards mediation in governmental officials.
- To support the public interest. (Public-policy support for negotiations between labor and management was embodied in executive initiatives and national legislation during the 1930's.)
- To establish a viable power base for all parties so that each will have "sufficient power to influence the action of each other party." (In the labor-management analogy, this power was conferred by a combination of legitimizing the strike and establishing legislative requirements to bargain.)
- To grant groups which oppose proposed projects a legitimate status as participants of the mediation process. (Labor originally achieved recognition through such economic actions as strikes, boycotts, or the physical takeover of production facilities.)
- To identify appropriate intervention opportunities in the decision-making process. (Typical *legislation* provided for mediation of labor-management

---

1 With respect to the German situation this statement is certainly true. Until today only few attempts have been made to implement mediation. In addition, most of these attempts were initiated by people who have used their personal connections to make mediation work.

conflicts identifies a point after which extensive negotiations should have occurred.)

- To support mediation services. (Here the labor-management analogy suggests the need for some *entity* independent and insulated from operating departments and agencies.)

The sociological and political impact of the mediation-talk becomes evident in Cormick's arguments. If these steps could be implemented, we certainly would have transformed our society into something like a "mediation-tailored society." The idea that we could change reality to accommodate a scientific result from the laboratory seems characteristic of modern scientific thinking of western cultures (cf. Bateson 1972). Bruno Latour (1983) tells the fascinating story of how Louis Pasteur managed to change French 19th century agriculture so that he could reproduce his laboratory results in the fields.

Cormick's suggestions for social change which would allow mediation "to evolve ... to a more broadly available option" are modeled with labor-management-disputes in mind. He uses this analogy to illustrate the suggested changes, but his analogy – unintentionally – sheds light on the essential differences between the domain of labor management dispute and the domain of environmental conflicts. To become an accepted procedure, mediation would require societal institutions similar to those currently in place for reconciling labor management disputes. Comparing the present state of environmental dispute resolution with the powerful, traditional, and well-reflected institutions for managing labor conflicts show without a doubt that we do not have comparable societal institutions for dealing with environmental disputes.

Reflecting on societal changes which might facilitate mediation, it may be misleading to use another historically evolved and culturally grounded model from a different societal domain. It seems more fruitful to focus on the role and function of environmental decisions in contemporary societies when we intend to conceptualize and design new models of environmental dispute resolution. It seems highly doubtful that we will be able to develop and establish social constructions and ideologies which will furnish the framework for a consistent and systematic use of mediation. Only such a framework would justify calling mediation a procedure in the legal-political sense. Not only is such a development unrealistic, it is also undesirable in my view. The non-procedural character of environmental mediation brings into question whether the technical terminology and the corresponding way of thinking (mediation as a tool, mediation as a useful instrument, mediation as a procedure) corresponds adequately with normatively reasonable conceptions of "the environment" or "nature."

Mediation defined as process rather than procedure would retain its distinctive charm, but also keep its problems: its basis in verbal nature, its ambivalence, its flexibility, and its paradoxical nature. The non-procedurality of mediation corresponds to features of the environment such as non-predictability, analog character, uncertainty, and complexity. This correspondence makes mediation a process more suitable for coping with environmental problems. It encourages the development of a new way of thinking about "the environment." It is not yet known what categories might be developed within such a framework, but I would like to present some admittedly speculative hints. It will be a cyclical way of thinking instead of a causal one and a way of thinking in categories which allow for ambivalence, paradox, flexibility, and non-deterministic considerations. Environmental mediation in this way will reinforce changes in our society already being developed. This new thinking can be described by the following aspects:

- It provokes new ways of thinking about environmental conflicts (cyclical instead of causal reasoning).
- It demands new criteria for judging environmental dispute resolution. These criteria are distinct from the usual criteria for traditional approaches such as "fair", "just", "competent."
- It questions the idea of "technique of conflict resolution" and forces the search for new conceptualizations and categories of thinking about ways to deal with conflicts.
- It stresses the relevance of the interactive process rather than the structural requirements for resolving environmental conflicts. It is significant how often the concept of 'discourse' is used in modern society when people talk about environmental problems.
- It changes the communicative patterns of our society, encourages new ways of dealing with conflict situations and changes societal standards of communication.

Within this framework of a new concept of environmental disputes, mediation as a process may play an important role in helping us cope with the new challenges of our environmental crisis.

## References

Bacow, L., and M. Wheeler, *Environmental Dispute Resolution.* (Plenum: New York 1984).

Bateson, Gregory, *Steps to an Ecology of Mind.* (Chandler Publ.: New York 1972).

Bingham, Gail, *Resolving Environmental Disputes: A Decade of Experience.* (The Conservation Foundation: Washington, D.C. 1986).

Bourdieu, Pierre, *Die feinen Unterschiede.* (Suhrkamp: Frankfurt/M. 1982).

Cicourel, Aaron, *Cognitive Sociology.* (Penguin: Harmondsworth 1973).

Cormick, Gerald, "The Myth, the Reality, and the Future of Environmental Mediation," *Environment,* 24 (1982), 14-17, 36-39.

Edelman, Murray, *The Symbolic Uses of Politics* (The University of Illinois Press: Urbana, 1964).

Goffman, Erving, *Interaction Ritual: Essays on Face-to-Face Behavior* (Anchor Books: Garden City 1967).

Kay, Paul, "Linguistic Competence and Folk Theories of Language: two English hedges," in: Dorothy Holland & Naomi Quinn (eds.), *Cultural Models in Language and Thought* (Cambridge University Press: Cambridge 1987), pp. 78-111.

Kessel, Kenneth, and Dean Pruitt, "Themes in the Mediation of Social Conflict," *Social Issues,* 41, (1985), 179-198.

Latour, Bruno, "Give me a Laboratory and I will Raise the World," in: Karin Knorr-Cetina & M. Mulkay (eds.), *Science Observed. Perspectives on the Social Study of Science* (Sage: London 1983), pp. 141-170.

Lind, Allen et al., "In the Eye of the Beholder: Tort Litigants' Evaluations of their Experiences in the Civil Justice System," *Law & Society Review,* 24, No. 4 (1990), 953-996.

Mnookin, Robert, and Lewis Kornhauser "Bargaining in the Shadow of the Law. The Case of Divorce," *Yale Law Review J.,* 88, (1979), 950-997.

Nader, Laura, and Harry Todd (eds.), *The Disputing Process. Law in Ten Societies* (New York 1978).

Nakamura, Robert, Thomas Church, and Phillip Cooper, "Environmental Dispute Resolution and Hazardous Waste Cleanups: A Cautionary Tale of Policy Implementation," *J. of Policy Analysis and Management,* 10 (1991), 204-221.

Nothdurft, Werner, *Konfliktstoff. Gesprächsanalyse der Konfliktbearbeitung in Schlichtungsgesprachen* (deGruyter: Berlin 1994a).

Nothdurft, Werner (ed.), *Streit schlichten* (deGruyter: Berlin 1994b).

Nothdurft, Werner, "How Procedural is Procedural Justice?" To appear in: Röhl,.K.F. (ed.), *Procedural Justice* (Proceedings of the 1993 Conference of the International Institute for the Sociology of Law, Onati, in press).

Nothdurft, Werner, *Muddling Through to Settlement: How Mediators Achieve Consensus* Paper Law & Society Conference, Manuscript (Amsterdam 1991).

Nothdurft Werner, "Kompetenz und Vertrauen in Beratungsgesprächen," in: Nothdurft, W., Reitemeier, U. & P. Schroder (eds.), *Beratungsgespräche* (Narr: Tübingen 1993).

Nothdurft, Werner, and Thomas Spranz-Fogasy, "Der kulturelle Kontext von Schlichten," *Z.f. Rechtssoziologie,* 86, No. 1 (1986). 31-52.

Ozawa, Connie, and Lawrence Susskind, "Mediating Science-Intensive Policy Disputes," *J. of Policy Analysis and Management,* 5 (1985), 23-39.

Pollner, Melvin, *Mundane Reason* (Cambridge University Press: Cambridge, MA 1987).

Schieffelin, Edward, "Performance and the Cultural Construction of Reality," *American Ethnologist,* 2 (1985), 707-724.

Stulberg, Gregory, "The Theory and Practice of Mediation: A Reply of Professor Susskind," *Vermont Law Review,* 6 (1981), 85-88.

Susskind, Lawrence, and Connie Ozawa, "Mediating Public Disputes: Obstacles and Possibilities," *J. of Social Issues,* 41 (1985), 45-159.

Tversky, Amos, and Daniel Kahnemann, "Judgement under Uncertainty: Heuristics and Biases," *Science,* 185 (1974), 1124-1131.

*Chapter 16*

# *Voluntary Siting of Noxious Facilities: The Role of Compensation*

*Howard Kunreuther*[1]

## Introduction

The siting of noxious facilities is an intrinsically difficult problem because the host region or community perceives that it absorbs most of the costs and risks while others accrue the benefits. There is normally little long-run employment created by facilities which store waste, and the tax revenues that may be forthcoming are generally not viewed as benefits if the residents perceive the facility as generating risks to themselves and future generations. Most people are *not* willing to trade off economic benefits for risk to themselves or others.

Until 1980 the traditional approach to siting was the *Decide Announce Defend (DAD)* method whereby the developer or contractor makes a series of technological choices as to the type of facility that should be built and where it should be located. The final decision is then announced to the concerned stakeholders and then defended. Under this process the proposed set of actions are not discussed with the concerned stakeholders in advance. The DAD approach has generally failed to locate facilities due to the adversarial nature of the process and the ability of those opposed to the facility to find legal ways of preventing it from being built (O'Hare et al. 1983).

### *Three examples*

Due to the failure of the DAD approach there have been legislative attempts to specify siting processes at the state, regional and federal levels in the United States. A *legislative congress* involves formal negotiations between the various parties with a direct interest in siting. The experience with such a

1 I wish to acknowledge very helpful discussions with Bruno Frey, Paul Kleindorfer, Roger Kasperson, Paul Slovic and the participants in the Humboldt Conference and the very useful suggestions by Tom Webler on an earlier version of the paper. Partial support for the paper is provided by the William and Flora Hewlett Foundation and NSF Grant #88-90299.

siting process in the United States has not been very positive to date as illustrated by the following three examples:

*Example 1:* The Massachusetts Hazardous Waste Siting Act of 1980 is designed to stimulate negotiation between a developer and host community, but to date there has not been a single successful siting experience and the law has recently been repealed.

*Example 2:* Those states that have used a legislated siting process to try and find homes for storing low level radioactive waste have met with fierce resistance from designated communities (English 1992). New York State has even suspended their search for a location and formally challenged the legislation. The Supreme Court ruled in June 1992 that the LLRW Policy Amendments Act of 1985, which required states to assume ownership and liability for wastes after January 1996, was unconstitutional (Easterling and Kunreuther 1994).

*Example 3:* The Nuclear Waste Policy Act of 1982 reflected a decision by Congress to find a site that would be technically feasible and politically acceptable. Many of the provisions of the Act have unraveled since 1982 and there has been strong resistance by the state of Nevada to having the wastes stored in Yucca Mountain, the site chosen by Congress.

## *Approach*

This paper develops prescriptive siting guidelines, shows the importance of compensation or benefit-sharing, and then proposes a voluntary approach which utilizes these prescriptive guidelines and compensation to facilitate siting. The concluding section proposes a set of questions for future research.

## *Rationale for procedure*

The voluntary approach places the burden on the developer or applicant to convince one or more communities that they should want to host the facility. For it to have a chance of succeeding there needs to be a feeling on the part of the potential candidate sites that the process is fair and the outcome is equitable. To achieve these objectives it is normally necessary for the developer to offer communities a compensation package. If more than one community volunteers to be a candidate then a lottery is held to select a tentative site. After this site is selected an auction is held whereby each of the interested communities (including the lottery "winner") indicates the maximum it would require to host a facility.

As shown below, the lottery reframes the problem by changing the reference point from "no site" to "tentative site with a compensation package." The foregone compensation is now viewed as a loss by those communities who were not selected by the lottery. If individuals in the community are loss averse as Kahneman and Tversky have suggested, then they are more likely to want to host the site if a lottery precedes an auction than if one engaged in a bidding process directly.

## Prescriptive Guidelines for the Siting Process[2]

Below are some common features of prescriptive guidelines that emerge from a review of the literature. These guidelines, coupled with compensation, form the backdrop for the proposed lottery-auction approach to facility siting.

*Feature 1: Differences in values and goals*

Interested parties often disagree about the desirability of a proposed facility because they have different goals and interests. In a study of technological conflicts and disputes about risky technologies, von Winterfeldt and Edwards (1984) point out that the debate between the relevant parties revolves primarily around value and moral issues where legitimate differences may exist. For example, should society develop technologies that we cannot fully control? Should society expose future generations to potential long-term costs and risks which are not fully understood?

*Prescription 1: Utilize a broad based participatory process*

Representatives of all affected groups should be involved in the siting process through broadly representative task forces or advisory committees. They should be given the resources needed for effective participation, and should have a chance to review the recommendations of facility proponents. Their values should be elicited so that similarities and differences between the different interested parties can be examined.

*Feature 2: Maintain the status quo*

There is a tendency to cling to the status quo even if there may be attractive alternatives (Samuelson and Zeckhauser 1988). In the case of the proposed siting of a new facility there will normally be strong objections by one or more groups unless all are convinced that there is a demonstrated need for the facility. In this connection, Rabe (1991) points out that one reason that Swan Hills, Alberta became interested in hosting a hazardous waste facility in their community is their local leaders emphasized the unsafe waste disposal practices that were being used in the community and the province.

*Prescription 2: Achieve agreement that status quo is unacceptable*

---

2 The prescriptive guidelines described below are part of a Facility Siting Credo that emerged from a National Workshop on Facility Siting which brought practitioners and researchers together to discuss ways of improving the siting process. For more details on the Facility Siting Credo see Kunreuther, Aarts, and Fitzgerald (1993).

The siting process must begin with an agreement that a facility is needed. The relevant parties need to understand the consequences of doing nothing – not just now, but in the future as well. Those who advocate building new facilities should be precise about the nature and scope of the problem that will result if the facility is *not* built.

*Feature 3: Concern with safety of facility*

Even if many residents of the area feel that a facility is needed it is still hard to induce change. The status quo still may prevail because the disutility induced by additional risks or losses is much larger than the utility induced by a comparable gain. Evidence provided by Zeiss (1991) on five siting cases in Canada including the Alberta experience, shows that there is a bias toward the status quo because residents are much more likely to be concerned with the potential negative impacts from hosting a waste facility than they are to be attracted by benefits of the same magnitude. This behavioral characteristic underscores the importance of undertaking preventive and mitigation measures for reducing the risks from any proposed facility before offering the community a benefits package to induce it to serve as a host site. By adding a set of design measures that isolate hazardous materials from the environment the proposed facility may appear safe.

Adding a second clay liner beneath a landfill to prevent wastes from leaching into the groundwater or incorporating a large buffer zone around the landfill to prevent hazardous or noxious materials from coming in contact with human populations are examples of mitigation measures that can be taken (Gregory et. al. 1991).

*Prescription 3: Guarantee safety standards are met*

No community should be asked to compromise its basic health or safety by building a facility. The proposed facility must meet all health, safety and environmental standards. Residents are unwilling to tradeoff off risk considerations for additional compensation.

*Feature 4: Lack of trust*

In recent years there has been a steady decline in the trust that the American public has placed in its scientists and institutions (Slovic 1993). Kasperson, Golding, and Tuler (1991) suggest that the NIMBY problem is due to a lack of confidence by the general public in scientists' ability to diagnose the relevant risks accurately.

In fact, there is a growing concern by laypersons that the risks associated with new technologies may not be well understood so that there is little reason to trust the experts. The different agendas of each of the interested parties in siting conflicts provides an additional reason why distrust exists.

The lack of trust is exacerbated by the discovery of health hazards from facilities that are declared relatively safe. For example, recent reports on the increased risks to individuals who resided near the nuclear testing facilities in Nevada and the public health hazards from the Hanford nuclear facility in

Washington State have exacerbated the problem. With respect to the siting of the high level nuclear waste repository in Nevada, there has been a loss of confidence by the public in the Department of Energy because of the absence of a trustworthy siting process (Slovic, Layman and Flynn 1991).

*Prescription 4: Work to develop trust*

One way to build trust between the public and technical experts is to relieve as many concerns as possible that the risk bearers and community have with respect to the safety of the facility.

## Role of Compensation or Benefit-Sharing

For residents to support a facility in their backyard, the benefits associated with having it need to be greater than maintaining the status quo. One way to satisfy this condition is to provide communities with compensation. However, compensation will be viewed as a bribe unless the affected groups feel that the facility satisfies rigorous, well enforced safety standards.

*Facilitating siting decisions.* The most common form of compensation is *direct monetary payments* which can take the form of guaranteed annual payments or tipping fees on the waste that is stored. In Charles City, Virginia, the developer of a landfill collects a tipping fee which it pays to the city, amounting to about $1 million in revenues annually. This has lowered property taxes and allowed for the rebuilding of the city's ailing school system.

*In-kind awards* takes the form of grants to communities or regions for improving health care facilities, housing, education or other services that enhance the citizens well being and reduces risks that they face. In Swans Hill, Alberta subsidized housing was provided for 35 housing units in conjunction with the siting of a hazardous waste facility (Rabe 1991). In Charles, City the operator collects the county's garbage free of charge. This is also the case in Grandview, Idaho where Wes-Con provides free garbage pickup to residents as part of a package associated with siting a waste-disposal facility.

*Contingency funds* are used to cover losses from an accident or other adverse effects of the facility. For example, trust funds could be established to cover the damages and health-related costs to victims of an accident. *Property value* guarantees protect residents of the host community and surrounding areas against any decline in the resale value of their home that would be due to the location of the facility. This type of compensation has been offered by Champion International Corp. as part of a program for siting an industrial landfill. The company monitors the changes in the sale prices of homes in the county over a ten year period and pays residents if there are any adverse changes in the property value due to the presence of a landfill. Kodak initiated a similar program to assure homeowners near its production facility that they would not lose equity.

*Benefit assurances* guarantees direct or indirect employment for community members, either during construction of the facility or during its operation phase. The hazardous waste treatment center in Swan Hills promised 55 new jobs and convinced town leaders that other new developments such as a new hospital would now be feasible. *Economic goodwill* refers to contributions to local organizations and expenditures for projects that are important to the community. The private corporation responsible for the Swans Hill facility planted 400 trees for town beautification, provided $65,000 to support local activities including golf course development and made charitable contributions such as sponsoring a hockey school and donating a bear rug to the town council chambers (Rabe 1991).

The importance of compensation or benefit-sharing as a part of the siting process can be summarized in terms of the following prescription:

*Prescription 5: Make the host community better off*

A package of benefits should be proposed by the applicant so that the host community feels that it is better off with the facility than without it.

## Elements of a Voluntary Siting Process

*Nature of Two-Stage Process.* This section develops a two-stage process for siting necessary noxious facilities. It relies on the five prescriptive guidelines developed above. The cornerstone of the approach is voluntary siting with a compensation package that makes the facility more attractive than the status quo to more than one community or region.

*Stage 1:* Determine a set of sites that meet acceptable criteria on technical grounds and that is an improvement over the status quo.

*Stage 2:* Develop a set of procedures which encourage one of the communities or regions that met the above technical criteria to consider hosting the siting. If more than one community volunteers then some type of lottery coupled with an auction would be utilized.

With respect to Stage 1 suppose a set of sites have been selected that are acceptable according to a set of technical, safety and physical criteria based on scientific expertise. In selecting these sites we have implicitly assumed that the first four prescriptive guidelines have been adhered to so that there is a broad-based participatory process, general agreement by the concerned interested parties that the status quo is unacceptable, guarantees that stringent safety standards will be imposed and an implicit trust in the process. No "best site'' in a technical sense will be specified. Rather the siting authority feels that the definition of "best" has to include an interest in hosting the facility.

Turning to Stage 2, the proposed procedure for voluntarily finding a site is a variation on one which has been successfully used in Alberta. Fourteen communities were initially interested in hosting the hazardous waste facility with nine communities subsequently eliminated on either environmental grounds or because of vocal public opposition.

The proposed procedure also is based on one that Illinois has utilized in trying to find a home for a LLRW facility and the process for finding a state, Indian reservation or territory willing to host a facility for the temporary storage of nuclear spent fuel.

In each of these situations planning grants are given to communities that express an interest in hosting the facility. These funds are for feasibility studies, public information efforts and other public outreach efforts. The acceptance of a grant does not imply a commitment to accept the facility. Rather the funds are designed to initiate a process so that the community or region has input into the process and can specify conditions, including compensation arrangements, that would make the site acceptable. In other words, the planning grants are designed to satisfy the five prescriptive criteria outlined in the previous section.

Depending on the type of facility that is being considered, there are different types of compensation arrangements that can be proposed. If a private developer is the applicant, the firm could offer a monetary payment that could be utilized by the community in any way it sees fit. Browning Ferris has operated in this manner in contacting communities who would be interested in hosting a landfill.

Compensation could be specified by a state or province in conjunction with the developer. Alberta utilized this approach in looking for a site that would host a hazardous waste facility. A benefits package could be specified by a representative of the country searching for a facility that would benefit the nation. The U.S. Nuclear Negotiator has an opportunity to play this role in searching for a MRS site for storing spent fuel.

*Lottery-auction procedure.* Suppose that planning grants were given to N communities and that some type of formal voting procedure had been conducted in each one to determine whether there was sufficient interest by the residents in hosting the facility if a prespecified set of conditions were met. For example, suppose that a referendum was held in each community with clearly defined safety standards, monitoring and control procedures, and a well-defined compensation package worth C* for the residents.

To be quite specific, suppose that any community where a majority of the residents favored the facility would be a viable candidate. Assume that $N > 1$ communities voted in favor of such a package. In the case of Alberta five communities held referenda in 1982 which drew heavy voter turnout and all supported having the facility in their backyard (McQuaid-Cook 1991).

We propose the following two step procedure whereby a lottery is first used to determine a tentative host site and then a competitive bidding system is introduced to determine the final selection.

*Step 1: Lottery to select tentative site:* Assume each of the N candidate sites has an equal chance of being chosen as the tentative site by lottery. By virtue of the positive vote in the referendum, a majority of the residents in the community that wins the lottery will feel better off having the facility in their backyard than maintaining the status quo.

*Step 2: Auction to determine final site:* Each of the N communities (including the one selected by lottery) now has an opportunity to host the facility by bidding an amount less than C*. One type of bidding process would be a sealed bid auction with the lowest bid hosting the facility. Alternatively, one could utilize a Dutch auction whereby the bidding proceeds downward from C* until there is only one community willing to accept the facility.

*Advantages of lotteries.* Historically lotteries have been utilized for selecting individuals to undertake onerous or burdensome tasks such as serving in the army. Draft lotteries were utilized to select men for the Army in the U.S. during World War II (1940) and in the Vietnam War (1970, 1971)[3]. In the Norwegian draft between WWI and WWII people were able to pay someone to take their place if selected by a draft lottery, an analog to the above proposed two step procedure used for siting the facility (Elster 1989)[4].

One reason for introducing a lottery as a first step in the siting process is to avoid discriminating against the poor communities[5]. At the end of this phase of the process everyone will know that a tentative site has been chosen to host the facility without any reference to income levels relative to the other candidate sites. Furthermore this site may well host the facility if the other

---

3 Lotteries need to be viewed as fair with each individual being given an equal chance of selection. In the 1940 and 1970 lotteries the balls indicating birthdates were not sufficiently shaken up so that a disproportionate number of people were selected from the later months.

4 In contrast to the successful Norwegian experience, the lottery with an option to buy out of the army was used during the U.S. Civil War and led to draft riots. This suggests the importance of making sure that every community understands the implications of the two-step lottery auction procedure before a referendum or plebiscite is conducted to determine whether residents want to participate in it.

5 Consider a compensation system where a community knows that if it is willing to accept t for hosting the facility it will have to be t/n in taxes should one of the other emerge as the host site. If the entire process was determined by a bidding procedure with the facility going to the lowest bidder, then it would most likely be the case that the poorest community would be the "winner". Because such a community could not afford to pay large sums of money to avoid having the facility, they would be forced to submit a lower bid than communities with higher incomes. For more detail on the assumptions and analysis of this model, see Kunreuther et. al., 1987.

communities do not want to make a counterbid for compensation lower than C*.

The equity argument for the lottery is a subtle but important one. By considering all communities as equally desirable for having the facility, there can be no complaints that poorer towns are being forced into taking the facility. Only if a poor community really wants to host the facility will it enter a bid below C*.

To illustrate this point, suppose one of the more affluent communities was chosen by lottery. A low-income community may decide not to enter the bidding even though it was willing to host the facility. For example it may have agreed to be a candidate site because residents felt that the status quo was unacceptable and a facility was needed somewhere in the region. By volunteering as one of the candidates, public-spirited citizens may have felt that they were satisfying a greater societal need. However, in offering themselves as a candidate they recognized that others were doing the same. In this case they would have little interest in bidding for the facility once another community was selected by lottery.

From a psychological perspective, there are good reasons to have a lottery as a first step in the siting process. Following the random drawing, the status quo has been changed from "no facility" to a "potential site" with the candidate having substantial benefits ($C^*$). Let $C_i$ represent the minimum amount of compensation that community would have to receive to be indifferent between hosting and not hosting the facility.

Any community not chosen by lottery may now feel that $C^*-C_i$ reflects a loss to them for not having hosted the facility. In other words the problem is now reframed in their minds so that if community i decides *not* to enter the bidding in Step 2, it is likely to feel that it has lost $C^*-C_i$ in potential benefits. In other words, loss aversion may stimulate an interest in bidding for the facility by those not chosen by lottery.

*Nature of bidding procedure:* Suppose that communities are ordered so that $C_1 < C_2 < \dots C_N$ One should note that the competitive bidding process by definition should lead to an efficient solution with community n as the "winner" demanding an amount less than $C_2$, unless there is collusion between communities.[6]

The low-bid auction should also satisfy the criteria of superfairness advocated by Baumol (1986). A distribution is considered to be "superfair" if each participant prefers its own share to that received by another group. In other words no group envies the other. By instituting a competitive process

6 Communities that agree to collude will demand an amount higher than their $t_i$, thus enabling to request an amount greater than $C_i$. In return community 1 will provide them with side payments after being designated as the host site.

with the site going to the lowest bidder, there should be no envy by either the winner or the other communities.[7]

## Suggestions for Future Research

*Alternative siting procedures:* One area for future research is the type of siting procedures that are likely to work in practice. A lottery-auction procedure is only one way that siting can take place if there is more than one candidate site that volunteers.

Other procedures are now being tried in which communities are provided with substantial benefits from hosting a facility in the form of tax revenues, in-kind compensation and grants to the community. Browning Ferris Inc. (BFI) has developed a Community Partnership Program in New York State in the hopes of finding volunteers for siting a landfill to store solid waste. It mailed a package to every local jurisdiction in New York State offering each of them the opportunity to host a solid waste landfill. Eagle, NY, a community of 1,300 residents responded positively in a public referendum. It will receive a benefits package worth between $1 and $2 million to go with the new landfill. Hence most residents feel that they are better of than under the status quo. (BFI 1993).

In Switzerland a novel procedure has been developed for siting a landfill but focusing on the default option. Each of the communities is legally responsible for managing its own solid waste. In most cantons, the cantonal government, however, has taken the initiative to develop waste management plans and to site regional waste disposal facilities that serve several communities and take advantage of the economies of scale. This effort has met substantial opposition by local governments and residents alike. Faced with the decision to site a landfill serving one third of all communities within the Canton of Aargau, the Cantonal government experimented with a novel approach to involve all the communities that could serve as potential sides for a regional landfill in a participatory voluntary selection process. If none of them volunteered to host the facility then each community would be responsible for its own solid waste (default option). As a result three communities have been selected by a unanimous votes of representatives of nine potential host communities (Renn and Webler 1994; Dienel and Renn in this volume). Landfills have a much better chance of being approved by the affected public than hazardous waste sites because the risks associated with solid waste are not perceived to be as great as those from hazardous or nuclear

---

7 A more detailed description of the lottery-auction procedure for different siting conditions (e.g., unequal construction costs, candidate sites directly benefit or do not benefit from the facility) appears in Kunreuther and Portney (1991).

waste. If one can site landfills through a voluntary system it may serve as a model for other projects.

*Transjurisdictional risks:* There are a number of open issues that are likely to be important in dealing with siting in the future. One key question relates to problems o£ transjurisdictional boundaries. Specifically, how does one address problems of risk as well as benefit-sharing for adjacent communities to the host site.

Consider the challenges associated with siting a hazardous waste incinerator near the border of a state, province, or country. What compensation should be provided to the neighboring state(s) or countries? What type of monitoring and control provisions should be instituted to protect these areas from any risks emanating from the facility? How will property values be guaranteed in the surrounding areas should they be affected by the facility?

In some cases the neighboring state may not be concerned with a proposed facility even through they may have reason to worry. The proposed HLNW repository at Yucca Mountain, Nevada has not attracted much concern in California even through it is less than 20 miles from the state border. An earthquake which registered 5.5 on the Richter scale occurred in nearby Yucca Valley, CA triggered some concern by geologists as to the safety of a repository should it be built at Yucca Mountain. It may also have generated concern by some Californians as to their future safety, should a repository be built in Nevada.

*Nature of a fair siting process:* More studies need to be undertaken to understand what the affected public considers to be an intrinsically fair process. For example, how important is it to institute a broad based participatory process in order to obtain a final solution that satisfies equity considerations? Are the costs and time required for this type of effort worth while given the benefits they are likely to produce? When can one locate a site without engaging in such a process and what are the costs and benefits of engaging in such effort?

How wide a net does one have to cast with respect to different stakeholders? What is the impact on the final solution when one introduces other interested parties (e.g. environmental groups) who have long-term objectives that may differ from the host community and the surrounding areas?

How does one incorporate concerns with future generations as an integral part of the analysis of the perceived risks? What types of data need to be collected for measuring these risks? How can the media play a creative role in providing data to the relevant interested party? How can one provide information to the public on the implications of maintaining the status quo so that they can better understand what it means *not* to take any action? How does one compare different alternatives to the status quo so that one can make an informed decision?

These questions and issues all appear to be important ones in examining ways to improve the siting process. It is my view that some type of voluntary process is an important starting point for addressing these questions. Voluntary procedures are viewed by the public and other interested parties as being fair and there is now increasing evidence that they are also successful in finding homes for facilities that previously were put in the Not In My Backyard (NIMBY) category.

## References

Atlanta Commissioner of City Planning. Personal communication with Leon Eplan, Commissioner of City Planning, City of Atlanta (December 1991).

Baumol, W., *Superfaimess* (Cambridge: MIT, 1986).

Browning-Ferris Industries, Inc. Personal communication with Philip Angell, BFI, Inc. (April, 1993).

Carnes, S. A., Copenhaver, E.D., Sorenson, J. H., Soderstrom, E. J., Reed, J. H., Bjornstad, D. J., and Peele, E., ''Inventives and Nuclear Waste Siting: Prospects and Constraints,'' *Energy Systems and Policy,* 7, No. 4 (1983), 324-351.

Easterling, D., and H. Kunreuther, *The Dilemma of Siting a Nuclear Waste Repository* (Kluwer, Boston, in press).

English, M., *Siting Low Level Radioactive Waste Disposal Facilities* (Quorum, *New York, 1992).*

Frey, B., "Financial Incentives for Environmental Protection Undermine Voluntary Efforts,'' *Joumal of Evnironmental Economics and Management* (in press).

Gregory, R., H. Kunreuther, D. Easterling, and K. Richards, "Incentives Policies to Site Hazardous Waste Facilities,'' *Risk Analysis,* 11, No. 4 (1991), 667-675.

Kasperson, R., D. Golding, and S. Tuler, *Siting Hazardous Facilities and Communicating Risks Under Conditions of High Social Distrust,* CENTED, Clark University, Worcester, MA (1991).

Keeney, R. L., *Siting Energy Facilities* (Academic Press: New York 1980).

Kunreuther, H., K. Fitzgerald and T. Aarts, "Siting Noxious Facilities: A Test of the Facility Credo," *Risk Analvsis,* 13 (1993), 301-315.

Kunreuther, H., and J., Linnerooth, Risk Analysis and Decision Processes: The Siting of Liquefied Energy Gas Facilities in Four Countries (Springer Verlag: Berlin 1982).

Kunreuther, H., Kleindorfer, P., Knez, P., and Yaksick, R., ''A Compensation Mechanism for Siting Noxious Facilities: Theory and Experimental Design,'' *Journal of Environmental Economics and Management* (1987).

Kunreuther, H., and P. Portney, ''Wheel of Fortune: A Lottery/Auction Mechanism for Siting of Noxious Facilities,'' *Journal of Energy Engineering,* 117, No.3 (1991), 125-132.

McQuaid-Cook, J., ''Case Study: Siting a Fully Integrated Hazardous Waste Management Facility with Incinerator and Landfill,'' in: *Innovative Approaches to Siting of Waste Management Facilities; Case Experiences and Perspectives:* (Auberge Montevilla: Montebello, Quebec, 1991).

Morell, D., and C. Magorian, Siting Hazardous Waste Facilities: Local Opposition and the Myth of Preemption: (Ballinger: Cambridge, MA, 1982).

O'Hare, M., L. Bacow, and D. Sanderson, *Facility Siting and Public Opposition* (Van Nostrand Reinhold: New York 1983).

Peelle, E., ''Hazardous Waste Management Outlook: Are There Ways Out of the 'Not-in-My-Backyard' Impasse?" *Forum for Applied Research and Public Policy,* 2 (Fall 1987), 68-88.

Rabe, B., ''Beyond my NIMBY Syndrome in Hazardous Waste Facility Siting: The Alberta Breakthrough and the Prospects for Cooperation in Canada and the United States," *Governance,* 4 (1993), 184-206.

Renn, O., and T. Webler, *Fairness and Competence in Siting a Landfill: A Case Study from Switzerland,* Paper for the Conference "Risk and Fairness" at the International Institute for Systems Analysis (IIASA), Laxenburg, Austria (August 1994).

Samuelson, W., and R. Zeckhauser, ''Status Quo Bias in Decision Making,'' *Journal of Risk and Uncertainty,* 1 (1988), 7-59.

Simon, H., ''Rationality as Process and as Product of Thought,'' *American Economic Review: Papers and Proceedings,* 68 (1978), 1-16.

Slovic, P., M. Layman, and J. Flynn, ''Risk Perception, Trust and Nuclear Waste: Lessons from Yucca Mountain," *Environment,* 33 (1991), 6-11, 28-30.

Tversky, A., and D. Kahneman, ''Loss Aversion and Riskless Choice: A Reference Dependent Model,'' *Quarterly Journal of Economics* (in press).

Von Winterfeldt, D., and W. Edwards, ''Patterns of Conflict about Risky Technologies," *Risk Analysis,* 4 (1984), 55-98.

Zeiss, C., ''Community Decision-Making and Impact Management Priorities for Siting Waste Facillties,'' *Environmental Inpact Assessment Review,* 11 (1991), 231-255.

*Chapter 17*

# *Voluntary Siting of Noxious Facilities: Additional Thoughts and Empirical Evidence*[1]

*Felix Oberholzer-Gee*
*Bruno S. Frey*

## Introduction

Failed attempts to find technical or political solutions to siting problems have provoked a profound change in the way siting procedures are structured. A first step was to introduce risk information programs as a means to improve the Defend Announce Defend approach. There is empirical evidence that information has a systematic effect on the way people form their risk perceptions and subsequently adjust behavior (Smith et al. 1990; Viscusi et al. 1987). Whether these processes strictly follow the expected utility model or can better be explained by taking framing effects and other anomalies into account is still an open question. The key role of information, however, is undisputed.

In a recent study, the Environmental Protection Agency found confirmation for the wide discrepancies between the risk assessments of its technical experts and the perceptions of the general public (Smith 1992). In addition, every accident and the discovery of hitherto unknown hazards brought about the question of fallible and possibly, biased experts (Otway 1987). As a result, it became clear that if communities were to accept noxious facilities on their grounds, information alone would not suffice to convince them. Information programs needed to be amended.

As a second step, public participation programs were included in siting procedures. The public was not only informed about the facilities it was expected to host, but people had a limited say in siting procedures and, to a lesser extent, in the design of the facilities. Governments and developers hoped that the use of more democratic processes would help to legitimize decisions to place these special burdens on citizens. As pointed out in Howard

1 This research was supported by Swiss National Fund, grant no. 12-25581.88, undertaken jointly with Werner W. Pommerehne. We particularly wish to thank Reiner Eichenberger for helpful comments.

Kunreuther's contribution to this volume, many of these participation processes have since unraveled.

Along with the frustration among developers and governments over siting difficulties, a tendency to reverse the democratization of siting processes is growing. An authoritative approach with the central government deciding where to build what type of facility is seen as the better and sometimes only solution. The Swiss government's decision in early 1993 to speed up the siting of a low level radioactive waste facility by overruling the cantons' say and possible popular referenda is but one example for this tendency (Neue Züricher Zeitung 40, 18.2.1993). We will argue that legislated siting programs did not fail because they were too democratic. On the contrary, clarifying the local opponents' entitlements by granting them the explicit right to deny the construction of any unwanted facility on their grounds seems to be the most promising way out of the siting dilemma.

## Entitlement Structure

At the heart of the siting problem is a dispute over the existing entitlement structure, i.e. the distribution of property rights. The entitlement structure determines if the local opponents have to bear the negative effects of a noxious facility or if they can call upon the state to protect their interests. Developers (private or state) expect the state's help in siting noxious facilities for it is the benefit of a larger group of citizens, and quite often public health in general, that they claim to pursue. Moreover, it is usually state agencies that had permitted the production processes which finally resulted in the toxic wastes under dispute. On the other side of the battle line, the status quo as defined by host communities unwilling to accept facilities consists of entitlements to physical safety and long-term health. As with the developers, they too turn to the state to protect their rights.

It is the prime task of the state to define and enforce property rights. Only when these are known can the parties to environmental disputes start to trade off their interests. If the state protects the communities from having to accept noxious facilities on their grounds, it is the developers who must offer a compensation package attractive enough to the prospective host community so that they accept the facility. If the state entitles the developer to site the project in a chosen community, it is the community who must compensate the developer if it forgoes its rights. The state's role is not an easy one. Given considerable transaction costs it cannot assign these rights arbitrarily (Coase 1960). While granted the power of eminent domain, governments up for re-election are reluctant to openly turn against host communities supported by environmental groups. Past incidents such as the Wackersdorf Nuclear Fuel Reprocessing Plant Project in Germany suggest that even the use of excessive force does not guarantee "successful" siting. If the government does not take action, the outcome is usually favorable for the host community which is in many cases able to gather legal support against the facility or to show at least enough determination to fight against the project to make it look economically unfeasible (O'Hare 1983; Samuels 1971). The government's reluctance or inability to take action does not improve the situation. As years go by and communities turn down project after project, siting problems become more and more pressing.

## Public Participation

By incorporating public participation schemes into siting procedures, legislators had hoped to find a way to acknowledge the host communities' entitlements without formally enforcing them and thereby making siting impossible. Legislated siting programs typically did not include the right of the community to the final decision, but sought ways to improve the chance of acceptance by establishing communication between the developer and the community. In our view, these types of public participation programs fail to solve the siting problem because they fail to clarify the underlying entitlement structure that is at the heart of the conflict.

This is not to say that risk communication is not important. Communication, as outlined in the facility siting Credo, described by Kunreuther et al. (1991), is itself a key essence in disputes over entitlement structures. New information regarding the risks of a facility, accidents at other facilities, and advances in technology determines the terms of trade under existing entitlement structures and provides powerful incentives to change them (Bromley 1992). However, when property rights are not clearly defined, there is no basis for trade. As a consequence, information designed to facilitate this trade cannot bring about a solution.

## The Lottery / Auction Mechanism

The Lottery/Auction mechanism proposed by Howard Kunreuther achieves the clarification of the entitlement structure that is essential to establish trade. It does not focus on the way negotiations and communication between the developer and the prospective host community should be structured. Instead, its focus is on decision making and, therefore, on the assignment of property rights. Taking the willingness of the communities to participate in the mechanism as a starting point, communities are fully entitled to decide whether or not a facility should be built on their ground. No doubt remains about the entitlement structure. It is the developers who have to offer a "benefit sharing package" attractive enough to make communities participate.

Having assigned the property rights, the lottery / auction mechanism seeks to combine considerations of efficiency with the requirement for fairness

(Kunreuther and Portney 1991). The latter is to be brought about by the lottery, the former by auctioning off any surpluses that might have accrued in the first round of decision making.

## *Fairness*

A test of the facility siting Credo that lists the most important elements of successful siting clearly shows the importance of fairness considerations (Kunreuther et al. 1992). Several variables such as trust, multiple options, geographic distribution and voluntariness are related to notions of fairness. While accepting the importance of fairness, however, we remain skeptical as to whether a lottery is generally seen as a fair mechanism. Our own research suggests that it is not.

To acquire insights into the perceived fairness of different decision making mechanisms, a study among 1,750 households in Zurich and Berlin was conducted by one of the present authors (Frey and Pommerehne 1993). Two situations of excess demand were presented: Respondents had to decide how to distribute 100 bottles of water to 200 thirsty hikers in the first situation, and the distribution of snow shovels the morning after a snow storm in the second. Four alternative decision making systems were given:

- a traditional procedure where a fixed rule was applied, in this case "first come, first served";
- a random mechanism where each person had the same chance to get a bottle of water or a snow shovel;
- a delegation of the decision to local authorities, who could decide on the distribution in accordance with their own principles; and finally
- the use of the price mechanism, i.e. a price increase for water and shovels that would clear the market.

In academic writings by economists the price as well as random mechanisms are advocated. The former will bring about efficient outcomes and is seen as an ideal tool to settle environmental disputes (Kneese and Schultze 1975; Baumol and Oates 1979). Random mechanisms have also been suggested as rational procedures, particularly for voting (Intriligator 1973; Mueller 1978). The view of the public, however, is quite different. The price as well as random mechanisms are not considered to be fair in situations of excess demand (see Table 1).

*Table 1:* Subjective Evaluations of Alternative Allocation Procedures (in percentage of all respondents)

| | Allocation procedure for water bottles | | | |
|---|---|---|---|---|
| Evaluation | price | tradition | random | administration |
| fair | 27% | 76% | 14% | 43% |
| unfair | 73% | 24% | 86% | 57% |

Subjective fairness clearly ranks the traditional procedure of "first come, first served" as the fairest. This is surprising since this decision making system does not take into account the thirstiness and thus the "needs" of individuals. The random mechanism is seen as the least fair. Only 14 percent judge it to be fair. The case for the lottery is not entirely lost, though. The ranking seems to depend on the specific situation. When asked to choose a fair rule to allocate snow shovels, the result was as exhibited in Table 2.

*Table 2:* Subjective Evaluations of Alternative Allocation Procedures (in percentage of all respondents)

| | Allocation procedure for snow shovels | | | |
|---|---|---|---|---|
| Evaluation | price | tradition | random | administration |
| fair | 23% | 93% | 27% | 48% |
| unfair | 77% | 7% | 73% | 52% |

In this case, a random mechanism is judged to be considerably fairer (27% instead of 14%). Compared with other decision making systems, however, random mechanisms such as lotteries still do not fare very well. Further empirical research will have to analyze whether lotteries are accepted as a decision making system in the context of siting problems. Based on our own study, however, people do not seem to attribute random mechanisms much fairness. Moreover, the use of a lottery seems to be in contradiction with another guideline of the facility siting Credo: "Choose the solution that best addresses the problem" (Kunreuther et al. 1991). If people truly believe that there is a "best solution" to the siting problem, be it technical or political, it must strike them as odd to select the starting point for the auction, and possibly the final outcome, by chance.

### *Auction*

To introduce an auction as the second stage of the siting process certainly appeals to economists and all those concerned with efficiency. In the absence of collusion the auction will bring about an efficient allocation of facilities. However, the results of our research suggest that the use of the price system does not seem to find much acceptance either. The preoccupation with distributional aspects of decisions is an important reason for the reluctance to use prices as a decision making process.

Distributional consequences of allocation decisions are often dismissed by economists on the grounds that (free) compensation is possible. The lottery/auction mechanism asks the communities to state the compensation demanded. The competitive structure of the auction then forces the communities to lower their compensation demands. Differences in demand for compensation essentially stem from differences in risk assessment, and differences in the evaluation of future income generated by compensation packages. The latter are the reason for the rejection of the price mechanism by the public in general and special interest groups in particular (Kearl 1979; Frey 1986; Kahneman et al. 1986). In this respect, to combine the lottery with the use of the price system bears the risk of destroying whatever credibility and fairness the lottery was attributed in the first place. Once the mechanism has been applied a few times with the result that the facility is built in relatively poor communities, the whole mechanism might be seen as an elegant way to take advantage of communities in need for funds.

## Conclusion

The lottery/auction mechanism developed by Howard Kunreuther combines important elements of successful siting procedures. Its major achievement is the clarification of the underlying entitlement structure which forms the basis for any trade in environmental disputes. In view of a de facto entitlement of prospective host communities to deny the siting of noxious facilities – the unwillingness or inability of governments to enforce siting decisions results in a de facto entitlement – we feel it is the right choice to formally acknowledge the communities' rights. The mechanism further addresses two key aspects of siting schemes: fairness and efficiency. While we agree on the importance of those aspects, empirical research does not support the view that lotteries and auctions are perceived as fair decision making systems. Further empirical research is needed to evaluate these mechanisms as parts of siting procedures or suggest other ways to reconcile demands for fairness with efficiency considerations.

Taking the prospective host communities' willingness to participate in a lottery/auction mechanism as a starting point provides the ultimate empirical test. If communities can be motivated to participate ex ante – before the facility's location is decided – and accept the results ex post – after the siting decision has been made – the fairness requirements are met. Drawing conclusions from other research, however, we fear that the mechanism will entice no more enthusiasm than the prospects of siting a facility in one's own backyard.

## References


Baumol, William A., and Wallace E. Oates, *Economics, Environmental Policy, and the Quality of Life.* (Prentice Hall: Englewood Cliffs, 1979).

Bromley, Daniel W., "Entitlements and Public Policy in Environmental Risks," in: Bromley, Daniel W. and Kathleen Segerson (eds.), *The Social Response to Environmental Risk.* (Kluwer: Boston 1992), pp. 1-22.

Coase, Ronald, "The Problem of Social Cost," *Journal of Law and Economics* 3: 1-44 (October 1960).

Frey, Bruno S., "Economists Favour the Price System – Who Else Does?" *Kyklos* 39: 537-563 (1986).

Frey, Bruno S., and Werner W. Pommerehne, "On the Fairness of Pricing – An Empirical Survey among the General Population," *Journal of Economic Behavior and Organization* 20: 295-307 (1993).

Intriligator, Michel D., "A Probabilistic Model of Social Choice," *Review of Economic Studies* 15: 553-560 (1973).

Kahneman, Daniel, Jack Knetsch, and Richard Thaler, "Fairness as a Constraint on Profit Seeking: Entitlements in the Marke,." *American Economic Review* 76: 728-741 (September 1986).

Kearl, James R., Clayne L. Pope, Gordon C. Whiting, and Larry T. Wimmer, "A Confusion of Economists?" *American Economic Review, Papers and Procedings* 69: 28-37 (May 1979).

Kneese, Allen V., and Charles Schultze, *Pollution, Prices, and Public Policy.* (Brookings Institution: Washington D.C. 1975).

Kunreuther, Howard, Thomas D. Aarts, and Kevin Fitzgerald, *Siting Noxious Facilities: A Test of the Facility Siting Credo.* Report 92-03-01 (Wharton School, Risk and Decision Processes Center: Philadelphia 1992).

Kunreuther, Howard, and Paul Portney, "Wheel of Fortune: A Lottery / Auction Mechanism for Siting of Noxious Facilities." *Journal of Energy Engineering* 117: 125-132 (December 1991.

Kunreuther, Howard, Lawrence Susskind, and Thomas D. Aarts, "*The Facility Siting Credo. Guidelines for an Effective Facility Siting Process.*" (University of Pennsylvania: Philadelphia 1991).

Mueller, Dennis C., "Voting by Veto," *Journal of Public Economics* 10: 57-75 (August 1978).

Neue Zürcher Zeitung, No. 40 from 18.2.1993. (Verlag für die Neue Zürcher Zeitung, Zurich 1993).

O'Hare, Michael, Lawrence Bacow, and Debra Sanderson, *Facility Siting and Public Opposition.* (Van Nostrand Reinhold: New York 1983).

Otway, H., "Experts, Risk Communication and Democracy," *Risk Analysis* 7: 125-129 (1987).

Samuels, Warren, "The Interrelations between Legal and Economic Processes," *Journal of Law and Economics* 14: 435-450 (October 1971).

Smith, V. Kerry, "Environmental Risk Perception and Valuation: Conventional Versus Prospective Reference Theory," in: Bromley, Daniel W. and Kathleen Segerson (eds.), *The Social Response to Environmental Risk.* (Kluwer: Boston 1992), pp. 23-56.

Smith, V. Kerry, William H. Desvousges, F. Reed Johnson, and Ann Fisher, "Can Public Information Programs Affect Risk Perceptions?" *Journal of Policy Analysis and Management* 9: 41-59 (Winter 1990).

Viscusi, W. Kip, Wesley A. Magat, and Joel Huber, "An Investigation of the Rationality of Consumer Valuations of Multiple Health Risks," *Rand Journal of Economics* 18: 465-479 (Winter 1987).

*Chapter 18*

# *Direct Participation in Macro-Issues: A Multiple Group Approach*
# *An Analysis and Critique of the Dutch National Debate on Energy Policy, Fairness, Competence, and Beyond*

*Cees J. H. Midden*

## Introduction

Energy policy has been a hotly debated issue in many countries for some decades. In contrast to the early debates, the role of the public has gained significant importance. Presently there are numerous examples of expensive facilities which had to be closed or were built and could not be used due to public opposition. These cases are not only expensive mistakes. They also show how technological developments can be thwarted by improper forms of decision making, which do not recognize the role of an involved public.

This chapter describes and analyzes how, in the Netherlands during the years 1981-1983, a large national debate on energy policy' was organized (in Dutch abbreviated as "BMD"). Due to its scale this national debate can be considered a rather unique form of public participation in policy and decision making. It can also be seen as a more general model of participation which can be used for other issues and in other contexts.

The Dutch model can be seen as related to approaches in various other countries such as in Sweden and Austria. The basic similarities in these cases are that the issue is large scale and decision making is on the national level. Secondly, a need for more elaborated public opinions is acknowledged leading to the organization of public information programs and small discussion groups. Thirdly these approaches have in common that they are advisory and that final decision making is left to politics. Nelkin (1977) in comparing a number of cases, distinguished between two types: the public advisory model and the elitist advisory model. The basic difference concerns the prime goals. The elitist model tries basically to inform the public about the issue at hand and the most desirable solution. This model tries to educate the public hoping to gain more acceptance.

This type of approach is mainly unidirectional, offers less opportunity to ask for information and is not really focused on dealing with value issues who are basically left to politicians. The problem is seen predominantly as a technical issue to be solved by technologists with the public in a passive role. It is hoped for that acceptance will grow when opinions are more based on sufficient knowledge to understand the technical issue. An example of this approach is the Austrian experiment in 1976 and 1977 concerning the siting of a new nuclear power plant. Strong opposition on the national level urged government to have a broad debate and more public participation. Characteristic for the approach was that there was not a strive for consensus. The basic objective was to promote informed public debate and to identify issues of agreement and disagreement. The whole process was aimed to analyze the facts and to bring these to the general public. For this purpose there were debates on television among experts, technical and economic dictionaries, possibilities for the public to put questions forward etc. A similar approach was also applied in France concerning industrial siting (see Nelkin & Pollak 1980). The Austrian approach did not get the outcome hoped for by industry: the information supply seemed to highlight the controversial nature of the conflict leading to more confusion instead of reassuring the public. As Nelkin & Pollak explain it contributed to public skepticism also because of the criticism on the elitist procedures which allowed no input on problem definition and the selection of debating experts. The eventual outcome was that the Austria nuclear program stagnated.

The participatory advisory model is more directed to advise decision makers about scientific opinion and about public concern. This implies that the issue is not primarily defined as technical, but that it is acknowledged that public opposition is not (only) based on lack of scientific literacy but is value-loaded. These value-differences need to be accounted for in decision making, although not in a direct way. An example of this approach has been applied in Sweden concerning energy policy and is called: Swedish Study Circles" (Nelkin 1977). Under the umbrella of a limited number of popular organizations like trade unions, political parties and religious organizations thousands numbers of small groups discuss the issue at hand trying to connect the facts to value choices and ideological perspectives. To enable realistic opinion formation, various types of information input are offered. Groups can write their own questions and ask for answers from government; also they can develop their own relevant expertise or consult self-selected credible experts. The results of these activities are summarized in reports and distributed among decision makers.

Clearly large-scale participation requires special qualities and has to take account of specific barriers. Questions that come to mind are of course: what were the effects on decision making and on public attitudes? Are these approaches considered legitimate within the existing political and administrative structures? What kind of issues can be addressed in this way? The Dutch national debate on energy policy has many similarities with the Swedish model. Following the conceptual framework used throughout this volume we will use the Dutch experience to contribute to the evaluation of this type of small group participation approaches to large scale technological decision making.

The chapter is composed as follows: firstly the embeddedness of the program will be discussed. This is followed by a discussion of the entire procedure. After which the model will be evaluated using the other standard criteria in this volume of fairness and competence of the discourse, legitimacy, applicability and transferability. Finally I will sum up and draw conclusions.

## Embeddedness

In 1974 the Dutch government made a principal decision on the expansion of the use of nuclear power for electricity generation. It was planned to build three large nuclear power stations. The following siting procedure was stagnated by strong public opposition against the plan itself, apart from the siting concerns. Energy policy became deadlocked. The conflict seemed deep and rooted in differences that went beyond energy policy and reflected basic disagreements about the need for economic growth, the role of large scale technologies in society, and issues of scale. The idea to solve this decision problem through a national debate to create a socially responsible and widely accepted decision was proposed by a number of social organizations including the Federation of Trade Unions (FNV), the Association of Dutch municipalities and a religious organization 'the Dutch Reformed Synod' (Nederlandse Hervormde Synode). Following this initiative a detailed proposal for this debate was presented by the so-called Initiative group Energy Discussion (IED), an independent volunteer group of energy experts and scientists from a variety of social and public organizations.

Parallel to this proposal the General Energy Council (in Dutch abbreviated as AER), a governmental advisory body, gave advise to the Minister of Economic Affairs on the organization of the debate. It differed in a number of ways from the IED-proposal. The IED proposed to discuss all relevant viewpoints, the AER wanted to discuss the plan of government; the IED proposed to discuss general energy policy, also in relation to socio-economic policy, AER proposed to restrict the discussion to general energy policy. The AER-proposal implied more freedom of choice to government and parliament. The IED-proposal aimed to present a conclusion which was already heavily discussed and evaluated by political decision makers. The Minister sent the official AER-proposal to parliament for approval. After debate parliament changed the proposal and shaped it more in accordance to the IED-plans. The only exception was the freedom of parliamentary choice which was safeguarded using the AER-model. One important element in this whole procedure was the democratic way in which problem definition was established. It was ultimately Parliament that decided on the charge of the debate and the problem definition. Both a procedure for achieving consensus and conflict resolution was available.

The model as designed by the IED and approved by Parliament had the following basic features:

1. *Charge definition*: energy policy as a whole was considered the problem to be discussed. There were no issues that could not be discussed. This was an important decision that emerged from a series of preliminary consultations of involved parties and that replaced the original plan of a debate on nuclear power.

2. *Model selection: p*articipation was aimed to be broad, which implied that all parties involved could contribute to the selection of procedures, time schedules, further organizational set up. Selecting the model was a participatory process.

3. *Public education and involvement functions: a* basic starting point was that participants should have available an information base on which to build their preferences and decisions. This information base should be as complete as possible. The · two goals were to make the information broad and comprehensive and to reach and involve as many participants as possible. Participation by involved and informed citizens was a major function.

4. *Value reconciliation:* the committee made no distinction between types of reactions, interests, norms, goals and the like. Equal treatment was given to reactions of a more emotional or rational nature. Calculated risks were important, but perceived risks as well. To achieve the information goals and a fair discourse funds were made available to enable parties to acquired missing information or develop viewpoints. The steering committee wanted to stimulate that parties take each other seriously, experts and laymen alike; pro's and cons of various sources and nuclear in particular should be able to participate equally; this approach to value reconciliation was an important step to address the conflict.

5. *Political impact:* As the whole effort was meant to stimulate decision making parliament had to play a central role from the start to ensure a good transfer of the results; however in the final model there was no formal link to political decision making; decision making was left to usual procedures through Parliament. To avoid vague and non-committing conclusions a strong effort was taken to achieve a reliable and valid way of outcome assessment. Several methods to measure preferences were applied.

6. *Organization and logistics*: these were assigned to a body with pre-agreed responsibilities and tasks, with a by government allocated budget and a fixed time schedule. This body needed a high level of transparency of decisions and activities. The members of the steering committee were not

supposed to be neutral. The composition however was meant to be balanced on what might be called an implicit pro-anti nuclear power dimension. The relation with government was meant to be strictly independent.

## Procedures

The full procedure included two main phases: the information phase and the discussion phase. It was preceded by a prologue for some preliminary actions and followed by a epilogue to make up the conclusions.

### *Prologue*

After parliamentary approval of the proposal and the allocation of a budget of 28 million Dutch Guilders the first step was the installation of the independent and evenly balanced steering committee. The steering committee was designed to be proficient in all the necessary topic areas. The committee was supported by a multi-disciplinary staff of about 30 people. Independence was formalized by giving the committee the legal status of a foundation.

To design the model the chairman consulted many representatives of political parties, social interest groups among which employers and employees organizations, environmental organizations, energy organizations and religious organizations. Although (of course) not every citizen was personally involved there was ample opportunity to have all viewpoints included in setting up the rules and the agenda for the debate and the final format appeared to be a fair refection of those diverse viewpoints. However access was limited to those groups that the steering committee selected.

### *Information phase*

The general goal of this phase was to conduct an inventory of opinions, beliefs, arguments, attitudes, and premises which might be relevant to a public judgment of the best energy policy. This phase also served to check information on validity. The end product of this phase was the Interim report, which was supposed to function as the common database for future discussions.

This phase was divided in three main activities: an inventory of judgments of informed citizens; hearings among organizations and controversy sessions. This phase was also used to launch a number of studies to shed more light on lacking or controversial information.

### *Inventory among informed citizens and organizations*

To gain insight into the different views on energy policy in society the steering committee (SC) used full page advertisements to invite individual citizens, groups, and organizations to send in soundly based and elaborated views on the energy issue. In addition, 260 organizations and pressure groups of all sorts were directly invited to present their views. The responses could be divided in 3600 rather short responses and 400 more extensive reactions.

The 400 extensive responses were thoroughly analyzed and summarized according to the viewpoints on a number of key aspects. The result offered a nice overview of available attitudes, arguments and conflicts. It also showed that different attitudes were often based on different types of knowledge, expressed as 'different schools.'

### *The hearings*

Following this inventory hearings were organized. Out of the 400 consulted organizations 93 were invited to participate in one of the 11 regional hearings. These organizations had very heterogeneous backgrounds. Among others there were: anti-nuclear groups, consumer organizations, energy companies, political parties, research institutes, employer associations, and labor unions. Participants were asked to prepare written reactions to a set of formulated questions that came out of the inventory. These reactions were prepared before the hearings and were explained and discussed during the sessions. Participation was very high (97 percent) and the meetings were carefully recorded and analyzed for use in the interim report.

### *The controversy sessions*

The foregoing actions enabled the steering committee to focus the debate on four issues:

> The costs of electricity with use of nuclear technology;
> the structure of the energy sector;
> the use of risk analysis and risk perception in decision making;
> options and risks of nuclear waste disposal.

The idea of controversy sessions was based on a number of motives: to reduce the segregation in camps of opposing parties and the lack of common knowledge and references; and to reduce the vagueness surrounding the nature of disagreements; i.e. to promote clarification of differences through confrontation.

Experts were invited to participate. The willingness to do so appeared to be great. The meetings were open to the public. Participants prepared written statements in advance to give opportunities for prepared reactions during the sessions.

The main conclusions were that participants were very well informed; that social attitudes were important factors behind specific viewpoints on energy policy; that government officials, utilities and technical experts had difficulties to understand more emotional and value driven judgments from other groups and found it difficult to discover the subjective elements in their own position; that many of the conflicting views were open to reach for more consensus.

### *Subsidies*

To achieve a fair and competent debate it was considered crucial to enable participants to present their views well prepared and argued. Many parties felt in a backward position in comparison to more wealthy groups which were able, often with public money, to organize expensive campaigns. To solve this problem participants could apply for subsidies for developing information programs, magazines, exhibitions and the training of moderators. Criteria for acknowledgement were a positive contribution to the debate, no doubling with other projects and the lack of own funds. In total the steering committee approved 77 project proposals and spent about 9 million guilders. Among the recipiants were environmental groups, youth organizations, women groups, political parties and religious organizations. Also subsidies were given to coalitions of various smaller groups.

### *Communication activities*

The whole debate encountered a lot of distrust and cynical reactions from many people. The growing nuclear controversy and distrust towards government were part of this attitude. Initial credibility was not very high. The steering committee had therefore an important communicative task to explain the goals of the debate, to explain the process and to make explicit the roles of experts and the general public. Communication was directed to reach all involved parties professionals and laymen alike.

This was done by different actions like advertisements, a newsletter, the publication of decision summaries, a brochure about how to participate, educational materials, lectures, posters and press releases.

### *The interim report*

The data collected in the information phase were transformed into the Interim report, a 170 page report, which was distributed on a large scale, mainly through public libraries. The report was positively received and served as the database for the discussion phase. In addition to the full size report there was a brochure meant to diffuse the essential information among the general public. This brochure contained information on the meaning of the debate, the role of citizens, basic information on energy, scenario conclusions and the announcements of the coming group discussions. More than a million of these brochures were distributed through post-offices and other channels.

## *Conclusions on phase 1*

It can safely be concluded that the information phase offered sufficient opportunities to everybody to speak up and influence the charge of the debate, the agenda and the selection of the participation model. A great effort was taken to inform the general public and to involve as many people as possible. Value reconciliation was served by giving fair treatment to various viewpoints and interests and to take all inputs seriously whether based on calculations or perceptions. Ample opportunities and financial support were available for groups and organizations to develop viewpoints and to put these forward. The resulting interim report formed a useful and competent starting point for the discussion phase.

## *The discussion phase*

The second year of the process was spent on the debate itself: the public, both individuals and organizations, was encouraged to express preferences in a number of ways.

To collect opinions in a valid and meaningful way the steering committee spent a lot of attention to the design of a good *questionnaire*. This was necessary considering all cynic reactions and mistrust around the debate. In the questionnaire preferences could be expressed on the use of the most important energy sources such as coal, nuclear, gas, oil, wind and solar. In addition there were more strategic questions on the goals of energy policy like keeping prices low, preventing pollution, reducing risks etc. Respondents were also asked whether the total consumption level should grow freely or be more restricted; how electricity supply should be organized according to scale and tariffs. The forms were distributed and used in huge numbers. It was surprising how little criticism was aroused by the design.

## *Local meetings*

Local discussions took place in a number of ways within organizations, in small groups on the local level, in schools. Furthermore individual citizens were allowed to fill in the forms and express their views.

The *local meetings* were organized in two phases: a first meeting in which participants could exchange information and a second meeting that was meant to discuss and complete the questionnaire. The meetings were moderated by about 1,000 locally hired and trained-on-the-job discussion leaders. Participants appeared to be very heterogeneous, about 19,000 people attended one or two meetings. Average group size varied from 19 in the first months to about 10 at the end. A number of 1,811 meetings actually took place. Evaluation indicated that discussions emerged without difficulties. The majority of participants appeared satisfied with the discourse. It was greatly appreciated that participants discussed with each other without any kind of pre-selection according to ideology or religious preference. On the whole discourses were characterized by tolerance towards each others viewpoints. The discussions did not have a strong effect on established attitudes, but had their strongest impact in the formation of viewpoints among participants which did not have elaborated viewpoints in advance.

In all kinds of *social organizations* like labor unions, political parties, religious organizations 1,120 meetings were organized with a total number of about 19,000 individual participants. About 9,000 *schools* were approached and more than 600 received materials. Twenty five schools organized special projects on the topic. Beside the organizational meetings on an individual basis there were also 344 social organizations such as political organizations, commercial companies and interest groups which were asked for an *institutional opinion.* The response consisted of 225 valid reactions. Apart from the institutional reactions about 42,000 people participated in the discussions, 25,600 of whom completed a questionnaire.

## Conclusions on Fairness and Competence of the Discourse

Although (of course) not every citizen was personally involved there was ample opportunity to have all viewpoints included in setting up the rules and the agenda for the debate. The final set up appeared also to be a fair reflection of those diverse viewpoints.

The composition and working style of the steering committee was widely discussed and the final set up was clearly a product of consensual approval. The choice of the style of moderation was based on advice from two different social organizations. They each advised the government, after which parliament finally approved. The steering committee worked hard for wide and universal participation. Understandable information was distributed on an enormous scale. The steering committee used five opinion rating instruments to get a fair representation of available opinions. The opportunities for participation on this scale seem unique.

Much effort was taken to allow validity claims. The steering committee aimed to test information on factual accuracy, premises on acceptability and arguments on consistency. The expert's consultation, ensured that all relevant issues were identified and especially the 'hot' issues to be debated. Schools of experts had to be distinguished, which were input for the controversy sessions with ample opportunity to make cognitive validity statements. The normative validity claims were already made during the agenda setting. In the participation phase the general public and also the social organizations could bring in normative and expressive validity claims. The public could make use of the special developed questionnaire which explicitly and systematically asked for preferences and wishes after the organized local discussion meetings. In addition peoples preferences were researched through studies using different techniques.

Disputes on cognitive validity could be put forward by large numbers of organizations. This was facilitated by offering subsidies to these groups which enabled them to conduct their own studies. The controversy sessions were excellent opportunities to discuss cognitive claims. Issues on normative validity and expressive validity were discussed in the discussion phase. Contrasting claims were recorded and reflected in the final report. There was no debate on unequal opportunities or non-representative claims in the conclusions. On the contrary there was strong agreement on the validity of the conclusions.

The steering committee put a lot of effort in collecting all the different viewpoints from social groups and experts. The controversy meetings were meant to discuss differences in assumptions and definitions. The conclusions from these sessions were reflected in the interim report which served as the information base for everybody. In this report the hot issues like the costs of electricity supply with different technologies, the use of risk analysis and the safety of nuclear waste storage were discussed on the basis of the preceding information phase. In addition the decision problem was summarized as a choice between four scenarios.

The steering committee received information from hundreds of organizations. It invited anybody with subjective expertise to react, which resulted in thousands of inputs. Organizations who needed so could apply for financial support to conduct studies, the controversy meetings allowed experts to discuss disagreements. Unresolved issues were discussed in the interim report.

## External Legitimacy

The final outcome did not have legal value. It was clear from the beginning that the outcome would function as advice to the government and to parliament. This damaged the legitimacy of the debate. Surveys showed that 50 percent of the population seriously doubted the political impact of the whole debate. The model was not flexible. There was not an attempt to achieve mutual agreement and compromise. The steering committee collected inputs and tried to draw conclusions without further interactions with involved parties.

The steering committee spent a lot of effort in assuring a reliable and complete *registration* of the effects of the debate on *public opinion.* To give a clear account of conclusions and effects the opinions registered had to be comparable, objective and quantitative. The data had to make clear how energy policy options were evaluated by the public. To judge representativeness, the registrations had to reveal also which people had participated in the process. A number of criteria may be addressed in looking for a proper registration (see also Vlek 1986).

### *Information content and elaboration*

Informed preferences may be considered more valuable than uninformed preferences. Part of this is the opportunity for discussion to develop more elaborated viewpoints. Offering information before registration may have as a negative effect that preferences are influenced by almost unavoidable information selectivity. An extensive information supply and discussion phase may also affect representativeness, because people who are less willing or able to process the information or participate in the discussions may be lost.

### *Elicitation of personal preferences and arguments*

The more respondents are allowed to give their own personal preferences the more valid and interesting the response may be. This supposes the opportunity for free responses. The drawback is that this approach affects reliability and also comparability, because response frames, wordings and forms may differ dramatically and responses are qualitative.

### *Quantitative results*

This criterion is obviously important because of objectivity and comparison. It demands a highly structured approach like in a multiple choice questionnaire. The drawback may be that answers are less personal and more suggestive to the respondent.

The steering committee acknowledged these trade-offs and developed a number of complementary methods: opinion registration after discussion, various types of randomly sampled surveys with pre-structured questions, hearing reports and personal reports (for organizations). There were two surveys: one rather traditional opinion study with measurements before and after the debate; the other one was a so-called choice-questionnaire, structured according to multi-attribute decision model, in which respondents had to make choices after relevant arguments were offered to improve the quality of decision making. Together all these methods offered a rather comprehensive picture of the effects and outcomes of the national debate.

## What Effects Did the Debate Have on Public Opinion?

Before discussing effects on opinions it may be worthwhile to consider the profile of participants. Compared to the national situation it can be concluded that the proportion of men compared to women was relatively large (62 percent vs 33 percent). The distribution of age was reasonably well with a slight under-representation of the youngest category (below 20) and the elderly (above 61); according to region there was an under-representation of the larger cities (16 percent among participants vs 30 percent nationally). According to education the citizens with primary school only were under-represented; the professionally and university educated citizens showed some over-representation. According to political preference social democrats appeared under-represented and some small left-wing parties were over-represented.

In general the differences are not very large. However the viewpoints of participants seem to be somewhat more consistent and pronounced. Participants were more in favor of energy conservation and less in favor of consumption growth than the population in general. Among participants supporters of growth are more in favor of nuclear power than among non participants. The latter are however more negative on nuclear power. Those who want to reduce consumption would like to invest much more in research. This relation is stronger for participants. Participants who want to conserve energy are more in favor of reduction of fossil sources. Non-participants are less negative about the use of fossil sources. The pre- and post test comparison indicated that in general attitudes towards energy issues had not changed significantly. Neither had the level of knowledge. The level of involvement showed a decrease. The awareness of the debate appeared to be very high. Attention in the form of reading information, participating in meetings etc. was low (although extremely high compared to other participatory activities on policy issues). Changes in viewpoints among the public were not related to level of participation, nor in level of knowledge or involvement. In sum the debate did not change the direction of viewpoints or the level of knowledge. Positive effects could be observed on the consistency and strength of viewpoints.

## Implementation

The *final report,* a 400 page volume, gave a complete overview of the total process of the debate. It describes the structure and process of the debate, all activities of evaluation and registration of opinions and policy preferences, the results with respect to energy consumption and conservation; the use of fossil fuels; the use of renewables and the use of nuclear power. It discusses the fuel packages for the next decades, the structure of the electricity sector, tariffs policy and international issues. The final section offers conclusions and specific policy recommendations.

Although not the focal point of attention in this chapter we summarize the central conclusions and recommendations of the debate:

- more energy conservation is desirable;
- renewable energy sources should be applied to a larger extent;
- natural gas should be used more;
- no new nuclear power for the moment;
- decentralized power generation systems should be promoted.

Given all the efforts taken and procedures used it is fair to say that the conclusions offered a representative picture of the preferences in Dutch society.

What happened after publication of the final report? Political parties reacted dogmatically. As if no debate had taken place the Christian Democrats and Conservative Liberals, at that time governmental parties, stated that the nuclear option should be kept open for environmental and economic reasons. The Social-Democrats continued demanding that existing nuclear plants should be closed. It was sad to observe that two years of extensive debate was washed away in one day. Two years later Parliament accepted a plan to build three large nuclear power stations under the condition that a proper solution for the radioactive waste was found. As government was bound to formal participation procedures in selecting sites, the debate was activated again. Experts and governmental representatives organized hearings at potential sites. These hearings offered an impression how badly treated people felt. All hearings ended in big turmoil and chaos and failed completely: the public was clearly not willing to listen to any new plans. The debate went on, but the Chernobyl accident heavily affected public opinion. After the accident almost 90 percent of the public opposed building new nuclear plants. The political basis for the expansion of nuclear power completely vanished and plans were adapted and partly frozen.

The odd situation now is that although the implementation of the debate conclusions was heavily frustrated the situation with regard to energy policy at this moment, in 1993, is strikingly consistent with the conclusions of the debate. The extent to which this should be attributed to the 'inescapable force' of public acceptance or the after effects of Chernobyl or other factors is of course difficult to assess.

## Applicability and Transferability

As could be observed in the Netherlands political decision making was in a deadlock situation with pro and con in fixed polarized positions. The debate managed to reopen discussions and to consider mutual viewpoints. The debate also ended in a clear conclusion which reflected the public opinion in the country fairly.

A participation project on this nation wide scale concerning a difficult abstract policy issue is without precedents. It is obvious that such an approach can only be reserved for some very basic strategic social issues and on a very low frequency. It is probably a nightmare for policy makers to become subject of such a political process too often. It would not be possible to activate so large numbers of people to participate on a too frequent base. The pattern of objectives demands a substantial organization, a considerable budget and a lot of commitment of many people. An open agenda and a pre-agreed style of moderation demand considerable preparation. Information content is crucial to give the conclusions external legitimacy. This makes a thorough and intensive information phase essential. To involve in this a large and very heterogeneous public requires a lot of effort. It is doubtful whether this kind of approach is feasible in other countries. Size only may be a become a problem to get the program organized. The political system may also be a factor. Dutch national policy has a strong consensus culture. In another policy culture this type of large scale participation might be less adequate.

Inputs from organizations formed an important part in the Dutch model. This was different in the Swedish case where some institutions were involved, but only in a managerial role. An advantage of discussions in newly formed groups is that the discussion is not heavily loaded with old conflicts and polarized positions. An advantage of involving existing organizations is that the link with the normal political process is more tight.

In the Dutch debate Parliament occupied a backseat position. It avoided commitment to the outcomes of the debate and deliberately created the opportunity to wait for the final conclusions and take distance if necessary. Although constitutionally it may be difficult to achieve direct decision making through a debate like described, alternatives may be looked for that create a stronger commitment of parliament to outcomes. A lack of commitment may have a negative effect on the perceived democratic quality of political decision making eventually leading to forms of political alienation. In that sense an in itself impressive participation process can be contra-productive. This was probably the greatest weakness of the Dutch approach.

Concerning transferability it should be clear that the model is typically meant for issues with long lasting strong implications; that demand fundamental discussions on strategic aspects; in which not only technical factors are relevant but also norms and values and worldviews play an important role. Value reconciliation should be an important objective.

The model is typically designed for large scale issues on the national or state level or similar scales. It is a macro-model. The model may be transferred to other fundamental issues of a comparable status and size as the energy problem, for example the issue of genetic engineering or basic issues of sustainable development. However we already referred to the low frequency that should be maintained. When problems are not well developed and too abstract it may also be very difficult to involve many people. It should be recognized that the whole process is a large investment for all involved parties.

The full Dutch program was a hybrid model. Parts which were added to the core model such as controversy sessions can be used in a flexible way.

## Conclusions

The model in this chapter has been characterized as macro model for broad strategic issues which are relevant for long term planning.

The core of the model is that large scale decision making is preceded by direct universal participation based on an elaborated opinion formation and registration process in small groups of citizens that are not selected according to any group membership. The core model can be enriched with various elements such as the direct participation of interest groups or expert debates. Essential is that participation is universal. More than the elitist models, like science courts or the Austrian model in which experts discuss the facts to educate the public, the present model asks for active public participation. Sufficient opportunities have to be offered to anybody to speak up and also to influence the charge of the debate, the agenda and the exact design of the participation model. Value reconciliation is an important motive. A strength of the model is that it allows an open confrontation of diverse viewpoints of experts and laymen with a rational or intuitive basis. The model allows interest groups to develop their viewpoints and to present these in the information phase. Fairness is served by compensating unequal starting points with regard to knowledge and budgets during the process.

A two phase model which starts with an information phase to establish a common base opinion formation and registrations has proved to be feasible and valuable. It reduces the risks that the debate is polluted by diverging levels of information and it enhances the focus on the real value issues which should be in the heart of the debate. A common information base reduces the risk of undesired confounding of facts and values.

Successful group debates are dependent on a carefully prepared agenda and structure and the availability of credible and capable moderators. Reliable and multiple registration of inputs is necessary to allow the debate to result in clear-cut and precise conclusions and to assess the representativeness and the validity of the outcomes.

Finally there is the issue of the link of the process to formal decision making. It is obvious from the Dutch case that ignoring the outcomes on the political level is detrimental to the solution of the issue at hand, but also, in a broader sense, to the confidence in democracy and the participation of the public in other political issues. In that sense it is short-sighted to consider public participation as a method to gain time or as a psychological trick to cool down a heated conflict. One solution to this problem might be the integration with a referendum. It is however questionable how this change might affect the opinion formation process. It would certainly not be an improvement if the debate would deteriorate into a polarizing political campaign. The greatest challenge to the type of participation described in this chapter is to connect an open participation process with an effective political commitment.

## References


Maatschappelijke Discussie Energiebeleid, *Het Tussenrapport, basis voor de Brede Maatschappijke Discussie.* (Stuurgroep Maatschappelijke Discussie Energiebeleid: Den Haag 1983).

Maatschappelijke Discussie Energiebeleid, *Het Eindrapport van de Brede Maatschappijke Discussie. Stuurgroep Maatschappelijke Discussie Energiebeleid.* (Stenfert Kroeze: Leiden 1983).

DeHoo, S. C., and C. Daey-Ouwens, "Forms of Public Participation in the Dutch 'Brede Maatschapijke Discussie' or BMD (Broad Social Debate) on Energy Policy," in: S. C. de Hoo, R. E. Smits, and R. Petrella (eds.), *Technology Assessment: An Opportunity for Europe,* Vol 2, ((Dutch Ministry of Education and Science: The Hague 1987), pp151-174.

Nelkin, D., *Technological Decisions and Democracy* (Sage Publications: Beverly Hills, CA, 1977).

Nelkin, D., and M. Pollak, "Problems and Procedures in the Regulation of Technological Risk," in: C. H. Weiss and A. F Barton (eds.), *Making Bureaucracies Work.* (Sage Publications: Beverly Hills, CA, 1980), pp. 233-253.

Vlek, C. A. J., *Rise, Decline and Aftermath of the Dutch Societal Discussion on (Nuclear) Energy Policy (1981-1983).* (Department of Psychology, University of Groningen: Groningen 1986).

*Chapter 19*

# *The Dutch Study Groups Revisited*

*Jeryl L. Mumpower*

## Introduction

The scale of the Dutch national debate on energy policy (or, BMD, to use its Dutch abbreviation) is remarkable. A brief review of a few key facts underscores its singularity (de Hoo and Ouwnes 1987; Midden 1995). The overall project was funded for over 29 million Dutch guilders (about US $15 million, in early 1980's currency). The Steering Committee that oversaw the project was legally incorporated as a foundation and was supported by a multi-disciplinary staff of 30 people.

The project was divided into two main phases: the *information phase*, which emphasized the collection and dissemination of data and information, and the *discussion phase*, which focused on preferences among the interested parties for various possible decision alternatives. During the information phase of the debate, over 4,000 written position statements on the energy issue were collected, and the 400 most extensive and elaborate of these were carefully analyzed and summarized. Eleven regional hearings were held for the purpose of collecting more in-depth information about the various points of view, and four so-called controversy sessions (analogous to Science Courts) were convened to consider key scientific and technical issues about which there was disagreement or contention. Research was initiated to develop several alternative scenarios defining various plausible future socioeconomic contexts for energy demand and supply, to serve as a backdrop for the discussions, and a number of subsidiary research efforts were supported. Subsidies for information programs and similar activities were given to 77 projects at a total cost of about nine million guilders. A 170-page interim report intended to serve as a data base for informing the ensuing debate was published and widely disseminated at nominal cost. Over one million copies of a brochure summarizing the key points of the report were distributed.

During the discussion phase of the debate, questionnaire data were collected from over 25,000 respondents, plus hundreds of institutions and organizations. Over 1,120 public meetings, facilitated by over a 1,000 trained and paid discussion leaders, were held throughout the country. These were attended by over 19,000 people. Additional activities were conducted in hundreds of schools. A 400-page final report was prepared and widely distributed.

The scale of the project is truly extraordinary, as a few extrapolations to some of the larger Western industrialized democracies helps to illustrate. A national survey that collected opinions from a similar percentage of United States citizens would involve approximately 375,000 respondents! A project of comparable scale in Germany would have involved the distribution of over five million brochures! Clearly, Midden (1994) makes little exaggeration when he concludes that "a participation project on this nation-wide scale concerning such a difficult abstract policy issue is without precedent," although the Swedish Study Circles and the Austrian Nuclear Open Debate of the mid 1970's also represent significant, broad-based national experiments in public education, consultation, and participation (for discussions of all three efforts, see Nelkin 1977).

The scale of the project complicates its evaluation in two significant ways. First, the BMD was multi-faceted, incorporating a number of distinctive components with different objectives. This multi-component character makes it difficult to distinguish the essence of the approach from the parts that are tangential or less central and, thus, makes it difficult to define clearly the object of evaluation. The BMD does not constitute a discrete, well-specified *model* for citizen participation in the same sense that most other techniques included in this volume, for instance the Planning Cells or Citizen Juries, do. Rather the BMD is a hybrid, which incorporates elements from multiple models. For this reason, I will refer to it as the BMD *approach*, instead of as the BMD model.

Second, and closely related, even if the overall evaluation of the BMD is quite favorable (mine is), it does not necessarily follow that such an approach is appropriate for every environmental policy decision. In its full-blown form, this approach is far too resource-intensive for most problems. Trying to evaluate how good the approach might be (or even what the approach would be like precisely) in a scaled-down version appropriate for problems of lesser scope is, at best, challenging and, at worst, speculative.

Because of the hybrid nature of the BMD approach, if the evaluation is to be reasonably unambiguous, it is important to be clear about which elements are regarded as key ones, and which are less central. Midden identifies several elements as vital to the approach:

"The heart of the model concerns the assumptions that the agenda was open for debate, as was the style of moderation. [Providing information] was a central characteristic [preceding] the discussions in phase 2. Also a key point was [...]that no distinction was made between types of reactions, emotional or rational, experts or laymen. Another important characteristic is the involvement of both individual citizens and organizations participating as organizations." (Midden, in this volume).

Midden appears to regard other elements of the approach, such as the subsidies for participants and the use of hearings and the controversy sessions, as important, but adaptable under different circumstances. Not only should he, as one of its developers, be given deference in defining the key components of the approach, his assessment strikes me to be fundamentally correct, and it is the above definition of the BMD approach to which the following evaluative comments pertain. In the balance of this paper, I will

- discuss the BMD approach in terms of the evaluative criteria based on Habermas's Theory of Communicative Action and proposed by Webler in Chapter 3;
- evaluate the approach in terms of several additional criteria that I believe are relevant for assessing public participation techniques; and
- offer some concluding observations about several further issues that the BMD approach raises.

## Fairness and Competence of the BMD Approach

Overall, the BMD rated high in terms of both fairness and competence. Its guiding philosophy emphasizes the right to attend, initiate, and debate – in all aspects of the process. In implementation, the BMD went to great lengths and devoted substantial levels of resources to encourage, recruit, facilitate and empower individuals and groups to participate. Likewise, it went to equally great lengths to ensure that the process was informed by the best possible information made widely available to all the various participants.

### *Fairness*

The evaluative criteria described in Chapter 3 emphasize giving everyone an opportunity to participate in placing concerns on the agenda and defining rules for the participation process. In this regard, the BMD approach scores high, although, it was not, of course, perfect.

Because the BMD was so expensive, it is not surprising that the Dutch Parliament made the final decisions about the charge of the debate and the problem definition. Practically speaking, it is hard to imagine a situation in which decisions involving financial implications of this magnitude would not be made by elected or appointed officials with fiduciary responsibility.

Clearly, the BMD did not provide everyone an *equal* chance to participate. Parliament members' deliberations and final decision were heavily influenced by proposals made by two policy elite groups (the so-called Initiative Group Energy Discussion and the General Advisory Council on Energy, which advises the Minister of Economic Affairs). No model can ever provide truly equal chances for participation, at least not for problems of substantial scope. Those who initially propose the agenda and procedural rules possess an inherent advantage in influencing their ultimate form. Perhaps it is possible to imagine a model of public participation that relied on public surveys or some similar polling mechanism to determine precisely which issues and concerns should be placed on the agenda and debated, but problem definition – a key element in determining eventual problem resolution – will normally be more heavily influenced by elites who have the resources, interest, and access to higher officials necessary to define the terms of the problem.

Most decisions about the final form and process of the BMD were made by the Steering Committee on Social Debate on Energy Policy, established by the Dutch Parliament. For large-scale complex debates, delegation of such decisions is probably inevitable. Under such circumstances, the appropriate question is probably not the extent to which such decisions are participatory, but whether they are indeed responsive to the interests and values of the broader polity.

Midden reports that, in the BMD, efforts were made to create a balanced, independent Steering Committee and staff. In formulating the final version of the agenda, the Steering Committee chairman held discussions with politicians, employers, labor, and environmental, energy, and religious organizations, among others. The agenda was further altered in response to the concerns and interests expressed in the national solicitation for views on the energy problem.

The approach emphasizes iteration and openness. In the BMD, decisions about agenda, process, and outcomes were made slowly and incrementally and were revisited numerous times before becoming final. The presumed subject of debate expanded over time. The initial question focused relatively narrowly on the necessity and desirability of nuclear energy in electricity generation in Dutch society. Over time, the BMD evolved into a much broader investigation of national energy policy, which considered what options were possible, what their impacts would be, and what societal support could be expected for those options.

The BMD approach represents a reasonable, practical accommodation to the inherent tensions between the need to overcome inertia in order to initiate any sort of discourse and the desire to be fair concerning agenda and process. By iteratively reconsidering decisions about agenda and process, the BMD came close to satisfying the spirit, if not the letter, of the fairness criteria, which emphasize participation in determining the basic content and ground rules of the discourse.

Even with substantial levels of resources, such as were enjoyed by the BMD, however, it is likely to prove impossible for public participation models to accommodate all the proposed agenda items anyone may want to discuss and provide everyone a chance to suggest changes to the agenda or rules. Although the BMD approach was quite permissive in these regards, limits were placed on the amount of time and number of iterations permitted before moving to the next stage of the process. Such a policy is likely to prove necessary for any discourse at a national or regional level. Attempting to accommodate everyone's desires concerning agenda is likely to make the process inefficient and potentially subject to manipulation by those who would use such a policy of openness as a vehicle to stall and delay.

According to the evaluative criteria of Chapter 3, public participation models should ideally provide everyone who is potentially affected by the decision proposal (positively or negatively) an equal chance to be present or represented at the discourse. In this regard, too, the BMD scores high, even if, again, and inevitably, it falls short of perfection.

The BMD went to extraordinary lengths to identify those who were potentially affected and to give them the opportunity and capability to participate. A number of steps were taken to protect against the possibility of omitting or neglecting significant points of view. The Steering Committee placed full page advertisements in virtually every newspaper throughout the nation requesting comments on future energy supply. As cited earlier, views were solicited from 260 organizations and public interest groups and about 4,000 replies were received and analyzed. Eleven regional hearings were held to discuss the opinions that were received from the solicitation. Four controversy sessions helped to clarify similarities and differences in the validity claims made about the facts at issue.

## *Competence*

The evaluative criteria emphasize the adequacy of the information and tools supporting the decision making process. The BMD invested substantial levels of resources to ensure that high-quality factual information was available from all relevant sources and that appropriate procedures were used for obtaining and validating that information. Financial subsidies were provided to participants without other means for effectively participating. These helped to ensure that all groups had access to relevant information and had adequate resources to make it possible for them to articulate their positions on the issues. The extensive series of opinion solicitations, hearings, controversy sessions, interim reports, and public opinion assessments contained in the BMD approach helped to eliminate or reduce misunderstandings about terms, definitions, and concepts.

The criteria also emphasize the development of access to an information base. In the BMD, in order to provide a good basis for discussion, considerable efforts were invested in scenario development, although these did not ultimately prove as helpful as originally hoped. The scenarios were intended to help to cope with the uncertainties involved in factors such as future energy needs, costs, and resource levels, among others, but the levels of uncertainty were so great that they did not prove very useful. Additional efforts aimed at developing widespread access to a useful and usable data base included release and distribution of the Interim Report, the widespread distribution of a popularized version of the report, the production of an information sheet, and regular media contacts.

In addition, the BMD provided financial support for supplementary research work in order to obtain what seemed to be important new information and to resolve critical controversies. Decisions regarding supplementary research were made by the Steering Committee on an expected benefit basis that weighed the results that could be expected within available time and their anticipated degree of relevance to the decision.

The criteria for competence in rational discourse emphasize that not only is it important for participants to possess access to a high-quality factual data base, but that public participation models should ideally promote the discovery and development of mutual understandings of values among all the participants. Not only should the participants understand the relevant facts of the matter, but also the value positions of one another.

The BMD invested substantial levels of resources toward this end. Extensive public opinion surveys supplemented the opportunities for learning that were available in the various public discussions and hearings sessions. Over 25,000 people filled out questionnaires regarding their preferences for the energy future.

The extensive use of public discussion and hearing sessions helped to encourage the likelihood that participants would engage in introspective reflection regarding their normative values, would consider the factual implications of their potential normative choices, and would reflect on the implications of their value positions. The numerous opportunities for discussion helped to create a circumstance that promoted the discovery and development of a mutual understanding of values that could serve as a basis for building compromise. Moreover, such discussions may contribute to the emergence of transcendent, collective interests and values. Even if the process does not lead to immediate resolution of the immediate problem, it may help to develop a more general level of societal competence involving broad-based participation and improved communication and understanding.

According to the evaluative criteria of Chapter 3, an ideal model of public participation should ensure that normative choices are not inconsistent with themselves, the general will, the law, or present expectations. The BMD does not address these issues specifically, but the extensive length of time and the substantial level of financial resources devoted to the process no doubt helped to weed out such inconsistencies and incompatibilities. On the other hand, the BMD application did not include provisions explicitly designed to structure values systematically (e.g., through decision or judgment analysis), which might have helped in this regard.

## Beyond Fairness and Competence

The fairness and competence evaluative criteria derived from Habermas's Theory of Communicative Action provide a basis for assessing the degree to which public participation models satisfy the ideal conditions for rational discourse. Within applied and political contexts, however, additional, practical concerns will also inevitably arise. These include, among others, cost, time, political support, and overall practicality. Moreover, the various criteria – both those enumerated in Chapter 3, plus the additional ones cited here – that we would like to see fulfilled by public participation methods will typically compete with one another. It will be impossible to satisfy all these criteria simultaneously. In order to achieve certain objectives, others must be foregone; tradeoffs are necessary.

In this section, I will discuss several additional criteria that seem germane to the BMD experience and consider the competition between various objectives and criteria.

### *Cost*

Public participation techniques should be affordable, as well as fair and competent. In its full-fledged form, the BMD is highly, perhaps prohibitively, expensive for all but the most important, broad-scale problems, as Midden himself notes. The Netherlands is a relatively small country by European standards, and even more so by American and Asian standards. Several American states are larger in population than the Netherlands, and a number of cities in the developing world are larger in population than the entire country. The Netherlands is also prosperous, by any measure. Its citizens enjoy one of the highest standards-of-living in the world.

Can a technique that works in the context of a small, relatively homogeneous country also work in much larger, more heterogeneous settings? Estimates of the costs of the BMD approach, if it were applied in more populous countries such as Germany, or the United States, approach truly formidable sums. Extrapolating on the basis of the relative sizes of the populations, a similar project would cost over 100 million DM in Germany and approximately $250 million in the United States. Perhaps certain economies of scale might be achieved, but additional costs would also likely be incurred as a result of difficulties and complications arising from implementation in a setting involving higher levels of social and cultural heterogeneity and vastly increased geographic scale.

Perhaps more to the point, does the BMD represent a sensible or even feasible approach in any countries other than the relatively small handful of prosperous industrialized nations? In anything resembling its full-blown form, I doubt it. Ironically, for an approach whose philosophy places great emphasis on promoting fairness in the public policy formation process, implementation of the BMD may itself raise questions of fairness, especially in poorer countries. Expenditures for promoting public participation compete with potential expenditures for other worthy public purposes, needs, and desires. Participation is an important public good, but there are many others, of course. Funds spent to support public participation cannot be spent for public health, improved medical care, risk reduction, environmental preservation, social welfare, national defense, or whatever.

## *Time*

Several of the evaluative criteria emphasize the need for sufficient time to allow everyone who wishes to do so to introduce items for the agenda, propose procedural rules, discuss and debate validity claims, and so forth. Satisfying these criteria is ordinarily accomplished at some cost in efficiency, however. Time is a finite commodity. Most legislative bodies exert considerable informal pressure, and sometimes still find it necessary to resort to formal rules (or at least the threat of such rules), to limit the amount of time that speakers are permitted on the floor and, thus, to circumscribe individual members' abilities to stall or prevent votes or other legislative action through tactics such as filibustering. If even small, representative bodies have found it necessary to impose time limits and to make other rules designed to keep the deliberative process moving forward, it seems that approaches that involve larger, more participatory bodies, such as the BMD, will likewise find a need to do so, especially after participants have gained sufficient experience with them so that they have learned strategies and tactics for manipulating the process to their advantage.

## *Political support*

Despite high marks with respect to the fairness and competence criteria, the BMD had relatively little direct political impact. It was rejected by some, endorsed by others, but more-or-less summarily evaluated by both proponents and detractors alike. Midden (1994) poignantly notes that "It was sad to observe that two years of really intensive debate was washed away in one day." Perhaps the failure of the BMD to have an immediate impact on policy (its ultimate impact remains harder to evaluate) lay in the failure of its architects to involve Parliament more directly in the design and implementation of the effort, but I suspect that the core of the problem is even more fundamental than that.

The evaluative criteria specified in Chapter 3, and the tradition of Habermasian thought from which they were derived, implicitly assume public participation to be part of a process that involves group decision making, problem solving, and cooperation. This presumes that participants share fundamental basic objectives and values, even though they may disagree about their relative importance or the appropriate levels of resources that should be invested to promote their achievement. Not every public policy problem can be framed this way. For some problems, the intended parties may have opposing or incompatible interests that must be resolved through more adversarial processes such as bargaining, negotiation, and confrontation. Such problems may eventually be successfully resolved, but the key to resolution lies in balancing interests, not in reconciling them or reaching consensus.

Certain problems, the Dutch national energy policy issue perhaps among them, may not be susceptible to resolution simply by the skillful application of a fair and competent procedure for managing discourse. The lack of political impact of the BMD exercise may be more reflective of the nature of the problem than the shortcomings of the approach. To expect more may be to expect too much. Public participation may play a role in helping to achieve a resolution to the problem, but it is perhaps unrealistic to expect that the process will lead directly to a consensually-based settlement.

Furthermore, whatever potential impact on the policy process can be expected for the BMD approach by virtue of its qualities of fairness and competence, it is unlikely that full impact will be achieved in early implementations. Not only are fairness and competence important to the success of public participation techniques, so too are the appearance of fairness and competence and, ultimately, trust in the fairness and competence of the process. The parties in a political debate are almost certain to be suspicious of a new technique and hesitant to endorse or accept it fully until repeated experience has demonstrated its ability to perform fairly and competently. Even then success is not guaranteed, because of the possibility of antagonistic efforts on the part of those whose primary objective is to advance their own individual interests, not to be fair.

## *Practicality*

The capacity for implementation is an important criterion for evaluating public participation techniques. Any other virtues of such techniques, no matter how substantial, become moot if the techniques cannot successfully be put into practice. The BMD is clearly more than simply "pie-in-the-sky," since it has been successfully implemented at least once, but it is not unfair to question how practical the BMD approach would be over the long run, with repeated applications. Are the ideas upon which the BMD approach is based practical or utopian?

In a recent review of several books dealing with environmental policy issues, Lynn (1992) posed a number of provocative questions applicable to the BMD and all other techniques that attempt to improve the quality of policy making through expanded involvement of a well-informed public:

"It is nothing less than utopian, this notion – one might say, conceit – that if members of policy elites could explain the relevance of their benefit-cost analyses and risk assessments to 'the public,' if 'the public' could be given a coherent account of the trade-offs they face, if self-interested actors and establishments who remain intransigent could be disempowered in favor of an inquiring and open-minded citizenry, public polices more clearly reflective of 'the public interest' would result.

"Those tempted by this view might perform the thought experiment of imagining that we could start with a fully-empowered citizenry of diverse value and ambitions, competing commitments to work, family, and community, varying levels of interest in the limitless number of issues on which they are in a position to deliberate and scarce resources of time, attention, and concern for larger questions. These free citizens just might do things like recruit agents to represent their interests, fail to reveal their preferences, behave in non-cooperative ways, seek to manipulate the terms of debate, become lackadaisical in monitoring their agents, ignore relevant information, become skeptical of expert claims, and allocate their scarce resources of time and attention toward those things they care about most. 'Then what would we do?'" (p. 137)

## *Competing values*

Whether attention is restricted to those criteria proposed in Chapter 3, or a broader range of criteria is considered, the various desiderata for public participation techniques are often in conflict with one another. A conceptual framework for thinking about public participation that explicitly recognizes the conflict between criteria is the competing values approach proposed by Quinn and Rohrbaugh (Quinn and Rohrbaugh 1981, 1983; Rohrbaugh 1987).

In the competing values approach, two basic value dimensions combine to define perspectives on effective decision making processes. The first value dimension is related to process structure; an emphasis on flexibility competes with an emphasis on control. The second value dimension is related to the focus of the process; an internally-focused emphasis on the needs and desires of those directly affected competes with an externally-focused emphasis on the needs and desires of the broader public. The relative emphasis on these competing values defines four distinct perspectives on effective decision making: the *rational* perspective, the *empirical* perspective, the *consensual* perspective, and the *political* perspective.

Each perspective emphasizes certain types of processes (means) that are intended to achieve desired outcomes (ends). For example, the consensual perspective emphasizes participatory processes, which lead toward supportable decisions, whereas the empirical perspective emphasizes data-based processes, which lead toward accountable decisions. The competing values approach is displayed graphically in Figure 1.

The competing values theory says that there is "no free lunch" with respect to decision making processes. Such processes naturally induce tradeoffs. Satisfying some objectives ordinarily occurs at the expense of others. More specifically, the theory identifies certain means-end relationships which are at odds with one another, as indicated by the diagonal lines in Figure 1. For instance, the theory suggests that adaptable processes do not typically lead to accountable decisions. Processes that maintain flexibility and which change or evolve over time do not yield the type of accountability associated with decisions reached via stable, constant, and consistent methods that have been clearly-specified in advance. Similarly, participatory processes do not lead to efficient ones. Processes that permit interested parties to participate intensively over extended periods of time do not typically lead to decisions that can be defended in terms of having been reached with minimal investments of resources in service of crisply articulated goals.

An intriguing question for future consideration is the degree of correspondence between the competing values theory proposed by Quinn and Rohrbaugh and the fairness and competence criteria proposed in Chapter 3 and inspired by Habermas' ideas. Are the elements above the horizontal midpoint of the competing values framework displayed in Figure 1 more-or-less the same as fairness? In other words, are the *fairness* criteria equivalent to achieving *legitimate* decisions through *adaptable* processes (the political perspective) and achieving *supportable* decisions through *participatory* processes (the consensual perspective) in the competing values scheme? Similarly, are the elements below the horizontal midpoint of the competing values framework as displayed in Figure 1 more-or-less the same as competence? In other words, are the *competence* criteria equivalent to achieving *accountable* decisions through *data-based* processes (the empirical perspective) and achieving *efficient* decisions through *goal-centered* processes (the rational perspective) in the competing values scheme?

If the proposed correspondence between the evaluative criteria of Chapter 3 and Quinn and Rohrbaugh's competing value theory is not too far-fetched, it suggests a still more intriguing question: Do fairness and competence compete with one another? This question requires more extensive treatment than can

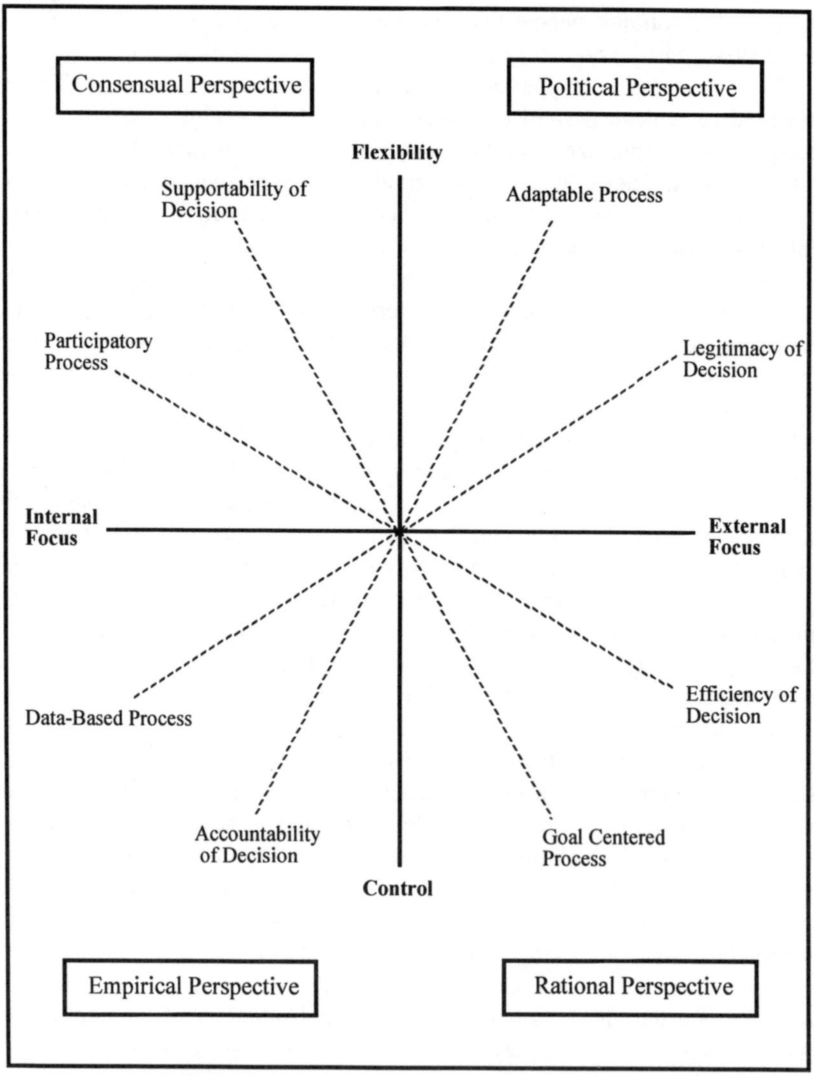


Figure 1: The Competing Values Model: Adapted from R.E. Quinn and J. Rohrbaugh (1983), "A spatial model of effectiveness criteria: Towards a competing values approach to organizational analysis," *Management Science*, 29:369.

be accorded it here, but I am inclined to think that the answer will generally be "yes," especially when resources are limited, which, of course, they always are. As the BMD experience demonstrates, resources devoted to ensuring widespread participation cannot be devoted to obtaining and disseminating the relevant data base. More fundamentally, Midden and his colleagues found it immensely challenging to try to supply lots of information, have organized discussion, elicit "free opinions," collect information in a structured quantitative form, and obtain a representative sample of the population, all at the same time.

## Concluding Remarks: Additional Issues and Concerns

The BMD experience raises several additional issues and concerns that are relevant to the evaluation of public participation techniques. These include decisions about whether or not to introduce public participation in policy making, a priori versus a posteriori standards of fairness, existing versus future preferences, the tension between democratic and representative processes, the issue of standing in disputes, and the role of science in public participation techniques.

### *Deciding on public participation*

Public participation is not achieved without cost. Because organized programs of public participation require budgetary support, some decision making power will fall to the hands of officials with budgetary powers. There is no way to avoid this, unless we institute public participation for purposes of governmental budgeting (perhaps not such a bad idea.) An underlying issue, then, is who decides what problems deserve the allocation of resources necessary to support public participation programs. In the BMD case, the budgetary decisions were made by the national legislature. Decisions about public participation (at least about specific public participation projects) will probably not often be made at such august levels, but, except for the most modest forms of public participation or spontaneous protests, programs of public participation will be dependent upon enabling decisions by those who control access to financial and other resources.

### *A priori and a posteriori fairness*

There are important differences between fairness before the fact and fairness after the fact, or, stated otherwise, between the fairness of the process and the fairness of the outcome. Keeney (1986) discussed this distinction with respect to the risk of fatalities. A risk that subjects each and every member of a population to a 1-in-10 chance of death is fair beforehand, but it is extremely inequitable after the fact to the individuals who make up the ten percent of the population who died from the risk.

A similar problem exists with respect to certain classes of public policy problems, such as siting. Citizens may endorse the siting procedure until it selects their backyard as the site. The fact that an individual approved of a participatory process for developing the siting criteria and, indeed, perhaps even participated in that process may affect his or her willingness to accept the verdict, but it certainly offers no guarantee. To willingly accept the result of a process that sites an undesirable facility in their backyard, many individuals are likely to require that local benefits outweigh local costs and risks, no matter how fair they think the process is.

## *Existing versus future preferences*

Like most public participation techniques, the BMD considered alternative decision options that were reflective of the range of existing social preferences. This raises interesting issues for both fairness and competence. For decisions that affect future generations – as many of the types of decisions important enough to warrant organized public participation programs do – the techniques are unfair because they focus solely on the concerns of those present and participating in the process. Future generations are effectively disenfranchised.

Even if we did try to anticipate the preferences of future generations and to incorporate them in the decision making process, it seems unlikely that we would be able to do so very well. In 1950's and 1960's, when many of the decisions that we now live with involving nuclear technology were made, could we possibly have anticipated the emergence of environmental concerns? Clearly we did not, but is it even possible that we might have anticipated the changes in values and technological impacts associated with the global climate change, ozone depletion, acid deposition, deforestation, and so forth? If we cannot reasonably hope to foresee the future development of social trends and values, fair and competent public participation may mean making decisions that are readily reversible or remediable by future generations.

## *Democratic versus representative participatory processes*

Techniques of public participation are ordinarily inspired, either explicitly or implicitly, by a strong belief in the value and superiority of participatory processes. All modern forms of democracy, however, are representative. In *The Federalist Papers*, written by Alexander Hamilton, James Madison, and John Jay to defend what was in their day a revolutionary charter – the United State Constitution, there is an extensive discussion of the tensions between representative and participatory forms of democracy. In Paper 63, Madison

(1787, 1961) argued for the need for a Senate – a relatively small body of citizens with relatively long tenure. He said:

"[T]here are particular moments in public affairs when the people, stimulated by some irregular passion, or some illicit advantage, or misled by the artful misrepresentations of interested men, may call for measures which they themselves will afterwards be the most ready to lament and condemn. In these critical moments, how salutary will be the interference of some temperate and respectable body of citizens, in order to check the misguided career and to suspend the blow mediated by the people against themselves, until reason, justice, and truth can regain their authority over the public mind?" (p. 384)

In the United States, the balance between participatory and representative forms of democracy once again became a salient issue during the 1992 presidential campaign, when one of the candidates (Ross Perot) repeatedly suggested that television and computer technology could provide an electronic means for much higher and more immediate levels of public participation in public policy making, via electronic polling or referenda. The BMD experience underlines the need to give serious attention to the relationship between the ideals of participatory democracy and those of representative democracy and to think hard about the relation between public participation and the powers and responsibilities of elected and appointed bodies.

## *Standing in disputes*

The evaluative criteria of Chapter 3 place a positive value on processes that permit participation by anyone who says they believe that they are affected by the decision. The BMD approach takes a similarly inclusive stand. The motivation for this principle is to ensure that people who are affected by decisions are not disenfranchised. The principle also runs the risk that certain individuals or groups will exert disproportionate influence on decisions that have a much greater influence on others, or will simply behave frivolously or irresponsibly. In most legal settings, individuals or groups must demonstrate that they are substantially and directly affected by the issues at question in order to enjoy legal standing. The question of where to draw the line between those who enjoy standing in disputes and those who do not has no definitive answer, of course, and must be constantly revisited. It does not seem self-evident, however, that public participation techniques should always permit the individual to adjudicate whether he or she should be permitted a voice in the deliberations, or how much weight should be accorded to that voice. Permitting anyone who wishes to participate to do so could conceivably lead to inequitable situations in which the interests of those only marginally affected by decisions outweigh the interests of those who are much more directly affected.

## *The role of science in public participation*

During the course of the BMD process, expectations steadily declined concerning science's ability to contribute to the solution of a variety of controversies in the area of energy supply. The scenarios developed on the basis of the best available scientific information and projections proved to be far less realistic, credible, and useful than had been hoped.

This lack of impact seems to be a frequent finding of those working on public policy problems that involve a substantial scientific or technical component (regardless of the degree of public participation that is involved.) A primary contributing factor appears to be the substantial levels of uncertainty – both real and apparent – regarding the relevant science. The BMD experience thus points to the critical need for further study and thought concerning the proper role of science in public participation processes.

## References

De Hoo, S. C., and C. Daey Ouwens, "Forms of Public Participation in the Dutch "Brede Maatschappelijke Diskussie" or BMD (Broad Social Debate) on Energy Policy: Results and Use," in: S. C. de Hoo, R. E. H. M. Smits, and R. Petrella (eds.), *Technology Assessment: An Opportunity for Europe* (Dutch Ministry of Education and Science: The Hague 1987), pp. 151-174.

Hamilton, Alexander, James Madison, and Jay, John, *The Federalist Papers*, (Mentor: New York, 1961; reprinted from 1786-1787).

Keeney, Ralph L., "The Analysis of Risks of Fatalities," in: Vincent T. Covello, Joshua Menkes, and Jeryl L. Mumpower (eds.), *Risk Evaluation and Management* (Plenum: New York 1986), pp. 233-248.

Lynn, Laurence E., Jr., "Book Reviews; *Site Unseen: The Politics of Siting a Nuclear Waste Repository*, Gerald Jacob; *Integrating Insurance and Risk Management for Hazardous Wastes*, Howard Kunreuther and Rajeev Gowda, Editors; *The Environmental Protection Agency: Asking the Wrong Questions*, Marc K. Landy, Marc J. Robers, and Stephen R. Thomas," *Journal of Policy Analysis and Management* 11. No. 1 (1992), 133-137.

Midden, Cees J. H., "Direct Participation in Macro-Issues: A Multiple Group Approach: An Analysis and Critique of the Dutch National Debate on Energy Policy, Fairness, Competence, and Beyond," (in this volume).

Nelkin, Dorothy., *Technological Decisions and Democracy*, (Sage: Beverly Hills 1977).

Quinn, Robert E., and John W. Rohrbaugh, "A Competing Values Approach to Organizational Analysis," *Public Productivity Review* 5 (1081), 141-159.

Quinn, Robert E., and John W. Rohrbaugh, "A Spatial Model of Effectiveness Criteria: Towards a Competing Values Approach to Organizational Effectiveness," *Management Science* 29 (1983), 363-377.

Webler, Thomas, "'Right' Discourse in Citizen Participation: An Evaluative Yardstick," (this volume).

Rohrbaugh, John W., "Assessing the Effectiveness of Expert Teams," in: Jeryl L. Mumpower, Lawrence D. Phillips, Ortwin Renn, and V.R.R. Uppuluri (eds.), *Expert Judgment and Expert Systems* (Springer Verlag: Berlin 1987), pp. 251-267.

*Chapter 20*

# *The Pursuit of Fair and Competent Citizen Participation*[1]

*Ortwin Renn*
*Thomas Webler*
*Peter Wiedemann*

## Comparative Evaluation of Eight Models for Environmental Discourse

In the previous sixteen chapters of this volume, eight models for public participation in environmental decision making have been proposed and reviewed using the discursive standard criteria laid out in Chapter 3. These evaluations are valid and insightful and go a long way toward providing a basis for positioning the eight models relative to each other. In this chapter we editors further pursue this objective by performing a comparative evaluation of all eight models on the discursive standard criteria. We then take this process one step further by proposing the foundations of a framework for classifying environmental decisions and public participation models. Finally, we close with comments and observations about the challenges still facing those who organize and take part in public participation.

### *Evaluation process and results*

Each of us privately evaluated all eight of the models on the discursive standard criteria. We then compared our evaluations and discussed points of disagreement. Most of our disagreements were based on misunderstandings and, with a little discussion, we were able to reach consensus on all of the evaluations.

Tables 1 and 2 present our summary evaluations of how each of the eight models perform on the main criteria. The criterion of fairness is separated

1 Views expressed in this chapter are those of the editors and do not necessarily reflect those of the other authors.

*Table 1.* Comparative evaluation of how each model restricts access to the discourse.

| Access to the Discourse | | | | | | |
|---|---|---|---|---|---|---|
| **Model** | **Highly restricted** | | | | | **Open** |
| CAC's | x | | | | | |
| RN | x | | | | | |
| Compensation | | x | | | | |
| Mediation | | | x | | | |
| Citizens Juries | | | | x | | |
| Planning Cells | | | | x | | |
| Iniatives | | | | | x | |
| Study Groups | | | | | | x |
| | | | | | | |

*Table 2.* Comparative evaluations of the eight models on the discursive standard criteria. (Four kinds of discourse are listed under competence: explicative, theoretical, practical, and therapeutic. Three activities are listed under fairness shape and agenda [Ag], choose rules and moderator [Mod], discussion [Disc]. Scoring is as follows: + = meets most criteria, S = meets many criteria, = meets few or none criteria.)

| | **Competence** | | | | **Fairness** | | |
|---|---|---|---|---|---|---|---|
| **Model** | **Expl.** | **Theo.** | **Prac.** | **Thera.** | **Ag.** | **Mod.** | **Disc.** |
| CAC | + | S/- | S/- | S | S | + | S/+ |
| NRM | + | + | - | - | - | - | + |
| Compensation | - | S | S/+ | - | S | - | - |
| Mediation | + | S | S | - | + | + | + |
| Citizen Juries | + | -/S | S | - | - | S/- | S |
| Planning Cells | S | S/+ | + | S | + | S | + |
| Cit. Initiatives | + | - | S/+ | S | S | S | + |
| Dutch Study | + | + | S | -/S | - | - | + |

into two sets of sub-criteria. In Table 1 the limitations of attendence of each model are compared. The assessments ranged from extremely constrained, as with regulatory negotiation (RN) where the regulatory agency picks the groups allowed to participate, to wide open, as with the Dutch Study Groups. Public hearings and public referendums are two additional models which place few restrictions on who may participate.

Table 2 contains the evaluations on the other sub-criteria of fairness (which have to do with access to liberties within the discourse) and the criteria for competence. Evaluations of competence are made on each of the four kinds of discourse. The main sub-criteria for competence – access to knowledge and use of the best selection procedures – are collapsed.

## *Citizen advisory committees (CAC)*

Attendance in CACs is highly restricted. Members are usually hand-picked by the institutional body seeking advice. Selection is based on the principal: to include a representative sample of the major interest positions. The full range of interests cannot, if only for logistic reasons, be included. Interested citizens may be able to attend meetings, but they cannot participate in structuring the agenda or the format, nor may they participate in discussion. Committee members are typically not involved in making the agenda, the sponsor establishes the charge and focus. Members are also discouraged from discussing or influencing the moderation, although they may convince the sponsor a change is needed if they are persistent.

Inside the process, CACs offer few opportunities for participants to discuss issues outside of the prearranged charge. Lynn and Kartez have described some of the more progressive CAC models, but these represent the potential that CACs could obtain rather than the norm. Within the confines of the agenda, discussions within CACs are fair. The committee has all the time it needs to discuss the allowed agenda items and a limited amount of freedom in deciding how to go about their work. They can set their own agenda (within limits of the charge), meeting times and places, rules for discourse (within limits), and may be able to select their preferred means to resolve disagreements. Nothing in the structure restricts members from raising explicative, factual, normative, or expressive issues, but the sponsor may not provide the resources to pursue certain subjects. "Unmanageable" CACs would likely be disbanded prematurely, starved to death, or simply ignored.

There is no guarantee that CACs will choose consensually approved methods for resolving disputes, but their structure certainly promotes consensual decision making. The group is expected to work out differences among themselves. There is no hierarchical relationship among the members, although there may be a chairperson appointed to serve as liaison to the sponsor. All of the discussions occur in a face-to-face communicative exchange. However, because the CAC model does not make an explicit commitment to the autonomy of the individual members, there is a danger that the consensus could be a false consensus, as some participants might feel pressured to conform. This danger could be alleviated by a style of moderation that welcomes minority viewpoints.

CACs are much less structured than Planning Cells or Citizens Juries. The lack of structural features such as a peer-reviewed educational program, an adversarial hearing on different points of view, or a value tree make the competence of theoretical, practical, and therapeutic discourse marginal at best. Access to the knowledge needed depends largely on the resources allocated by the sponsor. Explicative discourse has the best chance of being competent due to the education component. Theoretical discourse especially is promoted only insofar as the sponsor provides competent technical advisory staff. However, anecdotal evidence and alternative learning experiences are not likely to be valued. There are no structural properties of the model that force the CAC to consult with experts, seek out anecdotal experiences, or engage in theoretical discourse. Moreover, technical information provided to the CAC by the regulatory agency may not be peer-reviewed. This contrasts with the theoretical discourse occurring in the RN and mediation models, which is strongly structured.

CACs may never reach shared understandings about the factual or normative issues that are competent. Although their small group size, the face-to-face discussions over relatively long periods of time (regular monthly meetings), and the role of the participants as "value-cluster representatives" promote practical discourse, the membership of the committee is typically chosen from among the leaders of the community. They belong to the same class of elites as the governmental officials, experts, and stakeholders. They are more apt to rely on instrumental understandings of the problem and downplay the value of anecdotal evidence and competing normative argument. One illustration of this point is the resentment that "outsider" groups feel toward advisory committees.

To function more fairly and competently, CACs would need a budget, the freedom to allocate it, and more autonomy over specifying its charge. Of course the later would discount the likelihood of the sponsor adopting the CACs recommendation. CACs also work best when the problem under consideration is not wholly technical, but includes different types of tradeoffs spread over several interest positions and those positions are represented in the panel. Normative discourse would be promoted by CAC adopting some mechanism to solicit feedback from the non-participating public.

## *Planning Cells*

Planning Cells are a means to acquire informed recommendations about a specific policy or decision problem from a group of representative citizens. The group does not have any decision making authority, but its recommendations are intended as an aid to the legally-responsible or traditional decision making authority. The main challenges Planning Cells must overcome are (1) to gather a representative set of citizens and to bring them together in a structure conducive to dialogue and reflection, and (2) educate them about the problem at hand.

The fairness of Planning Cells hinges on the success of the random selection process. Random selection theoretically assures every individual and interest position in the affected population an equal chance to participate in the process. To be successful, an accurate picture of the relative distribution of interest positions in the targeted population is needed. In addition, there must be a popular awareness of the problem so that people whose interests are not primarily affected will choose to participate.

Another weakness in the fairness of Planning Cells is the tendency to force a moderator on the participants rather than inviting the participants to be more directly involved in designing the process right from the start. For the most part, the moderator is selected by the staff. People have a chance to speak out against the moderator, however, and they may persuade the staff to find a new one if they are persistent.

Planning Cells are highly competent in explicative, theoretical, and practical discourse. A distinguishing feature is their focus on bringing together technical knowledge, political and stakeholder interests, and citizen preferences into a coherent outcome. Technical information about the physical and social consequences of the decision options, political information about governmental agencys' and other political actors' reliability and credibility, and value information about stakeholders and citizens are gathered in systematic ways to produce a comprehensive understanding of the decision options.

Competence of Planning Cells is strong for a variety of reasons. First, seeking out knowledge and making it accessible to all participants is a major responsibility of the staff and the facilitator. The staff prepares the initial educational material (using peer review by uncommitted expert panels) and identifies potential expert witnesses for the testimony stage (the list can be amended by the citizens). Second, the structural conditions for discourse make certain that all participants have a chance to speak and that translations of expressive arguments are successful (although there is no formal procedure for this). If the citizens show no interest in interviewing certain interest groups or in attaining crucial pieces of information, the facilitator will encourage them to do so. Finally, it is the citizens who must conduct the exploration into the decision impacts on different value dimensions and incorporate those understandings into an holistic decision.

Planning Cells require a large investment of time and money and are not suitable for all types of problems and all contexts. If the problem is highly technical, it may be impossible to bring citizens up to the necessary level of understanding. Furthermore, if the decision options are too narrowly restricted and there is not enough room to allow tradeoffs on decision criteria, then the process will fail because everyone will find it easier to protest the entire process, rather than choose between two unwanted alternatives. This is especially true for siting issues. Planning Cells are relatively new. Unlike legal juries, they lack the cultural tradition that provides some degree of legitimacy and places them above immediate criticism. Even if the members are representative of all interests and values in the affected community, the novelty of the process leaves it open to criticism.

## *Citizens Juries*

Like Planning Cells, Citizens Juries rely on a randomly selected pool of citizens to evaluate policy alternatives. The major difference between the two is that Citizens Juries are much more focused than Planning Cells. Typically, Citizens Juries are asked to express a preference among three or four pre-selected policy options, Planning Cells are more engaged in designing policy options as well as making recommendations about what additional criteria might lead to policy acceptance.

There is some ambiguity with respect to fairness. The staff sets the charge and imposes the principal of majority vote as the means used to resolve conflicts. Majority vote does not guarantee the integrity of minority interest positions in the discourse. The new model Citizens Juries, which adds an oversight committee to set the agenda, the rules for discourse, and oversee the moderation adds fairness to the process, but the membership of the oversight committee is limited to those selected through random sampling procedures.

Explicative discourse is handled more competently by Citizens Juries than any other model discussed in this volume. Not only are the jurists free to raise comprehensibility claims at any time during the process, but they are provided with a draft glossary of terms and definitions (along with an educational packet) on the very first meeting. Disagreements or confusion about definitions or other comprehension questions can be addressed by the jury and the glossary revised.

Competence of discourse within Citizens Juries is encouraged by face-to-face mediated discussions in which everyone has an equal opportunity to make statements, challenge others' statements, and express opinions. One of the main differences with Planning Cells, and a weakness of the Citizens Jury model is that Citizens Juries do not promote critical inquiry into the factual issues and no systematic method is retained to reach the best possible understandings about facts and states of affairs. Crosby (this volume) recognizes a need to change this, but notes it is an issue of cost. On the other hand, anecdotal evidence is welcomed and respected in Citizens Juries and peer review of technical information packages may be used.

Participants of Citizens Juries have access to the knowledge needed to make normative judgments, largely by virtue of the random sample, which should include the plurality of interest positions and perspectives. They are also free to ask their own questions to "witnesses" who come before the jury, thus collecting knowledge they deem relevant and verifying understandings. The format of the discourse also promotes normative inquiry and debate among the jurists. However, the best methods to arrive at shared preferences about the available normative choices are not used. Citizen Juries do not structure values, make impact assessment profiles on the value dimensions, or evaluate the consistency of the outcome with pre-established norms and laws.

Therapeutic discourse is promoted in Citizens Juries, but only informally. There is no formal process for translating expressive statements or for verifying the sincerity of the speaker, but the jurists do develop a "safe" atmosphere for discussion, expression of feelings is not necessarily discouraged, and people are free to challenge the sincerity of the speaker.

As with all citizen panels approaches, both Citizens Juries and Planning Cells rely heavily upon: a clear mandate from the legal decision making institution to seriously consider implementing the recommendations of the panel; the involvement of randomly selected citizens; and legitimacy of the randomly selected citizens in the eyes of stakeholders and other citizens. It may be suspected that citizens in citizen panels would be too shy to ask critical questions of experts, regulatory officials, and interest group representatives. After all, these are the people who usually make the decisions. But in the experience of Renn et al. (1993), Crosby et al., (1986), and Kathlene and Martin (1991), citizens fulfill very well the task of "value consultants." They read the material, seek out needed information, and consider the subject matter very seriously – both reflexively and in groups. Experience has shown that citizens who participate take the process very seriously.

## Citizen Initiatives

The Citizen Initiative model provides an interesting mix of techniques to promote fair and competent environmental discourse. On the one hand, it strives to be fair by including in the process all individuals who are proven by objective means to be affected. These people are contacted, educated, and invited to participate to the fullest extent. At the same time, the initiative model works hard to be technically competent. The staff has expertise in the subject matter, and additional experts are hired or contacted to participate as needed. All technical matters are subjected to the scientific scrutiny of the staff and the governmental agency. Such theoretical discourse, however, takes place outside of the sight of citizens.

All people that the staff deems are adequately affected by the environmental problem are invited to participate in shaping the agenda (at "forum" meetings) and in the discussion sessions (public meetings). However, it appears that, despite ad hoc successes, the real agenda is imposed by the agency with the decision authority. Politicians and regulatory agency staff are also included, but little or no interest group outreach is done.

The Citizen Initiative process is heavily influenced by the moderator, who is paid by the governmental agency responsible for implementing the solution to the environmental problem. Changing moderators would not be easy. The moderator is explicitly committed to ensuring fairness in the discourse, encouraging less assertive individuals to speak out, translating technical jargon, and analyzing deficiencies in technical understandings. Consensus is the preferred mode of decision making.

Competence on the technical issues is mainly due to the skills of the staff and the available resources of the regulatory agency. If the polluter is identified, its involvement would provide even more resources. However, the lack of a external peer review, an external impact assessment, or public review of technical information, leaves open the possibility for errors or mistakes. The model also relies heavily on systematic scientific knowledge. It does not welcome anecdotal or intuitive contributions.

Practical discourse is enhanced by the fact that rules for discourse are explicitly drafted and signed by all participants. A diversity of individual interests are likely to be represented, since no one who is obviously affected is barred access. However, normative choices are more likely to represent a cost benefit calculation than the result of practical discourse, since values are never systematically structured or impacts evaluated.

Therapeutic discourse seems especially discouraged in this model, in part out of a necessity to strictly organize large public meetings and keep focused on the technical dimensions of the decision options. As with mediation, therapeutic discourse is more likely to take place in private discussions between individual participants and staff.

Citizen Initiatives work best when the problem is localized; the affected population is clearly distinguishable; citizens can benefit from having an outside group organize the initiative process; and there is external legal and time pressures that support the generation of a consensual decision (Claus, in this volume).

## *Regulatory Negotiation (RN)*

RN is a decision making process that is strongly shaped by the regulatory agency convening the event. The agency selects the participants, defines the charge, and hires the facilitator. Once the process begins, however, the agency takes part as an equal participant among the other members. All decisions are made consensually. Once organized, the committee has the power to add members, set the timetable, change the facilitator, pursue information needs, and terminate the process.

RN is committed to arriving at technically competent recommendations. Members either are experts or they become experts during the process. The competence of emerging agreements has been highly touted by past participants, and indeed, rules made by negotiation are less frequently challenged in court. But this may be because the parties that have the resources to go to court are those invited to participate in the negotiations. Unfortunately, negotiations are usually held out of the public eye. Therefore it is difficult to assure that the consensus is truly non coercive. There is little to prevent a negotiated compromise that is based on strategic maneuvering rather than the best technical understandings or normative debate. Since the negotiations are held in private, it is impossible to verify the motivations beneath the agreement. Access to information is a commonly-cited benefit to participants in negotiated rulemaking, but what guarantee is there that the information is not being used only toward the instrumental needs of the involved parties?[2]

One might consider that the technical competence of RN must be high, because the proposed rule will be scrutinized by public review. However, as Hadden (this volume) points out, the resulting rule of the discourse may indeed be subject to public review via the *Federal Register*, but the agency's commitment to the RN process and its outcome may clash with its commitment to considering public comments on the proposed rule.

RN has no explicit intention to address normative issues outside of a technical context in which cost benefit analysis is the dominant mode of reasoning. Discourse about norms and values (practical discourse) in RN suffers from a lack of representation. Thus, there is no formal effort to structure values or gage non-cost impacts. Although a committee of 25 people may be effective a reaching agreement, this committee cannot represent the full range of public interests, especially not unorganized interests (Amy 1987; Bingham 1986; Rushefsky 1984). Ad hoc groups or individual citizens are typically not permitted to participate. As a consequence, negotiated rule making lacks a legitimacy to make normative decisions for society, but they are very efficient at making technically-based decisions (Fiorino 1989; Susskind et al. 1983). The poor competence in practical discourse was pointed out by Susan Hadden in her review chapter. We fully agree with her argument that practical discourse in RN suffers in part because of the homogeneity of experiences among the participants. Regardless of the

---

2 This problem is in no way unique to RN. There has been much debate in the literature about what responsibility the facilitator bears toward bringing about an agreement that is not only fair, but also competent at resolving factual disputes. See Bacow and Wheeler (1984) for a summary of the debate between Susskind and Stulberg that occurred in The Vermont Law Journal in the early 1980s regarding the proper role for mediation facilitators.

structures of the discourse, it is the participants who make statements and judgments. Without a diversity of value positions, experiences, and perspectives represented, the full range of consideration cannot be given to normative issues.

Hadden points out that the confidentiality of RN and the likelihood that participants will encounter each other in future negotiations promotes honest behavior on the part of the participants. This is not the same as saying the RN promotes and encourages therapeutic discourse, which is about verifying that the speaker is being sincere and authentically representing his or her subjective impressions. But Hadden rightly points out that the high level of technical scrutiny does promote therapeutic discourse, because, in this technically-oriented discourse, insincerity would have to be disguised with technical argument, and RN is extremely competent at uncovering inaccuracies in factual understandings (theoretical discourse).

The appeal of RN is that a satisfactory solution is usually found in a reasonable amount of time and legal challenges are reduced or eliminated. RN is most appropriate for technical problems with clearly defined implications for interest positions. The affected interest groups must be few in number, able and willing to enter into negotiation and agreements, and have rather homogeneous backgrounds. Inviting a radical group that prides itself on not compromising would be fatal to the process. O'Hare, Bacow, and Sanderson (1983:154-157)[3] provide a detailed checklist for successful RN:

- Few parties are involved,
- Potential opponents are geographically defined,
- Potential opponents are cohesive,
- Mutually acceptable outcomes exist, (this criterion is crucial),
- Impacts are clear,
- Impacts can be mitigated or reversed,
- Parties can enter into binding agreement, and
- Parties are not initially hostile.

## *Mediation*

Mediation compares favorably to RN on fairness, because all aspects of the negotiation are open to discussion. Selection in mediation is still based on the power principal: included are interest positions that have enough political clout to interfere with decision implementation. The full range of interests cannot, if only for logistic reasons, be included. The agenda, rules, facilitation, and discussion are all subject to consensual agreement.

---

3 As quoted in Sigmon (1987:175).

Competence in reaching accurate understandings of states of affairs is only mediocre. Legitimacy of the agreement produced by mediation lies in the consensus among the committee members, not in its technical accuracy. If the a decision is proposed only on the basis of this agreement, rather than in accordance with criteria evaluated by analytical methods, then its competence is uncertain. There are several reasons to expect that the technical competence of a mediation discourse would be worse than that of RN. First, participants are not selected for their technical expertise on the subject matter. Second, arriving at a technically competent solution is not necessarily the main goal of mediation. Third, there is no technical assistance supplied to mediation discourse, no technical staff, no legal assistance. These problems are accompanied by an additional shortcoming, which is shared with RN: there is no attempt to seek out anecdotal knowledge or the local experiences of citizens and stakeholder group members.

Practical discourse in mediation is also not likely to be highly competent on normative issues. The main reasons are similar to those for RN. The model makes no commitment to normative rationality. There is no attempt to make certain all value positions or interests are represented. There is no formal value modeling or impact assessment done. And there is no intention to focus the discourse on normative argumentation. Mediation is oriented first to producing agreement. Understanding the factual and instrumental aspects of an issue is probably accomplished by mediation, but the interpretation of these factual issues with respect to normative goals and values is biased at best or missing at worst. Moral understanding is not seen as a prerequisite in this regard.

Only on explicative discourse is mediation likely to be competent. This is especially the case if the result of the discourse is prepared as a formal contract, in which terms and definitions would be clearly spelled out. Every member of the discourse would have the opportunity to review and approve of the definitions.

Mediation has proven to work well when the existing conflict has the following characteristics. First, a long-standing, bitter controversy has prevailed for some time with a belief that it is a win/loose situation. Second, the parties involved in the controversy no longer communicate with each other. They have predetermined their adversary's arguments and values and are opposing those conceptualized positions. Third, past decisions have been made based upon the political strength of the parties rather than on rational argument. Fourth, all have the perception that the war between the groups cannot be resolved in one battle, and no single group enjoys overwhelming superiority in this instance (Waldo 1987). Mediated solutions to these types of problems rely upon a willingness of all of the parties to "leave their well-oiled war machines on the sidelines" (Waldo 1987:7). Furthermore, all players must want a change in the status quo.

## *Compensation and benefit sharing*

Kunreuther's scheme to arrive at equitable outcomes in environmental decision making through developing compensation packages does not meet the criteria for fairness unless all affected individuals agree on a compensation package that they would accept. This condition is difficult to meet. First, there are theoretical reasons: Individuals show a large variance in assigning personal utilities to compensation and disutilities to the risk to which they might be exposed. Averaging over all individual compensation claims would violate the principle of Pareto optimality (nobody should be worse off after the transition than before). Negotiations about a compensation package that is supposed to be acceptable to all individuals will face the difficulty that the person with the highest perceived disutility due to the proposed transaction will always suffer a utility loss when a compromise is reached. In addition, it pays off to lie in these negotiations in order to receive a higher compensation sum than justified by the proposed utility loss. Kunreuther is aware of these problems and proposes several mechanisms to avoid strategic bargaining (such as competitive betting). But it appears almost impossible to prevent strategic behavior in spite of structural conditions forcing people to be honest about their preferences.

The second problem is that the model does not specify who should be participating in the compensation process. If individual utility is all that counts, then all affected people need to be included (even a representative sample would not be able to reflect the whole variance of potential utility losses). Organizing a bargaining process with everybody is certainly unrealistic. If at all, compensation can only be a topic within one of the other methods proposed in this book. .

To play the advocate for a moment, suppose we make the realistic assumption that the town council will handle the negotiations for the community. The council members are elected by popular vote, they are accountable to the town and represent the plurality of interests within the town. Citizens know their town council members and are accustomed to expressing wishes and desires to them. There is no guarantee that all interests will be represented fairly, but there is a degree of representation. If the council appoints a citizen board to negotiate, then the council will probably take pains to assure that the major interest groups in town are aptly represented.

Competence in Kunreuther's compensation scheme hangs on the individuals involved. It must be admitted that a small group of people, say three or four from each side, if driven by the desire to reach the best possible understandings and not merely the most advantageous political outcome, could engage in an extremely competent process. But this would depend greatly on their ability and willingness to solicit outside knowledge and advice. Kunreuther's model could be improved if he would define conditions under which the compensation package is designed. For instance, he could specify that each town produce a bid by a procedure that involved open debate and popular vote. A technical expert panel could help make certain the outcome is not corrupted with misunderstandings. Of course, such a description begins to look much more like a hybrid model than a model for compensation. A compensation process has potential for producing competent discourse on norms. This is because compensation is essentially a way to trade-off different values. As such, extensive conversation about the relative importance of values in consistant with arriving at a mutually acceptable package.

Informing the public may help to clarify the issues involved, but cannot resolve these more fundamental value conflicts. These conflicts are not caused by ignorance or misreading the relevant information, but by different value commitments and diverging interests between industry, regulators, stakeholders, cultural groups, and the affected public. One suggestion to cope with the problem is to compensate communities that are exposed to risk burden larger then their fair share. But compensation has not been able to win public acceptance of unwanted facilities. Aside from the equity issues, a problem with compensation is that people do not think that health risks can be offset by financial rewards. If there are no perceived benefits, only risks, associated with the facility, there is no reason for the public to accept it.

## *Dutch Study Groups*

Achieving fairness was one of the major considerations behind the design of the Dutch Study Groups. The project went to great lengths to ensure that (1) individuals and stakeholder groups throughout the country were informed about the process and the issue, and (2) had an opportunity to participate if they elected to do so. But the model did not live up to these hopes, according to the standard discursive criteria. The actual agenda was set by a select committee, which used an iterative process of soliciting public comment.[4] Inside the study groups themselves the discourse was quite fair. The non-hierarchical structure of the groups as well as their size and frequency of

4 In his review of the model, Jeryl Mumpower suggested that fair access to setting the agenda was not as important as making sure that all concerns were fairly represented on the agenda. This is precisely the tension between representative and participatory democracy. We would merely repeat the oft-made arguments that it is only through direct participation that individuals can be assured that their interests will be adequately protected and it is only through participating that individuals develop a political dimension to their being. That being said, we also recognize the legitimacy of the representation argument and suggest that an appropriate combination of representation and participation is probably the preferred path to take.

meeting all contributed to building an environment that encouraged people to make statements and raise questions.

Competence of the study group model was actually very high. Explicative discourse was especially competent, due to highly discursive and iterative processes to resolve ambiguities. Theoretical discourse was also high, mainly because the government allocated large financial grants to develop information packages and educational components, to do scenario analyses and modeling, to verify the known facts through peer review and intensive public scrutiny, and to assess uncertainties. A wide assortment of scientists and experts were involved, representing all imaginable points of view.

Practical discourse was enhanced by the open, well attended discussion groups and public meetings. A broad range of interest positions was present and there were no systematic barriers to participation. Meetings were held repeatedly at different times, in many parts of the country. Competence was also furthered by supplemental public surveys, which gathered inputs of those not at meetings. On the other hand, there were no systematic attempts to structure values or assess impacts, possibly because the scale of the project and the inherent uncertainties accompanying the data made this exercise unwieldy.

Therapeutic discourse was not specifically pursued. It is suggested that the large scale of the meetings prohibited substantial discussions about the sincerity or authenticity of individual expressions.

## Common Problems of Selection

For several of the models discussed in this volume, the number of people who can participate is very small. When more people want to attend the discourse than can be accommodated, some means of selection must be used. There are three possibilities to assign or select people to a participatory body:

- selecting representatives of groups or organizations that have shown interest in the issue.
- asking for volunteers within the affected population (for example through public announcements or advertisements)
- random selection or equivalent method to accomplish a statistical representation of a given population

Selection of any sort poses serious challenges in political cultures where volunteering for public service and participating on demand are considered basic rights. If societies favor an adversarial style of policy making, selection of participants by organized groups combined with volunteering is regarded as most legitimate, whereas random selection is often associated with an allegedly hidden agenda to keep the interested and knowledgeable persons out of the process. The experiences with CACs based on the first two selection methods tend to show, however, that deliberate selection of group representatives as well as volunteering lead to serious distortions of public values and interests. This bias can partially be compensated by making sessions open to public scrutiny. If such openness is not possible or undesired, control and feedback by the public are both in jeopardy. This might be the reason why in countries that do not share the U.S. policy of granting their citizens freedom of information, non-public discourses (such as negotiated rule making) are hardly discussed as instruments of participation, but rather are smooth elements of a streamlined decision process that enhances cognitive competence but excludes most people from the decision process.

Furthermore, if participation is left up to voluntary selection or appointment, onlookers may be cynical of the participants because either they appear privileged (by appointment) or too ambitious (through volunteering). Random selection – combined with a sense of citizenship and duty – is an attempt to gain the sympathy and support of outsiders, by virtue of the fact that they can picture themselves in that role. Random selection should create an entirely different impression of the motives of the citizens, thereby increasing the process's legitimacy.

Ideally, random selection assures that all values and preferences of the affected population are given an equal opportunity to be brought into the process. In theory, people who are not selected should be satisfied that their interests will be protected because there essentially is a guarantee that another person with similar interests will be selected. In practice, of course, such satisfaction is not forthcoming. People who are immediately affected and not selected in the random sampling feel deprived of a fundamental democratic right to protect their own interests. As a consequence, they may decide to seek other avenues to make their voices heard.

Another problem with random selection is: will enough of the randomly selected people agree to participate? The answer lies in how the problem is perceived in the community, how appropriate random selection is perceived, and how strong is the commitment to public responsibility among those chosen. It is also possible that, as a problem is brought to public attention, the attitudes of the general publics may shift during the time that elapses between the selection and the process's end. The randomly selected participants may not represent the greater community. Random selection also works only on the condition that the distribution of impacts is evenly distributed.

Although random selection theoretically guarantees everyone an equal chance to participate, there are additional conditions to be met: the sample size must be large enough, there should not be too many different value or interest positions, and the population needs to be informed of the problem and the possible impacts of alternative decisions. Achieving these conditions may be difficult, especially in cases of technological risk or environmental resource management, namely because the value differences are so broad, but also because the extent of local impacts is typically high.

Two models use some form of random selection to admit participants: Planning Cells and Citizens Juries. RN and CACs are by invitation only. The other models are generally open to anyone who decides to come. Public hearings, meetings, and workshops are widely accepting of participants, but they provide few means to engage in meaningful discourse and no opportunity to determine the final decision. Focus groups and workshops are open to the public – although they limit each event to small groups of people – and they do encourage more discourse, but they give the participants no more input to the decision than do public hearings. Influence over the final decision is provided to participants of small group models such as CACs, RN, citizen panel approaches, and, of course, referendums. But the public-at-large is excluded from all of these except for the latter. Referendum is the only model of participation that gives the decision making authority to a broad population, although in this case, no discourse occurs.

## A Preliminary Framework for Matching Problems with Models

Faced with these eight models – and a number of additional techniques – for citizen participation, how are those affiliated with a specific environmental problem to know which tool to use when? We expect that both affected citizens and decision makers will be involved in this selection. Clearly, the appropriate role of public involvement in making environmental decisions is contingent on the type of decision to be made and the type of conflict associated with the decisions. A classification scheme may help one to characterize problem and estimate the performance qualities of the available participation techniques.

Without doubt the most common classification has been that which reduces conflicts to being about facts or values. Frequently, this oversimplification spurs the simplistic prescription that conflicts diagnosed to be about facts are in need of "science," while those about values are tossed into a black box labeled "more politics." Such reductionism has dulled the potential of public participation by not recognizing that public value differences are tied to factual uncertainties and trust in public institutions.

### *Three levels of debate*

Functowicz and Ravetz have tried to address this deficiency by developing a taxonomy based on different types of evidence and argumentation (1985). Originally developed as a classification scheme for risk conflict, we have adapted it slightly to be more consistant with the theoretical approach taken here and applied it to the environmental context. According to this view, environmental debates fall into any of three spaces which are illustrated in Figure 1.

The three kinds of debate appear in a two dimensional space marked out by two axes, one labelled "degree of complexity" and another "intensity of conflict." In reality, a large number of dimensions are presumed important to accurately describe the features of the conflict (or decision) relevant to the performance of the public participation process. An assumption is made that these many dimensions, when collapsed into two, provide a reasonably meaningful characterization of environmental conflicts. It is important to emphasize that this taxonomy is a hypothesis and not the result of a factor analysis. With this caveat in mind, we suggest that the schema is useful in thinking about how to match participation methods with problems.

The first kind of debate involves factual arguments about probabilities, causal relations, and the extent of potential damage. Not only is it important to have clear understandings about facts, it is also important that estimates of uncertainty accompany this information. If expert authorities disagree about facts, a way of representing that disensus is needed. Factual debates become most problematic when experts hold opposing opinions about the validity of facts or when forecasts or projections are highly uncertain. The three examples of environmental decision making we have plotted in this area on the Figure are: turtle exclusion devices on shrimp fishing boats; appropriate fertilizer application rates (for crops); and the white-tailed deer hunting quotas in Southern Vermont. While values and issues of trust in governmental institutions do enter into each of these debates, they are *primarily* characterized by solutions driven by expertise. It should be noted that expertise means more than just science, we include local knowledge and personal experience in this term.

A second, more controversial level of debate concerns public confidence in institutions to deal with environmental threats. At this level the focus of the debate is on the trust or confidence that the public has in the decision making body (usually a governmental agency) to give adequate consideration to each party's concerns, to distribute costs and benefits equitably, and to fulfill promises and expectations. This type of debate does not rely on technical expertise, although reducing factual misunderstandings and reducing uncertainty may help. Instead, the emphasis is on achieving mutual awareness of each other's expectations and a commitment to the principle of reciprocity. Agreement in this discourse is gained by clarifying mutual expectations, demonstrating good will and a commitment to fulfill those expectations, and providing pathways for retribution or punishment if one party fails to live up to its promises. Government agencies have a responsibility to give well-intentioned, thorough consideration to each interest position. In response,

*Figure 1:* The three levels of environmental conflicts

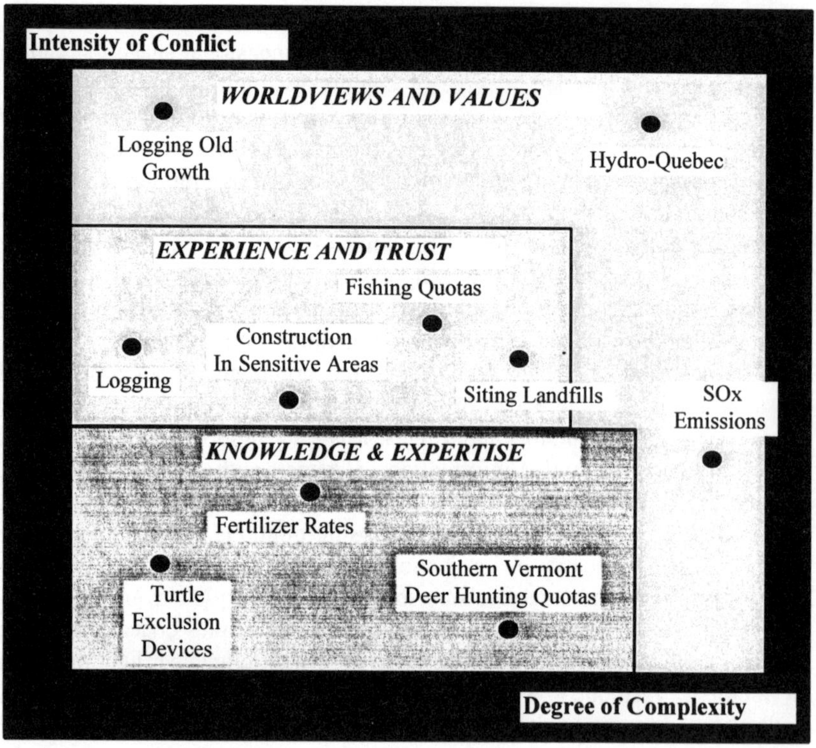


citizens and interest groups must not withdraw support if the consequences of the decision are within the range of expectations promised. We use the following as examples of debates on this level: non-extraordinary cases of logging; fishing quotas, landfill siting, and construction in sensitive areas.

In the third kind of debate the conflict is defined along competing social values, cultural lifestyles, and world views. In this case neither technical expertise nor institutional competence and openness are adequate conditions for reaching collective agreement. Decision making here requires a fundamental consensus on the issues that underlie the debate. The nuclear debate in the 1970's in Sweden leading to a national referendum on the future of nuclear power plants is an example of conflict resolution at this level. A referendum was the culmination of an extensive debate about the desired direction of technological development in which nuclear power served as a symbol for large centralized technologies and its impacts on economics and society (Olson 1984). The final vote to continue nuclear power for a limited period of time defined the legitimate role nuclear power was supposed to play within the larger technological scenario. The majority considered nuclear power plants as undesirable but necessary technologies that should be kept operating until alternative technologies could replace them. Replacement was estimated to be completed by the year 2010, after which all nuclear power plants were scheduled to be phased out. The agreement moved the issue from the third to the second level, where technical and organizational solutions could be discussed without the debate expanding into a conflict over the moral implications of nuclear power and its symbolic meanings. Other examples of this kind of debate are: logging in old-growth forests (Pacific Northwest of the USA), large-scale hydropower development (as in Northern Quebec, Canada), and sulfur oxides emissions controls (as under the USA Clean Air Act).

## *Employing the taxonomy*

Clearly not every environmental decision can be categorized into one of these three kinds of debates. Furthermore, the focus of a debate may shift over time, as with the Swedish debate on nuclear energy. The examples plotted on Figure 1 represent specific applications, not generalities that hold for all problems of that design. Each of these cases was clearly dominated by one kind of debate. To deal effectively with that problem, the public participation approach has to be effective at handling that kind of debate. This does not exclude the possibility that another debate may surface after agreement is reached on the first. Consider the case for arriving at hunting quotas for deer in Southern Vermont. That debate centered around wildlife biologists' calculations, so we have plotted that problem in the factual debate area in Figure 1. A similar problem – wolf hunting quotas in Alaska turned out to be a highly contentious value dispute. Reaching agreement in Vermont and Alaska on hunting quotas for deer and wolves, respectively, would take completely different forms of decision making techniques.

To complicate matters further, there is a strong tendency for reframing. Often business and government attempt to re-frame higher level conflicts into lower levels ones: third level conflicts are presented as first or second level conflicts, and second level conflicts as first level. This is an attempt to focus the discussion on technical evidence, in which the agency is fluent (Dietz et al. 1989). Citizens who participate are thus forced to use first level (factual) arguments to rationalize their value concerns. Unfortunately, this is often misunderstood by experts as "irrationality" on the part of the public. Frustrated, the public retreats to due process and routinization of the process, abscising it of substance, and departs with disillusion and distrust of the system.

## *Links with the discursive standard criteria*

We hypothesize a link between the discursive standard criteria (Chapter 3) and the taxonomy of environmental decisions described above. We suggest that each participation model has a domain in the space marked out in Figure 1 in which it is most capable of performing. Problems that lie within that domain can be aptly handled by that model.

Competence and fairness are needed in each of these kinds of debates. We want to stress not to relegate competence with first level debates and fairness with third level debates. The simple reason for this is that competence refers to discourse about values as much as it refers to scientific fact-finding. Likewise, fairness to contending factual interpretations is essential for establishing scientific validity.

That said, we do not want to reduce the message of this book to the slogan that all public participation must be as fair and competent as possible. What is lost in the basic truth of that statement are the sophisticated distinctions regarding the practical constraints of decision making and the performance of public participation models. We have already seen that some models rate better than others on certain criteria, however, *all* models have a role to play in solving our environmental problems.

The key is to recognize that, depending on the kind of problem and the social context in which the decision making takes place, *certain attributes* of fairness and competence will matter more than others. For example, few would disagree that there is a time and a place for public hearings, a model which is open to nearly all who wish to "participate," but highly constrains how they can take part in the discussions. Similarly, some problems demand a high degree of competence in practical discourse (discourse about norms and values) and lesser competence in theoretical discourse (about facts).

To locate each of the eight models studied in this volume on the two dimensional taxonomy of environmental problems, we sorted the discursive standard criteria into two sets, one relating to the y-axis, and one to the x-axis (see Table 3).

We reasoned that a model for participation would better be able to handle problems of high complexity if it allowed for fair participation in setting the agenda, and if it was extremely competent in explicative discourse (about terms) and theoretical discourse (about facts). Competence in practical discourse (about norms) is also helpful in these kinds of disputes. Fairness in all areas and competence in normative and therapeutic discourse are most important in resolving problems that have high conflict. Using this operationalization and the evaluative results from Tables 1 and 2, we plotted the eight models on Figure 2.

*Table 3.* The discursive standard criteria as indicators for intensity and degree of complexity.

Degree of Complexity

- Fairness to set agenda
- Competence in explicative discourse
- Competence in theoretical discourse
- Competence in practical discourse

Intensity of Conflict

- Open access to the discourse
- Fairness to set agenda
- Fairness to select moderation and rules for discourse
- Fairness to influence final decision
- Competence in practical discourse
- Competence in therapeutic discourse

For each of the three problem types, two models of participation seem well-suited. For problems that can be handled mainly through expertise, negotiated rulemaking and compensation are appropriate. For problems that involve more than disputes over facts, but deal directly with the legitimation problems of government, mediation and Citizens Juries are useful. When debates concern fundamental value differences, the Citizen Intiative and the Dutch Study Groups have the best potential to succeed. Finally, there are two models that lie on the boundary lines between two areas. Citizen advisory committees are appropriate for both disputes over facts (especially the technical advisory committees) and disputes over trust. Planning Cells also lie on a boundary. They have worked to solve environmental problems both about trust and value discrepencies.

## *Role of trust in environmental discourse*

Several of the authors in this volume have emphasized connections between public participation and trust. Governance suffers from a lack of public confidence in government. As so many experiences with these models for public participation have testified, trust must be a two-way street: administrative officials have to trust citizens and citizens have to trust government.

*Figure 2.* Locating the eight participatory models in the taxonomy.

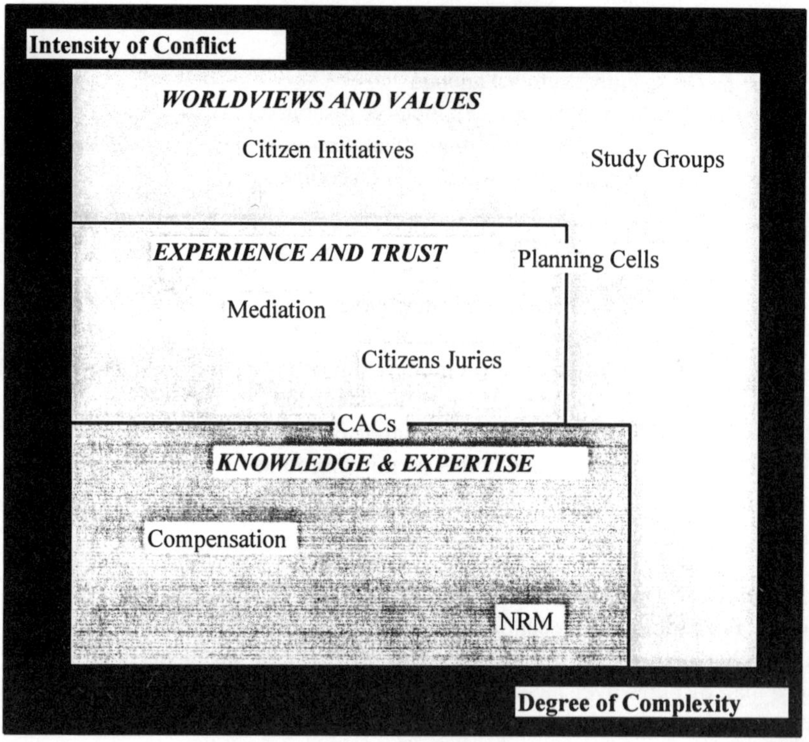


In the taxonomy of environmental problems, trust is the key to successfully solving problems residing in the second region. Problems of this kind are troublesome for decision making bodies because they cannot be solved merely through the application of expertise. They require the support of members of the public. The authors in this volume have suggested that trust is promoted when:

- there is a high likelihood that the participants will meet again in a similar setting;
- interaction takes place face-to-face in regular meetings over a reasonable period of time and people have a chance to get to know each other;
- participants are able to secure independent expert advice;

- participants are free to question the sincerity of the involved parties;
- citizens are involved early on in the decision making process;
- all available information is made freely accessible to all involved;
- the process of selecting options based on preferences is logical and transparent;
- the decision making body seriously considers or endorses the outcome of the participation process; and
- citizens are given some control of the format of the discourse (agenda, rules, moderation, and decision making procedure).

Not every one of these conditions is essential to building trust between the affected parties and the decision makers. The models that perform well on the issue of trust – namely mediation, Citizens Juries, Planning Cells, and to a lesser extent CACs – fulfill some of these conditions, but not all.

The discursive standard criteria do not explicitly mention the word trust. This is because the concept is built into the theoretical approach at a more fundamental level. Whether the treatment of trust needs to be made more explicit is something we will learn only through more experience with using these criteria. Certainly the authors of this volume found it necessary to make additional references to trust. Future evaluation work will have to pay special attention to how trust can be handled.

## Creative Solutions To Renewing Democracy

There is no ideal solution to the conflict between the legitimate demand for public participation, the need for technical and economic rationality, and the necessity for assuring accountability and responsibility of decision making bodies. We recognize that fair and competent participation will not come about through the rigid application of these models, but by creatively tailoring the models to be responsive to the problem setting. As Lynn and Kartez pointed out in their chapter, there are opportunities to improve upon these models and such reforms ought to be encouraged. Clearly, any participation exercise must combine technical expertise and rational decision making with public values and preferences (Stern 1991). Toward these ends, many creative approaches have been developed and some tested.[5]

We do not naively believe that public participation is appropriate for every decision. Representative democracy clearly has its merits. Continuity, practicability, accountability, and openness to public scrutiny deserve to be cherished and sustained. However, the representative model has limitations

---

5 Some examples include: Connor 1993, Nelkin and Pollak 1980, Burns and Überhorst 1988, Chen and Mathes 1989, Siting Task Force 1990, Kathlene and Martin 1991, Renn et al. 1991, 1993; and Webler 1992: 324ff.

and there are circumstances in which direct participation is an appropriate solution (Barber 1984). In particular, this is the case when citizens but not their representatives are affected by a political proposal. Despite the fact that representatives are supposed to act in the electorates' interests. The deficiencies of the traditional decision making process, especially the public hearing, were the main motivation to compile this book.

None of the eight models discussed in this volume were able to outperform all the others. Some criteria were never met, others we met only partially. In spite of these results, most of these models show promise. The message to take away from this analysis should not be to emphasize the deficiencies of these promising models for public participation, but to emphasize what steps can be taken to improve their performance. The challenge for public policy makers and planners is to tailor these new models to the circumstances and cases that require public input. We cannot offer a recipe for how to do this. Instead we mean for the evaluation of the models as performed in Table 2 and Figure 2 to provide guidance for interested users. We leave it to the reader to make his or her final judgment.

## Challenges to Participation from Rational Choice Theory

### *The before and after of public participation*

Public participation is an activity that encompasses much more than the actual meeting period. Each participation project has a history and is followed by social or political repercussions. Early on in this project, we decided to focus on the fairness or competence of the discourse. To put the discourse activity into perspective, it may be helpful to think of participation as having to meet three challenges. The second question, which deals with the structures and procedures used to facilitate discourse, has been the main focus of this book.

- How can the participants be brought into the process with a spirit of goodwill?
- How should the interaction among the participants be structured and carried out?
- How can the implementability of the decision outcome be assessed and the external aspects of the process be managed (monitoring decision impacts, issues of liability, etc.)?

### *"Is it rational to participate?"*

Rational choice theory claims that individuals have a tendency to resist participating in collective decision making. Mancur Olson described this as the free rider problem (1965). He reasoned that people will not participate in a process to deliver a collective good if they are reasonably assured that other people will participate and be successful in obtaining the good. In other words, people will reason that taking a free ride is the most rational decision. The free rider hypothesis poses a challenge for public participation. Not only must we explain why people do, despite the free rider tendency, choose to participate, but we also must consider what design features in our models for participation can enhance the willingness of people to participate.

There have been several answers to the free rider thesis, some from within rational choice theory, some from outside. One rational choice theory explanation is that people can be encouraged to participate if they are offered a positive payment or incentive to participate or a negative incentive to not participate. Free coffee and snacks are one widely used incentive. Citizens Juries and Planning Cells offer each participant a cash payment in exchange for their participation. Other positive incentives are the promise of an opportunity to voice grievances to a government official, or the promise to receive an information package or presentation about the proposed impacts. Incentives can also take the form of peer pressure, as when friends and neighbors entreat each other to participate. While the use of negative incentives by government agencies or project sponsors would certainly have negative reprocussions, we note that the trial jury system does rely on threats to get people to serve.

Another rational choice explanation is that people will participate if they conclude that other people will not. In other words the person reasons, "I had better do it, because no one else will." Motivation by perceived necessity leads, however, to a peculiar pattern of involvement. Neither need nor representativeness determine the strength of involvement, but the degree of exclusiveness of reaping the benefits of one's involvement. The less I perceive that others work for my interests and the more I believe that my own involvement does not spread the fruits of my work to all the others (free riders), the more I am willing to commit myself (Olson 1965, 1982). It is obvious that this reasoning leads to an unfair representation of public interests in the political arenas.

Typical non-rational choice responses to the free rider thesis are that people choose to participate not only on the basis of individual costs and benefits, but because they feel a social obligation or a desire to belong to a group. If everyone of your neighbors is going to the meeting tonight, you would be hard-pressed not to go along. There is both peer pressure and a voluntary desire to be included that drives participation. Organizers of citizen participation projects attempt to capitalize on this factor by framing the issue as a community or neighborhood problem and emphasizing how shared interests – such as community identity – are threatened.

Other people participate for moral or altruistic reasons. Perhaps they are strongly committed to environmental preservation, or perhaps they want to support the community against the powerful forces of big business or big government. In any case, their decision is not based on individual returns, but on a sense of duty or commitment to a higher value, which they perceive as threatened. Public participation organizers attempt to assess the existence of value commitment and frame the issue in a way that elicits the sympathy of these individuals or groups.

Once people have started to participate, they can be encouraged by making a formal agreement. Planning Cell organizers require the citizens sign a contract (in exchange for the money). This provides an additional motivating force of an agreement, which could be enforced in the courts. Of course, no one would expect Planning Cell staff to take a citizen to court for not attending a meeting, it is more of a symbolic measure that suggests the importance of fulfilling the agreement.

Solidarity is also a factor in perpetuating one's involvement. Especially in models of participation where people get to know each other on a first name basis, where they engage in lengthy conversations, have time to casually socialize, and develop a sense of responsibility to a shared goal, solidarity becomes a powerful force as a group identity forms. Frank Claus calls this the "crystallization point of citizen participation," when the process becomes self-maintaining. Staff no longer have to pester individuals to return. People telephone each other, share rides to the meetings, and reinforce each other's commitment to participate.

A final strategy that public participation organizers can take to improve motivation to participate is to re-frame the issue in such a way that individuals do not see free riding as a plausible alternative. To be successful, staff have to convince people that personal participation is essential. This is a difficult task, especially in national issues.

## *Developing a climate of cooperation*

Rational choice theory poses a second challenge for the field of public participation. It claims that people who act to enhance their personal interests are acting rationally. It also adopts the normative belief that policy making through interest group liberalism is the best way to serve the public good. The analogy used in the economic sphere of an "invisible hand" is also applicable to politics (Lindbloom 1959, 1977). The overall welfare of a society is then assured by the sum of the "selfish" actions of each group trying to maximize its influence and power. This is not the place to review the debate on pluralist democracies (cf. Dahl 1989, Bacharach 1967). But as many advocates of a pluralist view have pointed out, direct public participation is dysfunctional as long as a sufficient number of interest groups sustain the delicate balance of power. As long as individuals pursue their own interests, there is no incentive to cooperate other than joining or establishing a new interest group not yet represented in the respective policy arena.

This delegates moral action – which we interpret as action based on a judgment about what is best for "all" rather than what is best for "me" – as deviant or at least dysfunctional. This interpretation of moral behavior clashes with the widely held belief that solving today's environmental problems will not be easy if people refuse to surrender a portion of their individual autonomy for the collective good. Egocentric bargaining will work for some collective problems, but in absence of an outside force to drive people to compromise (such as legislation backed up by the power of the state), it cannot work for problems that have no win-win solutions. Without coordinated action, many problems will continue to deteriorate ecological sustainability. Most observers, at least of the environmental situation, agree with the diagnosis that the individual pursuit of self-interest coupled with the possibility of using the free-ride position has been the main cause for the continuing environmental damages.

If many of the contemporary problems such as environmental degradation require more than individual actions based on self-interest, three avenues are open to close the gap between the pursuit of personal interests and the need for a common good orientation. The first solution refers to internalization of social costs. It is the hope of rational choice theorists that using price incentives, such as subsidies and taxation, personal behavior may become more consistant with the public good. This market strategy is, without a doubt, necessary and worth while, but also insufficient. Not all desired or undesired behaviors can be monetarized and there is no convincing method to determine the right price for non-market goods.

The second alternative is to have the government prescribe the actions that are deemed necessary. In a democratic system that prides itself with upholding individual rights, however, heavy handed legislation is not seen as a legitimate solution, except in extraordinary cases. Most people reject the notion that decisions to improve the common good should be made by elitist governmental circles rather than by collective decision making.

One alternative remains: increased participation coupled with structural incentives to foster and promote communitarian values. Public participation needs to catalyze people into adopting an attitude that is oriented to cooperation rather than pursuit of individual interests. Adversarial decision making has its place in our societies, but it does not always produce the most fair or competent agreement. Consensual decision making is seen as more legitimate, especially when the problem is largely technical, impacts are uncertain and complex and not easily quantifiable, or values are in competition.

We like to emphasize the process of norm generation in public participation. By this we mean the creation of communitarian values and social preferences that transcend the egoistic perspectives many parties bring to the discourse process. This development of new values is a learning experience based on empathy (understanding the concerns of others) and quest for social rationality (common good perspective). The major accomplishment of a discourse is to create a set of shared values and goals that serve the common interest of *all* parties without placing undo burden on any one party. Such a learning experience is contingent on a group process in which all participants develop a common understanding of the problem and a desire to find a commonly acceptable solution. Acceptable solutions may include compensation if all parties agree that it is beneficial for the common good to have an inequitable solution. Without such an agreement compensation is unlikely to resolve distributional conflicts.

By adopting a moral attitude, people entertain suggestions that might not be in their own self interest, but which offer the best alternative for all. If we are to expect people to act morally and to cooperate, than we surely have to provide them with processes for participation that are both fair and competent.

## References

Amy, D. J., *The Politics of Environmental Mediation* (Cambridge University Press: Cambridge and New York 1987).

Bachrach, P., *The Theory of Democratic Elitism: A Critique* (Little Brown: Boston 1967).

Bacow, L. S., and M. Wheeler, *Environmental Dispute Resolution* (Plenum: New York 1984).

Barber, B. R., *Strong Democracy: Participatory Politics for a New Age* (University of California Press: Berkeley 1984).

Bingham, G., *Resolving Environmental Disputes: A Decade of Experience.* (The Conservation Foundation: Washington, D.C. 1984).

Burns, T. R., and R. Überhorst, *Creative Democracy: Systematic Conflict Resolution and Policymaking in a World of High Science and Technology* (Praeger: New York 1988).

Chen, K., and J. C. Mathes, "Value Oriented Social Decision Analysis: A Communication Tool for Public Decision Making on Technological Projects," in: C. Vlek and G. Cvetkovich (eds.), *Social Decision Methodology for Technological Projects* (Kluwer: Dordrecht 1989).

Connor, D., "A Generic Design for Public Involvement Programs," *Constructive Citizen Participation,* 21 (June & September 1993), 1 and 2.

Crosby, N., J. M. Kelly, and Schaefer, P., "Citizen Panels: A New Approach to Citizen Participation," *Public Administration Review,* 46 (1986), 170-178.

Dahl, R. A., *Democracy and its Critics* (Yale University Press: New Haven 1989).

Dietz, T., P. Stern, and R. Rycroft, "Definitions of Conflict and the Legitimation of Resources: The Case of Environmental Risk," *Sociological Forum* 4:47-70 (1989).

Etzioni, A., *The Spirit of Community. Rights, Responsibilities and the Communitarian Agenda* (Crown: New York 1993).

Fiorino, D., "Environmental Risk and Democratic Process: A Critical Review," *Columbia Journal of Environmental Law,* 14, No. 2 (1989), 501-547.

Funtowicz, S. and J. Ravetz, "Three Types of Risk Assessment: A Methodological Analysis," in: V.T. Covello, J.L. Mumpower, P.J.M. Stallen, V.R.R. Uppuluri, *Environmental Impact Assessment, Technology Assessment, and Risk Analysis* (Berlin: Springer-Verlag, 1985), pp. 831-848.

Kathlene, L., and Martin, J., "Enhancing Citizen Participation: Panel Designs, Perspectives, and Policy Formation," *Policy Analysis and Management,* 10, No. 1, (1991), 46-63.

Lindbloom, C., "The Science of Muddling Through," *Public Administration Review,* 19 (1959), 79-99.

Lindbloom, C., *Politics and Markets: The World's Political-Economic Systems* (Basic Books: New York 1977).

Nelkin, D., and M. Pollak, "Problems and Procedures in the Regulation of Technological Risk," in: C. H. Weiss and A. F. Burton (eds.), *Making Bureaucracies Work* (Sage: Beverly Hills 1980), pp. 233-253.

O'Hare, M., L. Bacow, D. Sanderson, *Facility Siting and Public Opposition* (Van Nostrand Reinholt: New York, 1983).

Olson, M. E., *The Logic of Collective Action. Public Goods and the Theory of Groups* (Harvard University Press: Cambridge, MA. 1965).

Olson, M., *Participatory Pluralism: Political Participation and Influence in the United States and Sweden* (Nelson-Hall: Chicago 1984).

Renn, O., T. Webler, and B. B. Johnson, "Public Participation in Hazard Management: The Use of Citizen Panels in the U.S.," *Risk: Issues in Health and Safety,* 197 (Summer 1991), 198-226.

Renn, O., T. Webler, H. Rakel, P. C. Dienel, and B. Johnson, "Public Participation in Decision Making: A Three-Step-Procedure," *Policy Sciences,* 26 (1993), 189-214.

Rushefsky, M., "Institutional Mechanisms for Resolving Risk Controversies," in: S.G. Hadden (ed.), *Risk Analysis, Institutions, and Public Policy* (Associated Faculty Press: Port Washington 1984), pp. 133-148.

Sigmon, E. "Achieving a Negotiated Compensation Agreement in Siting: The MRS Case," *Journal of Policy Analysis and Management* 6:170-179 (1987).

Siting Task Force, *Opting for Co-operation* (Siting Task Force: Ottawa 1987).

Stern, Paul C., "Learning through Conflict: A Realistic Stratgey for Risk Communication," *Policy Sciences,* 24 (1991), 99-119.

Susskind, L. E., L. Bacow, and M. Wheeler, *Resolving Environmental Regulatory Disputes* (Schenkman: Cambridge 1983).

Waldo, Jim, "Win/Win Does Work," *Timber/Fish/Wildlife: A Report From the Northwest Renewable Resources Center* 1(1):7 (1987).

Webler, T. *Modeling Public Participation as Discourse: An Application of Habermas's Theory of Communicative Action.* Doctoral Dissertation. Clark University, Worcester, MA. May 1992.

# *APPENDIX*

## *Biographies of Authors*

*Audrey Armour* is an Associate Professor in the Faculty of Environmental Studies, York University and President of Armour Environmental Consultants Inc. She has been engaged in the practice of environmental planning and related teaching and research since 1976. Dr. Armour is known internationally for her innovative work and research in siting locally unwanted facilities, assessing social impacts, and promoting an integrated approach to project planning and resource management. Dr. Armour has served on the Canadian Environmental Assessment Research Council, the Science Advisory Board of the International Joint Commission, and the Ontario Hazardous Waste Advisory Committee. She has also served as President of the International Association for Impact Assessment. She is currently a member of the Board of Directors of the Institute for Social Impact Assessment.

*Mike L. Baughman* is Founder and President of Intertech Services Corporation, a consulting firm specializing in empowerment of local government within multi-jurisdictional issue resolution contexts. Dr. Baughman has assisted in mediating disputes concerning hazardous waste management, water resource allocation, and community and economic development. He received his Ph.D. from Clark University, Worcester, MA. His thesis dealt with the experiences of siting hazardous waste facilities in the United States. He holds an advanced degree in Resource Economics from the University of Nevada, Reno.

*Frank Claus* is director of the Institute for Communication and Environmental Planning, a private consulting company that organizes citizen involvement projects. Dr. Claus holds a Ph.D in chemistry and worked as a lecturer for spatial planning at the University of Dortmund (Germany) from 1985-1989. He served as visiting professor at the University of Kassel before he founded his consulting company. Dr. Claus has authored several books and articles on environmental policies and citizen participation.

*Ned Crosby* has spent his professional career working on democratic reform at the Jefferson Center, which he founded in 1974. In the process, he has become familiar with agricultural, environmental, health and juvenile problems. His experience as an activist ranges from the streets of black neighborhoods of Minneapolis in the late 1960's through work with the peace movement in Washington D.C., concentrating on Africa in the early 1970's and on Central America in the 1980's (in the latter he worked closely with the Central American Working Group and with the office of David Bonior). In the early 1970's he was president of the Minneapolis Legal Aid Society; he also helped found and later served as chair of Operation De Novo, one of the nation's first pre-trial diversion projects. He received his Ph.D. in political science from the University of Minnesota in 1973, with extensive work done on the foundations of attitude measurements and on moral philosophy.

*Peter C. Dienel* is Professor for Sociology at the University of Wuppertal. After receiving a Master degree in theology and a Ph.D in sociology, Dienel worked as a director of a Protestant Academy and later within the State Government before assuming his position at the University of Wuppertal. In addition to his university affiliation, he is the director of the Center for Citizen Participation and Political Planning. He is the author of the book "Die Planungszelle" in which he develops and describes the Planning Cell procedure.

*Daniel J. Fiorino* is political scientist who has written several articles on democratic participation in environmental decisions. These have appeared in *Science, Technology, and Human Values, Risk Analysis, the Columbia Journal of Environmental Law and Public Administration Review.* He is also the author of Making Environmental Policy: An Introductory Analysis (University of California Press). He is currently the Director of the Waste and Chemical Policy Division at the U.S. Environmental Protection Agency in Washington, D.C.

*Bruno S. Frey* is professor of economics at the University of Zurich and author of various books, most recently *Economics as a Science of Human Behaviour* (Kluwer, 1992).

*Susan G. Hadden* is a Professor of Public Policy at the Lyndon B. Johnson School of Public Affairs, The University of Texas at Austin. She holds a B.A. degree from Harvard College and a Ph.D. in political science from the University of Chicago. She is the author of two books on lay use of technical information to reduce risk: *Read the Label: Providing Information to Reduce Risk* and *A Citizen's Right to Know: Risk Communication and Public Policy.* She has also published more than 60 articles on telecommunications policy, citizen participation, risk communication, and policies intended to reduce risks to human health or the environment. She has worked with a wide range of citizen and industry groups to develop programs for access to information about chemicals, and she has served on government advisory committees concerning labeling, risk communication, and a wide range of water policy and risk issues.

*Jack D. Kartez* is a Senior Fellow at Texas A & M University's Hazard Reduction & Recovery Center and an Associate Professor of Planning. His research on environmental hazards and consensual planning processes has been funded by the National Science Foundation, U.S. Environmental Protection Agency, and private foundations. He frequently serves as a planning process facilitator.

*Joanne-Linnerooth-Bayer* studied industrial management at Carnegie-Mellon University, Pittsburgh, USA, and received her Ph.D. in economics from the University of Maryland. Since 1972 she has worked at the International Institute for Applied Systems Analysis (IIASA) in Laxenburg, Austria, and has spent short periods at Resources for the Future in Washington D.C. She has published widely in the general area of environmental risk policy making, including major articles on the valuation of public programs that affect population mortality, the siting of hazardous facilities, the role of scientific expertise in the policy process, and the social control of transborder environmental risk. Her most recent interest is in "fairness" issues regarding the distribution of environmental risk burdens.

*Frances M. Lynn* is a Research Associate Professor in the Department of Environmental Sciences and Engineering in the University of North Carolina's School of Public Health. In addition, she founded and directs the Environmental Resource Program, an outreach and continuing education unit of the School of Public Health, which provides technical assistance to citizen groups, trains school teachers in environmental education, and facilitates collaborative decision making among citizens, industry and government.

*Jeryl L. Mumpower* is a faculty member in the Department of Public Administration and Policy at the Nelson A. Rockefeller College of Public Affairs and Policy, State University of New York at Albany. He also serves as Director of the Center for Policy Research. He was previously affiliated with the Division of Policy Research and Analysis, U.S. National Science Foundation. Professor Mumpower received his B.A. in philosophy from the College of William and Mary and his Ph.D. in social and quantitative psychology from the University of Colorado, Boulder. His primary research and teaching interests are in negotiation and bargaining, judgment and decision making processes, and environmental policy analysis.

*Felix Oberholzer-Gee* is research assistant in a project on siting financed by the Swiss National Fund. He is working in the Department of Economics at the University of Zurich. He received his MA degree at the University of Zürich and is presently Ph.D. candidate at the University.

*Ortwin Renn* serves on the board of directors of the Center of Technology Assessment in Stuttgart, Germany. He is also professor of environmental sociology at the University of Stuttgart. In 1992 he spent his sabbatical leave at the Swiss Institute of Technology in Zürich, Switzerland. From 1986 to 1992 he had taken the position of an Associate Professor for Environment, Technology, and Society at Clark University in Worcester (Massachusetts). Prior to joining Clark, he directed a research unit for technology assessment at the largest governmental research center in the Federal Republic of Germany (National Research Center KFA, Jülich). Renn received a MA degree in Sociology and Economics (European Diploma) and a doctorate degree in Social Psychology, both from the University of Cologne (School of

Economics and Social Sciences). Renn has published seven books and more than 50 articles in journals or chapters in books. Most of his publications focus on risk perception, the environmental movement, technology assessment, decision and value tree analysis, citizens' participation, and energy systems.

*Hans-Jörg Seiler* is director of the interdisciplinary research project on Risk and Safety of Technological Systems at the Swiss Federal Institute of Technology in Zürich, Switzerland. He studied law at the Legal & Economic Faculty University of Bern, Switzerland. In 1982 he was admitted as a legal attorney (Fürsprecher) for the canton of Bern. He completed his Ph.D. at the University of Bern and his habilitation for constitutional and administrative law. After being employed by the federal government (Office of Energy and Office of Justice), he joined the Swiss Institute of Technology. During his tenure at the Federal institutes, he worked among others on licensing procedures for nuclear installations, focusing on formalized participation of citizens.

*Thomas Webler* is currently a post-doctoral research associate with the Interdisciplinary Project on Risk and Safety at the Swiss Federal Institute of Technology in Zürich, Switzerland. He received his Ph.D. in Environment, Technology, and Society from Clark University in Worcester, MA in 1992 and holds an M.S. degree in Biomedical Engineering and a B.S. in Electrical Engineering. He has been involved with citizen participation in environmental decision making from the perspective of a citizen, a town official, and a research team staff. Together with Ortwin Renn, he has developed and implemented the Three-Step Citizen Panel Model and implemented it twice: in New Jersey in 1989 and in Aargau Canton, Switzerland in 1993. His dissertation served as the inspiration for this book. The theory and criteria developed in Chapter 3 were revised portions of his thesis.

# INDEX

# Technology, Risk, and Society

*An International Series in Risk Analysis*

1. J.D. Bentkover, V.T. Covello and J. Mumpower (eds.): *Benefits Assessment.* The State of the Art. 1986 ISBN 90-277-2022-3
2. M.W. Merkhofer: *Decision Science and Social Risk Management.* A Comparative Evaluation of Cost-Benefit Analysis, Decision Analysis, and other Formal Decision-Aiding Approaches. 1987 ISBN 90-277-2275-7
3. B.B. Johnson and V.T. Covello (eds.): *The Social and Cultural Construction of Risk.* Essays on Risk Selection and Perception. 1987 ISBN 1-55608-033-6
4. R.E. Kasperson and P.J.M. Stallen (eds.): *Communicating Risks to the Public.* International Perspectives. 1990 ISBN 0-7923-0601-5
5. D.P. McCaffrey: *The Politics of Nuclear Power.* A History of the Shoreham Nuclear Power Plant. 1991 ISBN 0-7923-1035-7
6. M. Waterstone (ed.): *Risk and Society.* The Interaction of Science, Technology and Public Policy. 1991 ISBN 0-7923-1370-4
7. A. Vari and P. Tamas (eds.): *Environment and Democratic Transition.* Policy and Politics in Central and Eastern Europe. 1993 ISBN 0-7923-2365-3
8. A. Vari, P. Reagan-Cirincione and J.L. Mumpower: *LLRW Disposal Facility Siting.* 1994 ISBN 0-7923-2743-8
9. O. Renn, T. Webler and P. Wiedemann (eds.): *Fairness and Competence in Citizen Participation.* Evaluating Models for Environmental Discourse. 1995 ISBN Hb 0-7923-3517-1; Pb 0-7923-3518-X

Kluwer Academic Publishers – Dordrecht / Boston / London